AF622659

Bioanalytical Aspects in Biological Therapeutics

Bioanalytical Aspects in Biological Therapeutics

Edited by

Xiaohui Xu
Daiichi-Sankyo, Inc., Basking Ridge, NJ, USA

Weifeng Xu
Merck & Co., Inc., Rahway, NJ, USA

WILEY

This edition first published 2022
©2022 John Wiley & Sons, Inc.

Published by John Wiley & Sons, Inc., Hoboken, New Jersey.
Published simultaneously in Canada.

No part of this publication may be reproduced, stored in a retrieval system, or transmitted in any form or by any means, electronic, mechanical, photocopying, recording, scanning, or otherwise, except as permitted under Section 107 or 108 of the 1976 United States Copyright Act, without either the prior written permission of the Publisher, or authorization through payment of the appropriate per-copy fee to the Copyright Clearance Center, Inc., 222 Rosewood Drive, Danvers, MA 01923, (978) 750-8400, fax (978) 750-4470, or on the web at www.copyright.com. Requests to the Publisher for permission should be addressed to the Permissions Department, John Wiley & Sons, Inc., 111 River Street, Hoboken, NJ 07030, (201) 748-6011, fax (201) 748-6008, or online at http://www.wiley.com/go/permission.

Limit of Liability/Disclaimer of Warranty
While the publisher and author have used their best efforts in preparing this book, they make no representations or warranties with respect to the accuracy or completeness of the contents of this book and specifically disclaim any implied warranties of merchantability or fitness for a particular purpose. No warranty may be created or extended by sales representatives or written sales materials. The advice and strategies contained herein may not be suitable for your situation. You should consult with a professional where appropriate. Neither the publisher nor author shall be liable for any loss of profit or any other commercial damages, including but not limited to special, incidental, consequential, or other damages. Further, readers should be aware that websites listed in this work may have changed or disappeared between when this work was written and when it is read. Neither the publisher nor authors shall be liable for any loss of profit or any other commercial damages, including but not limited to special, incidental, consequential, or other damages.

For general information on our other products and services or for technical support, please contact our Customer Care Department within the United States at (800) 762-2974, outside the United States at (317) 572-3993 or fax (317) 572-4002.

Wiley also publishes its books in a variety of electronic formats. Some content that appears in print may not be available in electronic formats. For more information about Wiley products, visit our web site at www.wiley.com.

Library of Congress Cataloging-in-Publication Data
Names: Xu, Xiaohui, Ph.D. editor. | Xu, Weifeng (Director at Merck) editor.
Title: Bioanalytical aspects in biological therapeutics / edited by Xiaohui Xu, Weifeng Xu.
Description: Hoboken, NJ : John Wiley & Sons, 2022. | Includes bibliographical references and index. | Contents: Overview of the development of biotherapeutics and the role of bioanalytical sciences / Robert Dodge -- Basic immunology for bioanalytical testing of biotherapeutics / Kang Chen and Weifeng Xu -- Platform and instrument considerations in bioanalytical testing / Dominic Warrino and Franklin Spriggs -- Pharmacokinetic assays / Tong-Yuan Yang and Eric Wakshull -- Recent progress in biomarker bioanalysis and target engagement assessment / Yan G. Ni, Lindsay E. King and Carmen Fernández-Metzler -- Immunogenicity risk assessment for biotherapeutics / Jochem Gokemeijer -- Bioanalytical strategy to support clinical immunogenicity assessment : antidrug antibodies / Ying Wang and Michael Luong -- Bioanalytical scheme for antidrug neutralizing antibody assays / Weifeng Xu, Bonnie Wu and Jim McNally -- Critical reagents in bioanalysis / Yang Xu, Agostinho Gomes Rocha, Shannon Chilewski, Krisna C. Duong-Ly, Kun Yang and Jonathan Haulenbeek -- Statistical aspects of bioanalytical testing / Arkady M. Gershteyn and Mark Ma -- Bioanalytical aspects in biological therapeutics : biotransformation / Wenying Jian, Cong Wei and Jinping Gan -- New modalities : multidomain therapeutics and gene therapy programs / Kelly Colletti and Mark Ma -- Regulatory aspects for assay development, validation, and sample analysis / Amy Lavelle and Megan Wiberg.
Identifiers: LCCN 2022012705 (print) | LCCN 2022012706 (ebook) | ISBN 9781119523215 (hardback) | ISBN 9781119523239 (pdf) | ISBN 9781119523284 (epub) | ISBN 9781119523246 (obook)
Subjects: MESH: Biological Products--analysis | Biological Products--therapeutic use | Biological Assay--methods | Drug Development--standards | Antibodies--pharmacology
Classification: LCC RM301.25 (print) | LCC RM301.25 (ebook) | NLM QV 241 | DDC 615.1/9--dc23/eng/20220513
LC record available at https://lccn.loc.gov/2022012705
LC ebook record available at https://lccn.loc.gov/2022012706

Cover image: Courtesy of Weifeng Xu
Cover design by Wiley

Set in 9.5/12.5pt STIXTwoText by Integra Software Services Pvt. Ltd, Pondicherry, India

Contents

Preface

With the 2011 approval of Yervoy, the first monoclonal antibody therapeutic targeting immune checkpoint inhibitor CTLA-4, the selection of cancer immunotherapy as the number one scientific breakthrough of the year by *Science Magazine* in 2013, and with the anti-PD1 monoclonal antibody therapeutics Opdivo and Keytruda approved for an ever-expanding list of oncologic indications, biologic therapeutics have entered their "Golden Age." This further pushes the expansion of the protein and monoclonal antibody biologics towards bispecifics, trispecifics, probody therapeutics, all kinds of conjugated proteins such as PEGylated cytokines, antibody–drug conjugates, and cell/gene therapies. Some of the associated new challenges with these booming and more complicated biologic therapeutics are the bioanalytic test methods for pharmacokinetics, pharmacodynamics, biomarkers, and immunogenicity assessments. While there have been many dedicated white papers and regulatory guidance, as well as conferences, focusing on different aspects of bioanalytic assays, we note that to date there has been no single book dedicated to covering the broad key aspects in the bioanalytical fields. With this in mind, when Wiley reached out to us for a potential book on bioanalytical aspects of biological therapeutics, we thought we could put together a wholistic and in-depth guide that would cover most of the important topics in the field of bioanalytics.

The book starts with an overview of the role of bioanalytical tests in the evolution of biotherapeutics (Chapter 1) and the basic immunological knowledge needed for bioanalytics (Chapter 2). We then introduce the platforms and instruments needed for bioanalytical methods (Chapter 3), and follow this with a discussion of the major bioanalytical assays: pharmacokinetics, biomarkers, immunogenicity predication, antidrug antibodies and neutralizing antibody assays (Chapters 4–8). In Chapters 9 and 10 we discuss the importance of critical reagents and statistical analysis, followed by biotransformation of proteins, monoclonal antibodies, antibody–drug conjugates, and PEGlyated biologics (Chapter 11), and ending with new modalities (Chapter 12) and regulatory guidance and compliance (Chapter 13).

We originally planned to have the book ready for readers much earlier. However, with the unprecedented COVID-19 pandemic, the lives of most of the contributing authors have been profoundly impacted. Luckily, all the chapters bar one were finally ready around the same time, when we saw the fall in new COVID-19 cases that coincided with the launch of the global vaccination program and with it the hope of returning to normal life. We deeply appreciate all the authors' dedication and Wiley's professional assistance with this project, as well as the vital help received from our families, friends, and colleagues. We hope both new and seasoned scientists find this book useful in one way or another, and we very much welcome readers' feedback. Together we will develop new therapeutics to make our lives healthier and happier.

Xiaohui Xu, PhD

Weifeng Xu, PhD

June 2022

Foreword

When I received the invitation to write the foreword to this book on *Bioanalytical Aspects in Biological Therapeutics*, I was honored and excited. As I reviewed the book outline, and subsequently read the draft versions of the chapters in order to write this Foreword, I am more than excited: I am elated, deeply honored, and humbled. The editors, Drs. Xiaohui Xu and Weifeng Xu, have recruited industry leaders, deep subject matter experts, and innovative thinkers to provide substrate for this book. The content is timely and practical as regards to the growing area of biotherapeutics and the advances being made with lifesaving therapeutics.

This book is a compilation of the most diverse scientific and technical information that pertains to the bioanalytical investigation of the current biotherapeutics under development. As far as I am aware, this is the first book that encompasses the breadth of biotherapeutics from simple peptides to the more complex cell and gene therapy products.

I started my industry career when proteins were introduced as therapeutics, namely the erythropoietin drugs, and it evolved during the explosion of the monoclonal antibody therapies. The primary techniques for quantifying a protein therapeutic were immunoassays: RIA (radioimmunoassay) or ELISA (enzyme-linked immunosorbent assay). After many decades, mass spectrometric methods and many formats of immunoassays with highly sensitive detection methods evolved. Since then we have seen the expansion of the field into bispecifics, and gene and cell therapies, enabling bioanalytical chemists to think beyond the use of a single technology to the use of complimentary assays to fully quantify these molecules. The deep understanding of the pharmacologically relevant species has led us to evaluate the data we generate in a multifaceted approach. As immunogenicity evaluations and risk-based approaches evolved, the criticality of statistical analysis and interpretation of data became important. Today, and in the future, nontraditional technologies that support regulated analysis such as flow cytometry, quantitative polymerase chain reaction (qPCR), and others are becoming more prevalent for gene and cell therapy bioanalysis. Multiple methods will be needed to fully characterize the pharmacokinetic (PK) and pharmacodynamic (PD) parameters. As described by the authors of these chapters, the complexity of the molecules with different functional moieties, and the delivery systems and their components, add to the complexity of bioanalysis. In addition to the complexity of the molecules themselves, the relevance of systemic exposure in the PK/PD analysis will be brought into question. The different sites of administration, the ADME (absorption, distribution, metabolism, excretion) properties of the disease state versus a healthy individual, the correlation between the animal models which may be the only avenue for predicting dose, and the biotransformation of these therapeutics will all be topics for discussion over the next decade. The recent experience of developing the vaccines for the novel COVID-19 virus also brings to light the close partnership that industry may need to pursue in order to solve the questions associated with these grave diseases.

As regards to cell and gene therapy products, precision medicine will be at the forefront. I remain humble in predicting this, based on my limited experience in comparison to many of the readers of this book. The efficacy of these products against the diseases that they are designed to treat is reliant on the translational and predictive biomarkers that we establish during the drug development process. Collaboration and cooperation of the vendors to translate these biomarkers into accurately measured endpoints will be critical. No cell or gene therapy program will be successful without a clearly defined pharmacodiagnostic path. The most critical component will be to have the regulators from many divisions cooperating when these products are approved.

During the past several decades, industry partnerships with the regulators have grown thanks to the efforts of many progressive and innovative leaders on both sides. Cooperation and open dialog at conferences has led to developing regulations and guidance to industry, with scientific rationale as the basis. The regulatory requirements for bioanalytical methods for quantitative measurement and immunogenicity assessments of biologics therapeutics were established within the last few years after many decades of scientific discussions. The validation of methods for gene and cell therapy and the relevance of quantifying the systemic exposure to correlate the PK and PD may be the next wave of intense regulatory discussions. For example, at the core of a quantitative measurement lies the procurement of a reference material that is relatively pure. In gene therapy what is the relevance of a "reference material" in a potency assay versus its clinical efficacy? In this new paradigm, academic institutions also play a significant role in developing new technologies to enable these nontraditional therapeutics. This industry/regulatory/academic cooper ation will need to evolve into more diverse and critical thinking.

There is also a growing engagement of the public interested in pharmaceutical drug development, patient advocacy groups, algorithms for monitoring large clinical trial data, modeling and simulation of the bioanalytical data, and so on, as part of the future of drug development. Most of the large pharmaceutical companies rely on academic laboratories and small biotechnology companies for innovations. As the industry becomes more global, patient diversity increases in clinical trials, and the bioanalytical data may not be the major driver for exposure or clinical endpoints, the demand for innovation in bioanalysis may be intensified. As was so eloquently articulated by Dr. Andrew Lo at the closing plenary of the 2021 National Biotechnology Conference (NBC), the power of gene therapy (or other novel therapies) does not necessarily rely on determining the quantification of the therapeutics and the determination of PK and PD but on the clinical endpoints from the patient outcomes. He described biomedicine as being at the inflection point, of "omics" and the importance of economics in addition to the scientific "omics" revolution. The growing data on monitoring and collecting patient outcomes' information, population effects, and other sources of patient outcomes may need to be carefully monitored, with data systems that can relate to efficacy endpoints. As an example, the mother of a 5-year-old with muscular dystrophy who was quoted as saying: "All I want is for my son to take five steps." While we as bioanalysts may be trying to measure the concentration of the gene therapy product driving that patient outcome, the regulators and industry may have to resort to more qualitative data. The question that remains unanswered is "How do we reconcile this dichotomy?"

In conclusion, I strongly believe that it is an exciting and exhilarating time to be part of the pharmaceutical and biotechnology industries as a bioanalytical scientist. The challenges that we may be presented with in order to understand the bioanalytical considerations for novel therapeutics will encompass a wide community with the patient being at the center guiding the innovation, transformation, and regulation of this industry.

Binodh DeSilva, Ph.D.
Vice President, Leads Discovery & Optimization
at Bristol Myers Squibb Company, Princeton, NJ

Acknowledgments

We thank Isaac Xu for helping on indexing of the book. We'd also like to thank AAPS PK and Immunogenicity working groups for providing the platforms to facilitate connection and discussion among chapter authors and other group members.

About the Editors

Xiaohui Xu, PhD, is Director at Daiichi-Sankyo Inc. She is actively serving in the American Association of Pharmaceutical Scientists (AAPS) PK Assay Group and IQ Consortium Metabolite Working Group. She has extensive experiences in regulated bioanalytical field including global submissions, covering small molecules, biologics and new modalities. She has extensive publicaitons and 2 book chapters. She received her Doctorate in Pharmaceutical Analysis from College of Pharmacy, the University of Georgia.

Weifeng Xu, PhD, is Director at Merck & Co., Inc. He is an active member and leader in the AAPS and the European Bioanalysis Forum's (EBF) immunogenicity community and has published multiple papers on immunogenicity assay development and validation. He is also co-leading the IQ Consortium Bioanalytical Considerations and Challenges for Cell Therapies subgroup. More recently Weifeng also became Bioanalytic lead to multiple vaccine projects within Merck including COVID-19 and pneumococcal to oversee all clinical immunogenicity assays.

About the Editors

List of Contributors

Kang Chen
Obstetrics & Gynecology
Wayne State University
Detroit, MI

Shannon Chilewski
Bristol Myers Squibb
Lawrenceville, Princeton, NJ

Kelly Colletti
Charles River Laboratories
Reno, NV

Binodh DeSilva
Bristol Myers Squibb
Lawrenceville, NJ

Robert Dodge
Novartis Institutes for BioMedical Research
East Hanover, NJ

Krisna C. Duong-Ly
PPDM, Merck & Co., Inc.
Rahway, NJ

Carmen Fernández-Metzler
PharmaCadence Analytical Services LLC
Hatfield, PA

Jinping Gan
HiFiBiO Therapeutics
Cambridge, MA

Arkady M. Gershteyn
Lantheus Medical Imaging Inc.
North Billerica, MA

Jochem Gokemeijer
Bristol Myers Squibb
Cambridge, MA

Jonathan Haulenbeek
Syneos Health
Princeton, NJ

Wenying Jian
Janssen R&D
Spring House, PA

Lindsay E. King
Pfizer Inc.
Andover, MA

Amy Lavelle
PPD
Richmond, VA

Michael Luong
Pfizer Inc.
Cambridge, MA

Mark Ma
Rallybio
New Haven, CT

Jim McNally
BioAgilytix
Durham, NC

Yan G. Ni
Regeneron Pharmaceuticals, Inc.
Tarrytown, NY

Agostinho Gomes Rocha
Smithers
Ewing, NJ

Franklin Spriggs
Bioanalytical Consulting LLC
Lenexa, KS

Eric Wakshull
Consulting
Santa Fe, NM

Ying Wang
Pfizer Inc.
Cambridge, MA

Dominic Warrino
KCAS Bioanalytical and Biomarker Services
Shawnee, KS

Cong Wei
Biogen Inc.
Cambridge, MA

Megan Wiberg
PPD
Richmond, VA

Bonnie Wu
Janssen R&D
Spring House, PA

Xiaohui Xu
Daiichi-Sankyo
Basking Ridge, NJ

Weifeng Xu
Merck & Co., Inc.
Rahway, NJ

Yang Xu
Merck & Co., Inc.
Rahway, NJ

Tong-Yuan Yang
Janssen R&D
Spring House, PA

Kun Yang
Regeneron Pharmaceuticals, Inc.
Tarrytown, NY

1

Overview of the Development of Biotherapeutics and the Role of Bioanalytical Sciences

Robert Dodge

The Novartis Institutes for BioMedical Research

1.1 Introduction

Bioanalysis, specifically the measurement of the concentration of biotherapeutics in biological matrices as well as the measurement of anti-therapeutic antibodies, has been critical to the development of biotherapeutics. In this chapter, a review of the critical role bioanalysis has played in advancing the development of biotherapeutics is outlined. While bioanalysis of small molecule or low molecular weight drugs has also played a critical role in development, this chapter is focused on biotherapeutics, defined here as protein-based therapeutics as well as novel modalities such as cell and gene therapies.

1.2 Pharmacokinetic Measurements

1.2.1 Initial Role of Bioanalysis in Pharmacokinetic Measurements

Starting with the introduction of the first biotherapeutics such as insulin, which involved extraction and purification of porcine or beef insulin for use in treating diabetes, bioanalysis has played a critical role in drug development. Some of the earliest and most critical aspects of the use of insulin as a biotherapeutic, such as formulation, were able to be addressed with accurate and

Bioanalytical Aspects in Biological Therapeutics, First Edition. Edited by Xiaohui Xu and Weifeng Xu.
© 2022 John Wiley & Sons, Inc. Published 2022 by John Wiley & Sons, Inc.

precise bioanalytical measurements to drive understanding of pharmacokinetic (PK) properties. For example, initial extracts of insulin had extremely short half-lives, requiring multiple injections per day. The need to develop formulations with longer half-lives was apparent, and methodologies to measure the PK properties were understood to be crucial in development (Yalow & Berson, 1960). The first robust insulin bioanalytical immunoassay used to measure insulin concentration was reported in 1960 (Yalow & Berson, 1960). In fact, a major advance in the use of extracted insulin as a useful biotherapeutic was brought about by formulations with protamine and zinc resulting in an increased half-life and a reduction in the number of injections a patient needed per day (Quianzon & Cheikh, 2012).

While animal-extracted biotherapeutics such as insulin and growth hormone were important drugs, the biotherapeutic field began to progress at a rapid rate with the advent of molecules produced using recombinant DNA methodologies. Recombinant production of biotherapeutics involves the production of a protein therapeutic from DNA exogenously added to a host, such as *Escherichia coli* or a mammalian cell line, resulting in production of a protein not constitutively produced by the host organism. Along with the increase in the number of biotherapeutics developed using recombinant technology, this was also the start of the need for very accurate, precise, and robust bioanalytical measurements. The resulting bioanalytical data played a determining role in allowing the assessment of safety and efficacy during the biotherapeutic development process.

For Humulin, the first insulin produced by recombinant DNA methodologies, bioanalysis was critical in assessing key aspects of development such as comparability, safety, and efficacy and dosing. Since the development of Humulin, the bioanalysis of biotherapeutics and the resulting concentration and immunogenicity data focused on understanding exposure, PK parameters such as half-life, and comparability (e.g., bioavailability and biocomparability) have consistently provided data to allow critical decisions and understanding during the drug development process. This has been true since the first recombinant DNA biotherapeutics were approved over three decades ago, and bioanalysis remains a small but crucial component in biotherapeutic drug development today.

The first recombinantly produced protein biotherapeutics designed for use in humans were insulin (Humulin) in 1982 and growth hormone (Protropin) in 1985 (Drugs@FDA, 2022). Prior to that time, all protein therapeutics, as well as vaccines, had been obtained exclusively via protein extraction from animals or human blood. Both of these recombinantly produced biotherapeutics produced by recombinant DNA methodology had historical counterpart protein drugs extracted from animal (insulin) or human (somatotropin) counterparts, although somatotropin had been removed from the market due to concerns about the transmission of Creuzfeldt–Jacob disease (Underwood et al., 1985).

Insulin as a purified extract had been used as a therapeutic for diabetes since 1922, and many aspects of the dosing and PK were well understood (Quianzon & Cheikh, 2012). Thus there was the need for accurate and precise bioanalytical measurements to determine and compare in vivo concentrations of the recombinant insulin to understand any differences between recombinant insulin produced in *E. coli* bacteria compared to the previous products extracted from porcine pancreas. While today we understand and can characterize post-translation modification, folding state, and correlate in vivo and in vitro activity of proteins, the understanding of any possible differences between recombinant and naturally produced proteins was not fully understood initially. Therefore, in order to ensure efficacy, dosing, and safety, bioanalytical measurements were a key component in verifying that recombinant insulin could be used in place of the extracted insulin products.

The bioanalytical assays put in place to understand insulin PK and immunogenicity were not so different to the concepts and methodologies that are still used today. Also since this first biotechnology-derived biotherapeutic insulin had been used in humans for decades as an animal-extracted product, research and a desire for accurate, sensitive, and selective methods for measuring and comparing the concentration of insulin were critical.

The procedure for developing the assay and the challenges to obtaining accurate, precise, specific measurements were not significantly different to those that the bioanalytical community faces today. Challenges such as the minimum required dilution, matrix interferences, and stability were all active areas of investigation in obtaining accurate and precise concentration measurements. Certainly examining the earliest development of bioanalytical assays for biotherapeutics concentration determination and contrasting this with current methodologies and challenges is an interesting and instructive exercise. It points to key areas for long-term solutions in biotherapeutic bioanalytical assay development, such that areas that were challenging in the 1960s and are still challenging today, six decades later, should lead to long-term changes in our approach to assay development today.

Yalow and Berson (1960) began development of their ligand binding concentration assay for insulin as would be done today for a ligand binding assay (LBA) for generation of antibodies for use as reagent or tools. They knew that high-quality reagents would be key to developing an effective LBA. They immunized guinea pigs with multiple rounds of beef insulin and isolated high-titer serum in responsive animals. Then, as today, this step of boosting animals with multiple doses of antigen in order to drive affinity maturation and high titer (concentration) of antibodies specific to the dosed antigen was well understood. The actual procedures followed to make polyclonal antibodies for LBAs today may vary in animal species (goat, monkey, rabbit, and even chicken), but the fundamental process and time required to generate polyclonal antibodies for assay reagents used by most bioanalytical scientists differs little from the original work. There are more modern methodologies for obtaining antibodies to analytes, for example, phage technologies such as HuCal (Knappik et al., 2000), that may significantly reduce the time needed to identify high-affinity antibodies, but this is not typically a default or widely used platform for reagent generation in the bioanalytical community and, as in many processes, biological generation of a tool reagent in animals is often more efficient than can be done ex vivo, or by human design.

The assay developed initially for insulin was a radioimmunoassay, with the same concept as a competition assay that may be used now with colorimetric or luminescence detection (Soeldner & Slone, 1965). However, some of the more sensitive methodologies such as fluorescence or luminescence probes were not available early in the development of sensitive LBAs, and so the use of radioisotopes, which are typically avoided in most bioanalytical laboratories today, was common in the late 20th century.

As with most LBA formats (ex sandwich format, competition format) in use in the modern bioanalytical laboratory, the next requirement in assay development is reagent labeling. This may involve antibody labeling with an enzyme, biotin, or, in the case of a competition assay, labeling the biotherapeutics. This is exactly the step taken in 1960, where the analyte, insulin, was labeled with ^{125}I. Not unlike issues encountered in labeling today, reagents in a modern assay, and adjustment of label level, were challenges faced in obtaining maximum sensitivity and specificity (Nowatzke et al., 2017). There are some technologies today that allow for site-specific or highly efficient labeling of reagents (e.g., azide–alkyne coupling [Breinbauer & Köhn, 2003]), but again these methods are not in general use. Despite advancement in ease of use through labeling kits with prepackaged chemicals and spin columns, novel chemistries for precise labeling of proteins, and for bioanalytical LBAs, the fundamental process and strategies and methods for developing LBAs are fundamentally unchanged from the initial LBAs. The fact that LBAs remain challenging is also driving newer methodologies to measure biotherapeutics in matrices such as those based on the mass spectrometry platform that is now routinely used (Yuan & Zhu, 2015).

1.2.2 Progression of Bioanalytical Testing for Pharmacokinetics

Following the success of insulin and growth hormone products from extracted animal products to recombinant protein biotherapeutics, numerous other products, including the first monoclonal antibody directed against CD3 (muromonab), have been approved (Kozlowski, 2007). Since that time, the majority of biotherapeutics approved have been monoclonal antibodies. While no health authority or industry guidance specific to the bioanalysis requirements were initially available, the vast majority of these biotherapeutics used the enzyme-linked immunosorbent assay (ELISA) format for analysis (Damen et al., 2009). ELISA, more generally termed LBAs, thus became the standard for biotherapeutic bioanalytical quantification for the purpose of PK analysis over the next several decades of biotherapeutic development.

While the United States Food and Drug Administration (US FDA) did issue general Bioanalytical Method Validation Guidance in 2001, it was specifically focused on low molecular weight xenobiotic drugs and did not address issues or challenges unique to protein biotherapeutic bioanalysis. However, in 2003, a Ligand Binding Assay Bioanalytical Focus Group of the American Association of Pharmaceutical Scientists published a consensus document designed to give guidance specifically on validation of bioanalytical methods for macromolecules or protein biotherapeutics (DeSilva et al., 2003). This paper became the basis for the required parameters for evaluation and acceptance criteria until 2012, when the European Medicines Agency (EMA) published a final guidance bioanalytical method validation, subdivided into two sections: one section specific to small molecule therapeutics with a mass spectrometry platform and the other specific to LBAs. Shortly afterwards the US FDA issued as a draft an update to the original Bioanalytical Method Validation Guidance which included expectations as regards LBAs. These health authority guidelines and consensus white paper recommendations ultimately resulted in LBAs for protein biotherapeutics being consistently accurate, precise, and free of interference across the industry.

In addition to the expectations and resulting accurate assays, protein biotherapeutics assays faced additional challenges over small molecule drug assays in providing relevant data for pharmacology models (Mahmood, 2005). Most schemes for measuring small molecule drugs involve extracting the drug from proteins or precipitating proteins that may be interacting with the drug. Therefore, assays of drug concentrations are independent of protein or target concentrations. However, protein drug concentrations are typically measured directly in the matrix (ex serum), thus resulting in additional parameters to be understood, namely whether the protein drug is free in the serum or is bound to a target in the serum. This is a particularly important factor to consider because most LBAs, as the assay format name suggests, use a ligand-binding capture or detection format. Hence free drug and total drug may be detected differently in an assay, depending on the assay format (Lee et al., 2011). In addition to the natural ligand present in the matrix, there may also be the confounding factor of immunogenicity. Formation of antidrug antibodies (ADAs) can have the effect of blocking detection of the drug in the assay, meaning that the drug is present, but is not detected in the LBA. These facts usually make the concentration data obtained from an LBA subject to interpretation, and are why large molecule bioanalytical scientists are typically key contributors to pharmacology teams' understanding of the PK and why they are often asked to write regulatory filing documents, answer health authority questions, and help with PK model building and interpretation. This is in contrast to small molecule bioanalytical scientists, whereby the concentration data they generate are typically much more straightforward and are not affected by numerous confounding factors (exogenous ligand, immunogenicity, and so on) and consequently they have a more limited role in providing any particular insights into the pharmacology or in the overall drug development programs.

With the development of robust, sensitive, accurate, and precise large molecule bioanalytical assays, the understanding of the PK of protein therapeutics could progress to the state of understanding we have today. For example, the understanding of the now routinely observed and sometimes expected target-mediated drug disposition was only possible with the ability of sensitive bioanalytical assays to detect the subtle changes in PK profiles (Levy, 1994). For example, understanding the PK characteristics of a biotherapeutic with first-order or higher elimination versus apparent higher-order multi-exponential elimination caused by the binding of a biotherapeutic to its target was only possible with the advent of sensitive bioanalytical assays capable of detecting the entire range of a PK profile.

In addition, the full understanding of routes of delivery versus exposure, such as intravenous injection compared to the bioavailability of a subcutaneous injection of biotherapeutics, has only been possible with precise and sensitive bioanalytical assays (Charman et al., 2000). Finally, of the stated three pillars (exposure at site of action, target binding, and expression of functional pharmacological activity) of therapeutic survival in the drug development process to phase II (Morgan et al., 2012) both "exposure" and "target binding" are measurements typically made by bioanalytical scientists, ensuring and validating the bioanalytical measurements' crucial role in the success of the drug development process.

1.2.3 Recent Bioanalytical Contributions of Bioanalytical Measurements to Complex Molecules

As noted earlier, bioanalytical sciences has been key in providing both PK and immunogenicity information in the course of biotherapeutics prevalence in the marketplace today. Historically, bioanalysis for PK assessment has been a single measurement of a biotherapeutic concentration; or, in terms of immunogenicity, determining whether there is an immune response to the biotherapeutic. However, over the last decade, both the complexity of the biotherapeutics in development, and in concordance with the sophistication of the bioanalytical methodologies, has allowed the bioanalyst to provide much more in-depth information about in vivo biotherapeutic characterization than single measurements at each time point.

An example of how the bioanalytical scientist is now providing more detailed information to assist in the development of a compound is the antibody drug conjugate (ADC) class of molecules. ADCs are monoclonal antibodies linked to a cytotoxic drug. These molecules take advantage of the specificity of the monoclonal antibody to target the drug to a specific cell type for killing, ameliorating nonspecific cell killing by the drug. However, there is no single measurement that will characterize the PK of an ADC completely (Wang, 2017). Since the monoclonal antibody biotherapeutic is covalently bound to the linker and then to a drug, and the linker may be cleavable in vivo, there will be the possibility of free drug present in a subject. In addition, depending on the methodology used to make the ADC, one could expect multiple species of antibody biotherapeutics, having different quantities of drug linked to them, resulting in different drug to antibody ratios (DARs) being present in vivo. The complexity of these molecules and the amount of information required to fully characterize the PK, optimize the dose, and understand the mechanism of action has resulted in a recommendation that up to four bioanalytical assays may be needed to provide sufficient information to drive development (Gorovits et al., 2013). These include an assay for all drug-conjugated antibody, the total amount of antibody (conjugated and unconjugated), and the amount of drug that is conjugated to the antibody and existing as free cleaved drug. Then the concentrations determined from the four bioanalytical assays that describe the PK profiles of each of the species allow for the best understanding of exposure–response relationships in the PK models.

1.3 Immunogenicity Evaluation

1.3.1 Initial Role of Bioanalysis for Immunogenicity Evaluation

As immunogenicity evolved as a concern for biotherapeutics capable of initiating a humoral immune response, the large molecule bioanalytical scientist evolved to include immunogenicity assessment as part of their role. Initial humoral immunogenicity concerns were initially addressed by bioanalytical scientists for extracted biologics (e.g., for insulin) as it was known that proteins from foreign source (ex porcine) had the potential to be recognized as foreign by the human immune system and elicit an ADA response. The interest in understanding the differences between pork and beef insulin immunogenicity was the first instance of humoral immunogenicity evaluation by bioanalysis. In this case, the drive was to understand why insulin from the two species and the processes of insulin extraction from the animals were notably different even though the primary amino acid sequences were mostly conserved between porcine, beef, and human insulin. These observations and bioanalytical assessment of immunogenicity led to the eventual understanding of the role and impact that contaminating proteins from the extraction process could play in causing unwanted immunogenicity. In this case, these impurities induced immune responses in humans preferentially by beef insulin. This resulting knowledge led to additional emphasis on purification methodologies in these early biotherapeutics (Fineberg et al., 1983a). Interestingly, this understanding is similar to concerns one has today, even with recombinantly produced protein therapeutics where host cells such as *E. coli* and Chinese hamster ovary (CHO) cells produce foreign proteins and may result in an immune response not to the biotherapeutic but to copurified host cell proteins (Jawa et al., 2016). Or, as was the case with beef insulin, it was shown with detailed bioanalytical testing that host cell proteins copurifying with a biotherapeutic can enhance an ADA response as was shown more recently for the recombinantly produced growth hormone, somatropin (Romer et al., 2007). These case studies illustrate the importance of understanding the formulation- and process-related impurities that may be in a biotherapeutic drug product, and this knowledge, although well-known and understood today, was an important contribution of bioanalytical sciences to the protein drug development process.

The concerns and bioanalytical techniques developed for investigating differences between pork and beef insulin were then extended to characterize and compare immunogenicity between extracted biologics (pork insulin) and recombinant human insulin. (Fineberg et al., 1983b). The bioanalytical assays developed for these early immunogenicity studies were radioimmunoassays. These immunoassays measured the polyethylene glycol (PEG)-precipitated antibody–^{125}I-labeled insulin complexes. Determination of positivity for anti-insulin antibodies was done using a cut-point approach whereby an increase in an insulin-treated subject's background binding level above an insulin-näive subject was based on the coefficient of variation of the subject's background binding and a two-fold increase would result in classifying a subject as positive. This statistical parametric approach to classifying subjects as positive or negative to anti-insulin antibodies is equivalent to methodologies outlined in current health authority guidelines using the concept of subject-specific cut-points in these studies. Results and interpretation of results were based on titer-like reporting of a percent antibody-bound insulin using insulin levels present in a subject.

In addition to the same concept of parametric cut-point, the methodologies used in these first immunogenicity evaluation studies were similar to the study design of today's studies. In these initial assessments drug-näive subjects with diabetes treated with either recombinant human insulin or purified pork insulin were monitored for the presence and levels of anti-insulin antibodies over a treatment period of 1 year. An analysis of the potential for immunogenicity from pork insulin versus recombinant human insulin was determined. It was clear from these early studies that human insulin had a lower immunogenicity rate than purified pork insulin, lower titer, or percent-bound anti-insulin antibodies, and that pork insulin had a lower immunogenicity rate than mixed beef/pork insulin products (where essentially all subjects developed anti-insulin antibodies). However, human recombinant insulin still induced an anti-human insulin antibody response in approximately 40% of treated subjects, which were surprising results as the initial thoughts might have been that there would be a much lower propensity for a humoral immune response with human insulin in human subjects. However, in these studies, laboratory tests for ADAs did not appear to correlate to any clinical efficacy, PK, or safety endpoints. Thus, many of the questions about the immune response to biotherapeutics raised during these initial insulin immunogenicity studies remain today, nearly four decades later, and as to why some humans develop antibodies to biotherapeutics and some do not. Questions such as the route of injection (intravenous), formulation components, drug aggregation, and propensity for the human immune system to mount an immune response to self-proteins or non-endogenous counterpart proteins still remains an active area of investigation aided by ever more sensitive and descriptive bioanalytical assays. In addition, a key point from these early studies remains important today, namely evaluating the presence or absence of a correlation between laboratory-based bioanalytical assays with clinically observable effects. In the case of insulin immunogenicity, no clear effect or correlation with anti-insulin antibody development or titer was seen in the patients in these studies. Yet, as in the case of insulin immunogenicity, having bioanalytical assays available to test and measure PK and immunogenicity responses during the investigation of the first recombinant protein therapeutics enabled researchers and regulators to understand how these molecules would behave in humans, and also highlighted some of the benefits, such as a somewhat lower immunogenicity response in human recombinant insulin versus extracted animal insulin. Finally, as is the case today, bioanalytical assays were able to provide critical information to aid understanding of the immunogenicity of the earliest recombinant protein therapeutics.

1.3.2 Historical Progression of Immunogenicity Testing Schemes and Methodologies

As described in the previous section, the role of the human immune system in potentially recognizing protein therapeutics as foreign entities and the understanding that this may not be desirable was recognized at the start of biotherapeutics drug development. However, even though a large number of patients in the relevant insulin studies developed anti-insulin antibodies during treatment, the clinical consequences of the response were not clear and development of anti-insulin antibodies was not classified as a safety risk or relevant to the efficacy of the product.

The first biotherapeutic that resulted in a serious clinical consequence when anti-therapeutic antibodies developed was recombinant erythropoietin (EPO) (Casadevall et al., 2002). EPO is a glycoprotein hormone that controls erythropoiesis, or red blood cell production. It is a cytokine for red blood cell precursors in the bone marrow and recombinantly produced exogenous EPO is used to treat anemia. In the initial study of the phenomena of clinical consequences from the development of anti-EPO antibodies, 13 patients were studied that had developed chronic disease pure red cell aplasia (PRCA). All the patients studied were under treatment with EPO for anemia from chronic renal failure. All the patients studied also had anti-EPO (ADAs) present. The antibodies from these patients to EPO did not just bind to the cytokine, they blocked its function (i.e., they neutralized the activity of EPO). Hence in the patients these antibodies neutralized both the drug as well as the patient's endogenously produced EPO. With the neutralizing of EPO, the patients no longer had the ability to produce red blood cells, resulting in PRCA, and thus the patients required either transfusions to obtain red blood cells or immunosuppressive therapy to repress the anti-EPO neutralizing antibodies, thus allowing EPO to complete the precursor cells' transformation to red blood cells.

Because PRCA is such a serious condition, and it was caused by immunogenicity to a biotherapeutic, industry and health authorities began a series of improvements in required testing schemes and data evaluation for protein drugs having the potential to elicit a neutralizing humoral immune response. In addition, enormous efforts in the bioanalytical science field were undertaken to advance the understanding and monitoring of subjects receiving proteins or other modalities capable of inducing an antibody response to exogenously administrated therapeutics. In addition, health authorities developed general and even specific immunogenicity testing guidelines like the EMA's specific guideline on the development of recombinant EPO products which specifically outlines timing, types of immunogenicity assays, detection requirements (ex low-affinity IgM antibodies), and neutralizing potential as part of the clinical safety requirements (see Health Authority Guidelines: EMA, 2010).

1.3.3 Risk Assessment Immunogenicity Testing

It was recognized early on that there was a potential disparity between incidence and consequence of immunogenicity between different biotherapeutics, for example, in the two cases described, insulin had a large number of patients develop anti-insulin antibodies with few apparent clinical effects, whereas EPO had an extremely low incidence of patient immunogenicity but with extremely severe clinical effects. Understanding both the risk of developing ADAs and comparing this with the consequences of developing ADAs was clearly important and resulted in a drive to formalize risk-based approaches to immunogenicity testing.

In 2004, Worobec and Rosenberg from the FDA published a series of papers (Worobec & Rosenberg, 2004a, 2004b, 2004c) giving recommendations on using a risk-based approach to evaluating and assessing immunogenic concerns from consequences to factors in the manufacturing process that generate risk. For immunogenic consequences, they focused on hypersensitivity reactions and neutralizing antibodies to endogenous counterpart proteins, both areas of high safety concern today also. For potential hypersensitivity to protein therapeutics, their recommendation was to assess hypersensitivity with tests that were standard at the time, for example, skin prick tests or radioallergosorbent test (RAST) assays. While it is now known that most protein therapeutics do not elicit a dangerous hypersensitivity reaction, some do as in the case of cetuximab, and testing for IgE antibodies, which are responsible for Type I hypersensitivity reactions, is recommended by FDA guidance when there is a risk of this type of reaction (see Health Authority Guidelines: FDA, 2014).

Cetuximab (Chung et al., 2008) is a chimeric mouse–human IgG1 monoclonal antibody against epidermal growth factor receptor and is approved for several cancer indications. It was observed that there was a high prevalence of hypersensitivity reactions to cetuximab in the USA mid-Atlantic region. In this case, testing for the presence of anti-IgE antibodies to cetuximab in subjects having a hypersensitivity reaction was carried out using an ImmunoCap assay, which is a variation of the RAST test to detect specific IgE antibodies to cetuximab. Many subjects having the hypersensitivity reaction to cetuximab were found to have specific preexisting IgE antibodies to the galactose-α-1,3-galactose (α-gal) carbohydrate group present on cetuximab. Thus the underlying mechanism for the hypersensitivity reaction was determined, and the implications for treating an individual with preexisting α-gal antibodies could be assessed and mediated as appropriate.

The FDA authors defined the highest risk as the case where a biotherapeutic has an endogenous counterpart involved in mediating an important biological function which could result in the neutralizing of the drug by an immune response. For example, in the case of thrombopoietin (TPO), which is a pegylated recombinant human megakaryocyte growth and development factor (PEG-rHuMGDF), some subjects developed severe thrombocytopenia which was attributed to neutralizing antibodies to the biotherapeutics as well as the endogenous MGDF (Li et al., 2001). For types of biotherapeutics such as TPO or EPO, the FDA authors recommended the most sensitive assays, at the earliest point in development. Thus even though the recommendations in the series of papers given by the FDA authors was not guidance or requirements, they outlined the health authority thinking at the time and the acceptance and encouragement of risk-based approaches for and the importance of bioanalytical testing immunogenicity in human clinical trials of biotherapeutics.

At the same time as the initial risk-based approach was being developed, the industry and the FDA also began to reach consensus on assay design, formats, and critical parameters for the bioanalytical assays that should be used for the detection of antibodies against biotechnology products (Mire-Sluis et al., 2004). Prior to this time, as in the initial radioimmunoassays described for insulin, most testing was carried out as a single test, designed to give a positive or negative result with no clear standard procedures or methodologies to determine positivity. This paper also introduced descriptions for what is now known as the first tier of immunogenicity testing, the screening assay, and although not explicitly stated, the concept of a second tier of testing for non-specific binding using immunodepletion in a confirmatory assay was also introduced at this time. Further advancement of the concept of tiered testing or a "multistep" approach was elaborated and introduced (Geng et al., 2005). Then a joint industry–FDA white paper proposed an "algorithm immunoassay based testing" process with the tiered step to "confirm" ADAs with specificity tests (Shankar et al., 2006). In addition an additional tier, a titer assay, was also proposed. Finally, the concept of tiered testing was formalized in the EMA guidance of 2008 giving the bioanalytical sciences community clear tools and methods to use in evaluating biotherapeutic immunogenicity (see Health Authority Guidelines: EMA, 2008). While further refinements to the risk-based approach, reporting nomenclature (Shankar et al., 2014), and updated health authority documents have been introduced, the overall concepts concerning the importance of bioanalytical assessment of immunogenicity have remained clear and essential to biotherapeutics development.

In addition to bioanalytical testing to provide data for PK parameters and models evolution, the testing and interpretation that bioanalytical scientists provide towards characterizing immunogenicity is becoming more complex. After an initial report of impact on immunogenicity of pegylated biotherapeutics proteins with PEG polymer linkages, the question arose as to not just whether the biotherapeutics were generating a humoral immune response, but what part of the biotherapeutic molecules were antibodies binding to, the protein portion or the PEG portion (Garay et al., 2012). This phenomenon resulted in the additional recommendation from the FDA to design bioanalytical assays that could detect both the antiprotein antibodies and antibodies against the PEG moiety or any nonprotein high molecular weight entities (e.g., hydroxyethyl starch) in the composition of the biotherapeutic (see Health Authority Guidelines: FDA, 2014).

1.3.4 Immunogenicity Assessments Beyond Safety Testing

The key driver for the exponential increase in the interest and testing for humoral responses to biotherapeutics was driven by safety, primarily from the concern about serious consequences if an ADA neutralized an endogenous counterpart. In addition to the role of safety assessment, and simple positive or negative testing results, immunogenicity information provided by the bioanalytical scientist can provide much more detailed information to aid the drug development process. In the case of GSK1995057, it was noted that certain subjects had preexisting antibodies to the biotherapeutic and those autoantibody patients may have a less than optimum response to treatment (Holland et al., 2013). In order to fully understand what the reason was for these preexisting antibodies, there was an effort to extract patient preexisting antibodies and characterize the specific epitopes within the biotherapeutic they bound to. In the course of the bioanalytical characterization, a new class of autoantibodies existing in a large percentage of the human population was discovered that will greatly assist the drug development process for future molecules having these reactive epitopes. Other work (Steenholdt et al., 2013), although additional investigation may be required (Allocca et al., 2013), suggests preexisting antibodies to infliximab may have an effect on long-term efficacy, and careful analysis and understanding of preexisting antibodies to biotherapeutics remains an active area of bioanalytical research (Gorovits et al., 2016).

Finally, as ADA testing and reporting continues to evolve and adds more value to the interpretation of clinical results, we see the bioanalytical scientists and bioanalysis in general having the ability to affect drug development progression. The recent decision about the developability of bococizumab and the cancelation of clinical development was based on the examination of the rates of how immunogenicity to the biotherapeutic in patients correlated with efficacy (Nature Biotechnology, 2016). Specifically, researchers plotted the level of low-density lipoprotein cholesterol levels against negative ADA, low-titer ADA, medium-titer ADA, and high-titer ADA patients (Ridker, 2017). The dramatic and clear correlation between cholesterol levels and ADA titer were conclusive for the lack of efficacy based on ADA response in patients.

1.4 Future of New Biotherapeutics Entities and Evolving Role of Bioanalysis

The pharmaceutical industry is constantly evolving and developing beneficial therapeutics to treat disease. Especially in the area of oncology and genetic disease, the now traditional low molecular weight and biologic therapeutics may not be sufficient and so new cell and gene therapies are coming in to use. With these new types of therapeutics, bioanalysis is evolving to measure and characterize the PK of these new molecules.

Gene therapies or gene therapy medicinal products (GTMPs) generally consist of a vector or delivery formulation/system containing a genetic construct engineered to express a specific therapeutic sequence or protein responsible for the regulation, repair, addition, or deletion of a genetic sequence (see Health Authority Guidelines: EMA, 2018). As such, the biotherapeutic does not consist of a single covalently linked molecule but instead is a complex entity consisting of a nucleic acid and a delivery system. The delivery system is generally complex, consisting of, for example, a lipid nanoparticle made of multiple molecules, or even a virus consisting of numerous proteins assembled into a particle capable of delivering the nucleic acid to cell targets.

With the complexity of a GTMP the bioanalysis becomes complex. Questions arise as to how much of the intact product (delivery system and vector) is in circulation or in tissues, concentrations of the nucleic acid component (Tremblay & Oldfield, 2009), concentrations of the delivery system (Hawthorne et al., 2018), and whether there are degradation products, similar to catabolites or metabolites, present. Typically the complete bioanalysis will involve a variety of traditional platforms, such as liquid chromatography–mass spectrometry and LBA; however, new platforms for the bioanalytical scientist, such as quantitative polymerase chain reaction (qPCR), may also be of use for PK measurements of the GTMPs (Ma et al., 2017).

In addition to PK measurements, bioanalytical testing for GTMPs will also involve immunogenicity evaluation. While a humoral response to delivery systems consisting of organic compounds such as lipid nanoparticles will be less prevalent, preexisting antibodies to viral delivery systems are well known and testing is recommended by health authority guidelines (see Health Authority Guidelines: EMA, 2018; FDA, 2014). This testing expectation exists because of the known direct impact that preexisting neutralizing antibodies can have on efficacy (Falese et al., 2017). In addition, cellular immunity and potential innate immune response may need to be evaluated, thus expanding the bioanalytical immunogenicity evaluation well past the historical ADA assay testing for traditional protein therapeutics.

Knowledge of the PK and immune system responses of GTMPs (distribution, bioanalysis) will continue to drive the understanding of safety, and the dosing levels that correlate to efficacy, allowing the development of these novel therapies to progress.

The most recently approved biotherapeutic modality comprises cell therapy products, currently with several chimeric antigen receptor T cell (CAR-T) products approved for use. CAR-T products are biotherapeutics specific to each patient. The production of these therapeutics begins with the patient donating blood cells which are sent to a manufacturing facility. The patient's T cells are transduced with a viral vector that integrates protein with a receptor targeting a specific antigen (e.g., CD19) and an intercellular region capability of activating the T cells. These cells are expanded and then given back to the patient. As in gene therapy, understanding the PK (Larson et al., 2009) and immunogenicity (Norellia et al., 2016) of these complex biotherapeutics is and will continue to be critical in understanding this novel treatment, and bioanalytical methodologies will evolve to gain this understanding. For example, PK interpretation will require not just the understanding of the concentration of a single molecule over time, but of an entire cell expressing multiple CAR-T molecules, and traditional definitions such as adsorption, bioavailability, metabolism, and excretion will need to be redefined. In addition, immunogenicity of cell therapy products will need to incorporate novel techniques as well as a broader scope than traditional ADA testing where typically only a humoral (antibodies) immune response is tested for.

In addition to concepts surrounding PK measurements of cells, platforms such as mass spectrometry and LBAs will no longer be appropriate for the bioanalytical scientist measuring cell types and concentrations. Adaption of platforms such as flow cytometry for whole-cell counting or qPCR to determine total CAR gene concentrations will be required as well. While both of these platforms have been used extensively for biomarker testing, using them for validated concentration measurements used for PK modeling may need to be developed (Kakkanaiah et al., 2018). Finally, as several health authority guidelines have been published, additional challenges in the interpretation and implementation of the required bioanalytical testing remain (Rousseau et al., 2018).

1.5 Summary

From the start of biotherapeutic development using extracted animal proteins to the novel cell therapy products that have recently been approved, bioanalytical testing has played an integral role in the development of biotherapeutics. Starting with initial formulation studies over half a century ago, bioanalysis started to play a critical role in understanding biotherapeutic PK. Then with the advent of recombinant DNA methodologies, bioanalysis was critical in the comparison of existing animal- and human-extracted protein therapeutics to newer and safer recombinantly produced biotherapeutics. With the focus on improving the robustness, matrix interference elimination, accuracy, and sensitivity of bioanalytical assays, new PK phenomena such as target-mediated drug disposition and route of injection bioavailability understanding progressed.

Now the focus of bioanalysis to aid more complex biotherapeutic development is moving past single-point concentration measurements to more complex assays and/or multiple measurements for detailed PK models to fully describe in vivo behavior and correlation with clinical observations. In addition, new biotherapeutic entities, such as cell and gene therapies, are expanding the need for more complex bioanalysis, requiring the bioanalytical scientist to adapt to new testing platforms and paradigms and to continue to play a critical role in the drug development process.

Unique to biotherapeutics is the possibility that foreign proteins in humans may elicit a humoral immune response, causing specific ADAs to develop. This was recognized even for the earliest animal-extracted proteins such as insulin. As the potential for serious consequences to patient safety resulting from neutralizing antibodies to endogenous biotherapeutic counterparts were fully recognized, bioanalytical testing for ADAs for high-risk molecules moved to the forefront of the development process. Even in cases where safety is of less concern, immunogenicity bioanalysis is driving other decisions in the drug development process, such as the long-term efficacy prospects of treatments as well as potentially stratifying the clinical eligibility of treatments based on pre-existing antibodies. From PK measurements to immunogenicity evaluation, bioanalysis has played and continues to play an extremely critical role in the biotherapeutic development process.

References

Allocca, M., Fiorino, G., & Danese, S. (2013). Commentary: Antibodies reacting with the infliximab Fab portion – Something new? *Alimentary Pharmacology & Therapeutics*, 38(5), 552. https://doi.org/10.1111/apt.12401

Breinbauer, R., & Köhn, M. (2003). Azide–Alkyne coupling: A powerful reaction for bioconjugate chemistry. *ChemBioChem*, 4(11), 1147–1149. https://doi.org/10.1002/cbic.200300705

Casadevall, N., Nataf, J., Viron, B., Kolta, A., Kiladjian, J-J., Martin-Dupont, P., Michaud, P., Papo, T., Ugo, V., Teyssandier, I., Varet, B., & Mayeux P., (2002, Feb 14). Pure red-cell aplasia and antierythropoietin antibodies in patients treated with recombinant erythropoietin. *New England Journal of Medicine*, 346(7), 469. https://doi.org/10.1056/NEJMoa011931

Charman, S. A., Segrave, A. M., Edwards, G. A., & Porter, C. J. H. (2000). Systemic availability and lymphatic transport of human growth hormone administered by subcutaneous injection. *Journal of Pharmaceutical Sciences*, 89(2), 168–177. https://doi.org/10.1002/(SICI)1520-6017(200002)89:2<168::AID-JPS4>3.0.CO;2-Q

Chung, C. H., Mirakhur, B., Chan, E., Le, Q.-T., Berlin, J., Morse, M., Murphy, B. A., Satinover, S. M., Hosen, J., Mauro, D., Slebos, R. J., Zhou, Q., Gold, D., Hatley, T., Hicklin, D. J., & Platts-Mills, T. A. E. (2008). Cetuximab-induced anaphylaxis and IgE specific for galactose-α-1,3-galactose. *New England Journal of Medicine*, 358(11), 1109–1117. https://doi.org/10.1056/NEJMoa074943

Damen CW, Schellens JH, & Beijnen JH.(2009). Bioanalytical methods for the quantification of therapeutic monoclonal antibodies and their application in clinical pharmacokinetic studies. *Human Antibodies*, 18(3), 47–73. https://doi.org/10.3233/HAB-2009-0206

DeSilva, B., Smith, W., Weiner, R., Kelley, M., Smolec, J., Lee, B., Khan, M., Tacey, R., Hill, H., & Celniker, A. (2003, November). Recommendations for the bioanalytical method validation of ligand-binding assays to support pharmacokinetic assessments of macromolecules. *Pharmaceutical Research*, 20(11), 1885–1900.https://doi.org/10.1023/b:pham.0000003390.51761.3d

Drugs@FDA: FDA-Approved Drugs (2022). https://www.accessdata.fda.gov/scripts/cder/daf

Falese, L., Sandza, K., Yates, B., Triffault, S., Gangar, S., Long, B., Tsuruda, L., Carter, B., Vettermann, C., Zoog, S. J., & Fong, S. (2017). Strategy to detect pre-existing immunity to AAV gene therapy. *Gene Therapy*, 24(12), 768–778. https://doi.org/10.1038/gt.2017.95

Fineberg, S. E., Galloway, J. A., Fineberg, N. S., & Goldman, J. (1983a). Effects of species of origin, purification levels, and formulation on insulin immunogenicity. *Diabetes*, 32(7), 592–599. https://doi.org/10.2337/diab.32.7.592

Fineberg, S. E., Galloway, J. A., Fineberg, N. S., Rathbun, M. J., & Hufferd, S. (1983b). Immunogenicity of recombinant DNA human insulin. *Diabetologia*, 25(6), 465–469. https://doi.org/10.1007/BF00284452

Garay, R. P., El-Gewely, R., Armstrong, J. K., Garratty, G., & Richette, P. (2012). Antibodies against polyethylene glycol in healthy subjects and in patients treated with PEG-conjugated agents. *Expert Opinion on Drug Delivery*, 9(11), 1319–1323. https://doi.org/10.1517/17425247.2012.720969

Geng, D., Shankar, G., Schantz, A., Rajadhyaksha, M., Davis, H., & Wagner, C. (2005, 15 September). Validation of immunoassays used to assess immunogenicity to therapeutic monoclonal antibodies. *Journal of Pharmaceutical and Biomedical Analysis*, 39(3–4), 364–375. https://doi.org/10.1016/j.jpba.2005.04.045

Gorovits, B., Alley, S., Bilic, S., Booth, B., Kaur, S., Oldfield, P., Purushothama, S., Rao, C., Shord, S., & Siguenza, P. (2013, May). Bioanalysis of antibody-drug conjugates: American Association of Pharmaceutical Scientists Antibody-Drug Conjugate Working Group position paper. *Bioanalysis*, 5(9), 997–1006. https://doi.org/10.4155/bio.13.38

Gorovits, B., Clements-Egan, A., Birchler, M., Liang, M., Myler, H., Peng, K., Purushothama, S., Rajadhyaksha, M., Salazar-Fontana, L., Sung, C., & Xue, L. (2016, March). Pre-existing antibody: biotherapeutic modality-based review. *The AAPS Journal*, 18(2), 311–320. https://doi.org/10.1208/s12248-016-9878-1

Hawthorne, G., Harbach, S., & Wilson, A. (2018). Cationic lipid bioanalysis: Understanding distribution of lipid nanoparticles for delivery of RNA therapeutics. *Bioanalysis*, 10(5), 275–277. https://doi.org/10.4155/bio-2017-0279

Holland, M. C., Wurthner, J. U., Morley, P. J., Birchler, M. A., Lambert, J., Albayaty, M., Serone, A. P., Wilson, R., Chen, Y., Forrest, R. M., Cordy, J. C., Lipson, D. A., & Bayliffe, A. I. (2013). Autoantibodies to variable heavy (VH) chain Ig sequences in humans impact the safety and clinical pharmacology of a VH domain antibody antagonist of TNF-α receptor 1. *Journal of Clinical Immunology*, 33(7), 1192–1203. https://doi.org/10.1007/s10875-013-9915-0

Jawa, V., Joubert, M., Zhang, Q., Deshpande, M., Hapuarachchi, S., Hall, M., & Flynn, G. C. (2016). Evaluating immunogenicity risk due to host cell protein impurities in antibody-based biotherapeutics. *The AAPS Journal*, 18(6), 1439–1452. https://doi.org/10.1208/s12248-016-9948-4

Kakkanaiah, V. N., Lang, K. R., & Bennett, P. K. (2018)Flow cytometry in cell-based pharmacokinetics or cellular kinetics in adoptive cell therapy. *Bioanalysis*. 10(18), 1457–1459. https://doi.org/10.4155/bio-2018-0203

Knappik, A., Ge, L., Honegger, A., Pack, P., Fischer, M., Wellnhofer, G., Hoess, A., Wölle, J., Plückthun, A., & Virnekäs, B. (2000). Fully synthetic human combinatorial antibody libraries (HuCAL) based on modular consensus frameworks and CDRs randomized with trinucleotides. *Journal of Molecular Biology*, 296(1), 57–86. https://doi.org/10.1006/jmbi.1999.3444

Kozlowski, S. (2007). Protein therapeutics and the regulation of quality: A brief history from an OBP perspective. *Biopharm International*, 20(10), 37–55, 40–45. https://www.biopharminternational.com/view/protein-therapeutics-and-regulation-quality-brief-history

Lee, J. W., Kelley, M., King, L. E., Yang, J., Salimi-Moosavi, H., Tang, M. T., Lu, J-F., Kamerud, J., Ahene, A., Myler, H., & Rogers, C. (2011). Bioanalytical approaches to quantify "total" and "free" therapeutic antibodies and their targets: Technical challenges and PK/PD applications over the course of drug development. *The AAPS Journal*, 13(1), 99–110. https://doi.org/10.1208/s12248-011-9251-3

Levy, G. (1994). Pharmacologic target-mediated drug disposition. *Clinical Pharmacology and Therapeutics.* 56(3), 248248–252. https://doi.org/10.1038/clpt.1994.134

Li, J., Yang, C., Xia, Y., Bertino, A., Glaspy, J., Roberts, M., & Kuter, D. J. (2001). Thrombocytopenia caused by the development of antibodies to thrombopoietin. *Blood*, 98(12), 3241–3248. https://doi.org/10.1182/blood.v98.12.3241

Ma, M., Balasubramanian, N., Dodge, R., & Zhang, Y. (2017). Challenges and opportunities in bioanalytical support for gene therapy medicinal product development. *Bioanalysis*, 9(18), 1423–1430. https://doi.org/10.4155/bio-2017-0116

Mahmood, I. (2005, April). Pharmacokinetic and pharmacodynamic considerations in the development of therapeutic proteins. *Clinical Pharmacokinetics*, 44(4), 331–347. https://doi.org/10.2165/00003088-2005 44040-00001

Mire-Sluis, A. R., Barrett, Y. C., Devanarayan, V., Koren, E., Liu, H., Maia, M., Parish, T., Scott, G., Shankar, G., Shores, E., Swanson, S. J., Taniguch, G., Wierdal, D., & Zuckermanm, L. A. (2004, June). Recommendations for the design and optimization of immunoassays used in the detection of host antibodies against biotechnology products. *Journal of Immunological Methods*, 289(1–2), 1–16. https://doi.org/10.1016/j.jim.2004.06.002

Morgan, P., Van Der Graaf, P. H., Arrowsmith, J., Feltner, D. E., Drummond, K. S., Wegner, C. D., & Street, S. D. A. (2012, May). Can the flow of medicines be improved? Fundamental pharmacokinetic and pharmacological principles toward improving Phase II survival. *Drug Discovery Today*, 17(9–10), 419–424. https://doi.org/10.1016/j.drudis.2011.12.020

Nature Biotechnology Pfizer drops phase 3 lipid-lowering antibody, 2016/12/07 online, 34, 1214. https://www.nature.com/articles/nbt1216-1214

Norellia, M., Casucci, M., Bonini, C., & Bondanz, A. (2016, January). Clinical pharmacology of CAR-T cells: Linking cellular pharmacodynamics to pharmacokinetics and antitumor effects. *Biochimica et Biophysica Acta (BBA) – Reviews on Cancer*, 1865(1), 90–100. https://doi.org/10.1016/j.bbcan.2015.12.001

Nowatzke, W., Spriggs, F., Fitzgerald, V., Davis, A., & Bowsher, R. R. (2017). Development and maintenance of critical reagents for ligand binding assays to support regulatory-compliant bioanalysis. In M. Rocci Jr & S. Lowes (Eds.), *Regulated Bioanalysis: Fundamentals and Practice. AAPS Advances in the Pharmaceutical Sciences Series* (Vol. 26). Springer.

Quianzon, C. C., & Cheikh, I. (2012). History of insulin. *Journal of Community Hospital Internal Medicine Perspectives*, 2(2), 10.3402. https://doi.org/10.3402/jchimp.v2i2.18701

Ridker, P.M., Tardif, J-C., Amarenco, P., Duggan, W., Glynn, R.J., Jukema, J.W., Kastelein, J.J.P., Kim, A.M., Koenig, W., Nissen, S., Revkin, J., Rose, L.M., Santos, R.D., Schwartz, P.F., Shear, C.L., & Yunis, C. (2017). Lipid-reduction variability and antidrug-antibody formation with bococizumab. *New England Journal of Medicine*, 376(16), 1517–1526. doi: 10.1056/NEJMoa1614062

Romer, T., Peter, F., Saenger, P., Starzyk, J., Koehler, B., Korman, E., Walczak, M., Wasik, R., Ginalska-Malinowska, M., Solyom, E., & Berghout, A. (2007). Efficacy and safety of a new ready-to-use recombinant growth hormone solution. *Journal of Endocrinological Investigation*, 30(7), 578–589. https://doi.org/10.1007/BF03346352

Rousseau, C. F., Macluaitis, R., Sladowski, D., & Narayanan, G. (2018, 28 May). Cell and gene therapies: European view on challenges in translation and how to address them. *Frontiers in Medicine*, 5, 158. https://doi.org/10.3389/fmed.2018.00158

Shankar, G., Arkin, S., Cocea, L., Devanarayan, V., Kirshner, S., Kromminga, Q. V., Richards, S., Schneider, C. K., Subramanyam, M., Swanson, S., Verthelyi, D., & Yim, S. (2014, July). Assessment and reporting of the clinical immunogenicity of therapeutic proteins and peptides-harmonized terminology and tactical recommendations. *The AAPS Journal*, 16(4), 658–673. https://doi.org/10.1208/s12248-014-9599-2

Shankar, G., Shores, E., Wagner, C., & Mire-Sluis, A. (2006, June). Scientific and regulatory considerations on the immunogenicity of biologics. *Trends in Biotechnology*, 24(6), 274–280. https://doi.org/10.1016/j.tibtech.2006.04.001

Soeldner, J., & Slone, D. (1965). Critical variables in the radioimmunoassay of serum insulin using the double antibody technic. *Diabetes*, 14(12), 771–779. https://doi.org/10.2337/diab.14.12.771

Steenholdt, C., Palarasah, Y., Bendtzen, K., Teisner, A., Brynskov, J., Teisner, B., & Nielsen, C. H. (2013). Pre-existing IgG antibodies cross-reacting with the Fab region of infliximab predict efficacy and safety of infliximab therapy in inflammatory bowel disease. *Alimentary Pharmacology & Therapeutics*, 37(12), 1172–1183. https://doi.org/10.1111/apt.12330

Tremblay, G. A., & Oldfield, P. R. (2009). Bioanalysis of siRNA and oligonucleotide therapeutics in biological fluids and tissues. *Bioanalysis*, 1(3), 595–609. https://doi.org/10.4155/bio.09.66

Underwood, L. E., (Chair), Fisher, D. A., Frasier, S. D., Gertner, J. M., Kaplan, S. L., Kirkland, R. T., Lippe, B. M., & Raiti, S. (1985). Degenerative neurologic disease in patients formerly treated with human growth hormone: Report of the Committee on Growth Hormone Use of the Lawson Wilkins Pediatric Endocrine Society, May 1985. *The Journal of Pediatrics*, 107(1), 10–12. https://doi.org/10.1016/S0022-3476(85)80604-1

Wang, J. (2017, April). Current status of antibody-drug conjugates bioanalysis. *Journal of Applied Bioanalysis*, 3(2), 16–30. https://doi.org/10.17145/jab.17.004

Worobec, A., & Rosenberg, A. S. (2004a). A risk-based approach to immunogenicity concerns of therapeutic protein products, Part 1: Considering consequences of the immune response to a protein. *Biopharm International*, 17(11). https://www.biopharminternational.com/view/risk-based-approach-immunogenicity-concerns-therapeutic-protein-products-part-1-considering-conseque

Worobec, A., & Rosenberg, A. S. (2004b). A risk-based approach to immunogenicity concerns of therapeutic protein products, Part 2: Considering host-specific and product-specific factors impacting immunogenicity. *Biopharm International*, 17(12). https://www.biopharminternational.com/view/risk-based-approach-immunogenicity-concerns-therapeutic-protein-products-part-2-considering-host-spe

Worobec, A., & Rosenberg, A. S. (2004c). A risk-based approach to immunogenicity concerns of therapeutic protein products, Part 3: Effects of manufacturing changes in immunogenicity and the utility of animal immunogenicity studies. *Biopharm International*, 18(1). https://www.biopharminternational.com/view/risk-based-approach-immunogenicity-concerns-therapeutic-protein-products-part-3-effects-manufacturin

Yalow, R. S., & Berson, S. A. (1960). Immunoassay of endogenous plasma insulin in man. *Journal of Clinical Investigation*, 39(7), 1157–1175. https://doi.org/10.1172/JCI104130

Yuan, L., & Zhu, M. (2015). Quantitative bioanalysis of proteins by mass spectrometry. *Materials and Methods*, 5, 1332. https://dx.doi.org/10.13070/mm.en.5.1332. https://www.labome.com/method/Quantitative-Bioanalysis-of-Proteins-by-Mass-Spectrometry.html

Health Authority Guidelines

Guideline on Non-Clinical and Clinical Development of Similar Biological Medicinal Products Containing Recombinant Erythropoietins (Revision) 1 October 2010 (https://www.ema.europa.eu/documents/scientific-guideline/guideline-non-clinical-clinical-development-similar-biological-medicinal-products-containing_en-2.pdf)

Guideline on Immunogenicity Assessment of Biotechnology-Derived Therapeutic Proteins, Effective April 2008, European Medicines Agency, Doc. Ref. EMEA/CHMP/BMWP/14327/2006 (https://www.ema.europa.eu/en/documents/scientific-guideline/guideline-immunogenicity-assessment-biotechnology-derived-therapeutic-proteins-first-version_en.pdf)

Guidance for Industry Immunogenicity Assessment for Therapeutic Protein Products Final (US Department of Health and Human Services; Food and Drug Administration, August 2014) (https://www.fda.gov/regulatory-information/search-fda-guidance-documents/immunogenicity-assessment-therapeutic-protein-products)

Biotherapeutics, or biotechnology derived products, are defined for the purposes of this chapter as preparations of biological products from genetically engineered bacteria, yeast, fungi, or cells (http://www.who.int/biologicals/vaccines/biotherapeutic-products/en/).

European Medicines Agency, Committee for Advanced Therapies (CAT). Guideline on the Quality, Nonclinical and Clinical Aspects of Gene Therapy Medicinal Products (2018). (https://www.ema.europa.eu/en/documents/scientific-guideline/guideline-quality-non-clinical-clinical-aspects-gene-therapy-medicinal-products_en.pdf)

2

Basic Immunology for Bioanalytical Testing of Biotherapeutics

Kang Chen[a] *and Weifeng Xu*[b]

[a]*Wayne State University*
[b]*PCD, Bioanalytics, Merck & Co., Inc. (corresponding author)*

2.1 Introduction

The fundamental function of our immune system is to distinguish non-self- or altered self-antigens from self-antigens in order to protect the host from potential harmful challenges. Despite the goal of treating patients with more sophisticated biotherapeutics to better mimic self-antigens and hence minimize host immune responses, all biological therapeutics are still "foreign" antigens and thus have the potential to induce immune responses and unwanted side effects that are sometimes life-threatening. While factors impacting the generation of anti-biotherapeutics antibodies such as similarity to self-antigens, impurity, and patient status will be discussed in greater detail in Chapter 6, this chapter starts with an introduction to the cells and organs of our immune system, and follows this with an outline of the different administration sites and routes of biotherapeutics to facilitate an understanding of how foreign antigens such as biotherapeutics elicit immune responses and finally induce the production of antidrug antibodies (ADAs).

Bioanalytical Aspects in Biological Therapeutics, First Edition. Edited by Xiaohui Xu and Weifeng Xu.
© 2022 John Wiley & Sons, Inc. Published 2022 by John Wiley & Sons, Inc.

2.2 Innate Immune Responses to Biologic Drugs

The physiological function of the immune system is defense against infectious microbes or cancerous host cells, but noninfectious or noncancerous foreign substances, including biotherapeutic peptides, proteins, nucleic acids, or cells, can also elicit immune responses. Innate immunity is the first line of response against infection or perturbation of tissue homeostasis. Similar to pathogens, immunogenic biologic drugs administered to the body are first detected and captured by components of the innate immune system. Cells of the innate immune system, such as dendritic cells, macrophages, and granulocytes, use germline-encoded sensors called pattern recognition receptors (PRRs) to detect pathogen-associated molecular patterns (PAMPs). PAMPs are unique to or more abundant in microbes but not the host. Recognition of PAMPs by PRRs triggers inflammation and the recruitment and activation of phagocytes and other immune cells to control and/or eradicate the microbe, or blockade of viral replication and killing of virus-infected cells. Multiple PRRs have been identified and classified thus far on the basis of their structure and/or cellular location (Kawai & Akira, 2010; Rathinam et al., 2012; Wu & Chen, 2014). One well-known example of PRRs is the family of Toll-like receptors (TLRs), which includes many functional members that recognize a broad range of PAMP ligands such as virus double-strand RNA (recognized by TLR3), bacterial lipopolysaccharides (LPS) (recognized by TLR4), flagellin (recognized by TLR5), or CpG DNA (recognized by TLR9) (Takeda et al., 2003). Due to their ability to boost the innate immune response, some of these TLR ligands have been approved or are being tested as immune adjuvants (Steinhagen et al., 2011). PAMPs from different microbes may differ in their detailed structures but always share a common molecular pattern. The various PAMPs detected by PRRs are not structurally related but share features that reflect the evolutionary strategy of innate immune recognition. PAMPs are unique or typical to microbes but not host cells. They are invariant among pathogens of a given class and often have essential or important functions in microbial survival. Importantly, PAMPs are unique to microbes but not unique to pathogens; hence the innate immune system needs to distinguish noxious pathogens from normally innocuous microbes, such as commensal microbes. More relevant to the context of the administration of biologic therapeutics such as monoclonal antibodies (mAbs), PRRs also detect danger-associated molecular patterns (DAMPs), which are self-antigens expressed or released during tissue damage, often as a result of stress or injury such as injection, infection, or tumorigenesis, that signify these "danger" events. Furthermore, certain cells of the innate immune system, such as natural killer (NK) cells, are kept in check by the balance between the activating and inhibitory PRRs, with the inhibitory PRRs recognizing molecules expressed on healthy host cells. Once such self-molecules are downregulated as a result of infection or transformation, NK cells are activated. Therefore, the innate immune system is also triggered by the recognition of "missing self". The incorporation of PAMP or DAMP structures in biologic drugs has emerged as a promising strategy to optimize the therapeutic efficacy of these agents (Demaria et al., 2019).

Classically, several cardinal features distinguish innate immunity from adaptive immunity. Innate immune responses are often rapid, typically within hours and days, and precede the development of adaptive immune responses, which usually takes 2 to 3 weeks. Innate immune recognition relies on germline-encoded PRRs for a limited number of antigenic PAMPs whereas adaptive immune recognition uses antigen receptors generated by somatic recombination that have enormous diversity and precise specificity. Due to the absence of clonal selection of the cells bearing antigen-specific receptors and the differentiation of these cells into highly specific and potent memory cells, innate immunity has been considered devoid of memory. However, discoveries in the past two decades have given a more complex yet nuanced picture of the innate immune system and the blurring of the boundary between innate and adaptive immunity. Certain adaptive immune cells, such as γδ T cells, invariant natural killer T (iNKT) cells, mucosal-associated invariant T cells (MAITs), and marginal zone (MZ) B cells, exhibit features of innate immunity with limited antigen receptor diversity and rapid responses (Lanier, 2013). In addition, following initial microbial exposure, macrophages and NK cells can expand and contract their populations, upregulate genes involved in pathogen recognition and antigen presentation through epigenetic and metabolic reprogramming, and secrete cytokines that augment the antimicrobial activity of bystander immune cells. This increases host resistance to reinfection and offers cross-protection against other infectious agents, leading to enhanced innate immune responses that have given rise to the concept of "trained innate immunity" or "innate immune memory" (Netea et al., 2011; Song & Colonna, 2018). Of note, innate memory-induced decreased responsiveness has also been observed (i.e., "trained innate tolerance") (Boraschi & Italiani, 2018; Domínguez-Andrés et al., 2019). Trained innate immunity and tolerance conceivably have critical implications for the design and the monitoring of the immune responses to biologic drugs, especially if such drugs are repeatedly administered.

Innate immunity has the crucial role of instructing adaptive immunity. First, the specific structural patterns that PRRs detect carry the information on the type of organisms from which the antigens originate (i.e., viruses, bacteria, fungi, or parasites, or even foreign or self-tissues) (Kawai & Akira, 2010; Rathinam et al., 2012; Wu & Chen, 2014). Second, PRRs have tissue-specific expression as well as localization in distinct cellular compartments, including the plasma membrane, endosomes, lysosomes, and cytosol (Barton & Kagan, 2009). Pattern recognition therefore further encodes the information on the tissue and cellular location of the infection or insult. Such requisite information from the innate immune system instructs the activation of the relevant adaptive effector responses (Iwasaki & Medzhitov, 2015).

2.2.1 Components of the Innate Immune System

The innate immune system encompasses physical and chemical barriers such as epithelia and various antimicrobial molecules, phagocytic cells (neutrophils, macrophages, monocytes), dendritic cells (DCs), mast cells, innate lymphocytes, as well as blood proteins including the complement system and other mediators of inflammation. Many innate immune cells, such as macrophages, DCs, and mast cells, are constantly present in tissues and function as sentinels to survey for invading microbes and other tissue-damaging substances.

2.2.1.1 Neutrophils

Neutrophils are the most abundant leukocytes in the blood and the principal cell type in acute inflammatory responses. Their cytoplasm contains two types of membrane-bound granules. Most of these granules, called specific granules, are filled with enzymes, such as lysozyme, collagenase, and elastase. The remainder of the cytoplasmic granules, called azurophilic granules, contain enzymes and microbicidal molecules. Neutrophils differentiate from bone marrow precursors after stimulation by granulocyte colony-stimulating factor (G-CSF) and granulocyte-macrophage colony-stimulating factor (GM-CSF). Neutrophils can migrate rapidly out of the circulation into sites of infection or tissue injury, which is called extravasation, where they function for 1 to 2 days to phagocytose microbes, especially opsonized microbes and products of cell necrosis, degrade them in phagolysosomes, and then undergo apoptosis. In addition, neutrophils can release inflammatory cytokines, granular contents, antimicrobial molecules, and chromatins to trap and kill extracellular microbes and recruit other immune cells to amplify the immune defense. However, overaction of this response can result in inflammatory tissue damage. Recent studies have shown that neutrophils can modulate adaptive immune responses by expressing cytokines, chemokines, and surface molecules as well as carrying and presenting antigens to support the function of effector or regulatory B and T cells.

2.2.1.2 Mononuclear Phagocytes

The mononuclear phagocyte system includes circulating monocytes, which can differentiate into macrophages upon migration into tissues, and tissue-resident macrophages, which are derived mostly from hematopoietic precursors during prenatal life with minor blood monocyte input under steady state in many tissues. In adults, monocytes arise from bone marrow precursors with the stimulation of macrophage colony-stimulating factor (M-CSF). Human monocytes have been grouped into three major populations: classical ($CD14^{+}CD16^{-}$), intermediate ($CD14^{+}CD16^{+}$), and non-classical ($CD14^{dim}CD16^{+}$) monocytes. Each population is also distinguished from one another by their functions in homeostasis and diseases. Upon activation, classical monocytes have high phagocytic and antimicrobial activity, but low production of inflammatory cytokines, whereas intermediate monocytes can produce high levels of proinflammatory cytokines, such as tumor necrosis factor (TNF), interleukin-1β (IL-1β), and IL-6. Non-classical monocytes are recruited into tissues after infection or injury and may contribute to tissue repair. In mice, blood monocytes have been classified into two major subsets with distinct phenotypes and nonoverlapping functions. One subset expresses high levels of Ly-6C and low levels of CX3CR1, referred to as Ly-$6C^{hi}$ inflammatory monocytes, are rapidly recruited to sites of infection and inflammation, and are functionally similar to human classical monocytes. Another subset expresses high levels of CX_3CR1 and low levels of CCR2 and Ly-6C, often called CX_3CR1^{hi} monocytes, and usually adhere to, migrate, and survey the luminal surface of small blood vessels in a process referred to as patrolling (Shi & Pamer, 2011).

Under homeostatic conditions and after an infection or tissue damage, macrophages can perform diverse functions depending on the stimuli they receive. During the development, growth, and renewal of healthy tissues of an organism, macrophages recognize, engulf, and eliminate apoptotic host cells before the dead cells can release their contents and induce inflammation. During an infection, macrophages are activated by proinflammatory PAMPs (such as LPS) or DAMPs and the cytokines interferon gamma (IFN-γ) and TNF to contain and destroy microbes using multiple cellular and biochemical mechanisms, such as ingestion of microbes by phagocytosis or opsonization via complement receptors or Fc receptors, formation and acidification of phagosomes to contain and damage the ingested microbes, fusion of phagosomes with lysosomes, production of reactive oxygen and nitrogen species in lysosomes that are toxic to the microbes, digestion of microbial proteins and carbohydrates by hydrolytic lysosomal enzymes, and the formation of granulomas which are structures that contain the microbes or indigestible foreign bodies surrounded by multinucleated giant cells resulting from the fusion of activated macrophages as a protective mechanism. Macrophages also ingest necrotic host cells and neutrophils at the site of infection or sterile tissue injury as part of the clean-up process. Besides these classical functions, macrophages can assume alternative functions upon receiving other stimuli to promote tissue remodeling and repair by stimulating angiogenesis and fibrosis. Macrophages also produce a myriad of cytokines and chemokines to recruit and stimulate other innate and adaptive immune cells and can serve as antigen-presenting cells (APCs) to promote the differentiation of effector T cells and support the production of antibodies by B cells.

2.2.1.3 Dendritic Cells

DCs are either tissue-resident or blood-derived cells that are widely distributed in lymphoid, mucosal, and organ parenchymal tissues. DCs have long membranous projections and phagocytic capabilities. Most DCs arise from a precursor that can also differentiate into monocytes. Maturation of DCs is dependent on the cytokine Flt3 ligand (Flt3L), which signals through the Flt3 tyrosine kinase receptor on the precursor cells. Langerhans cells (LCs), a type of DC in the epidermal layer of the skin, develop from embryonic precursors in the yolk sac or fetal liver early during development, and reside in the skin prior to birth. DCs sense the presence of antigens to initiate the innate immune defense. They are also efficient at capturing protein antigens of microbial origin, degrading these antigens, and presenting these antigens for recognition by T cells. All DCs express both major histocompatibility complex class I (MHC-I) and class II (MHC-II) molecules, which present antigens to $CD8^+$ and $CD4^+$ T cells, respectively. DCs are considered the most important professional APCs because of their unique capability of activating näive T cells. DCs express multiple classes of PRRs to recognize microbial molecules, and can upregulate MHC-II, costimulatory molecules, and certain components of the proteasome for more efficient antigen processing and presentation through a process called maturation. Many biologic drugs that are under development or in clinical use target various populations of DCs to treat cancer and autoimmune diseases (Saxena & Bhardwaj, 2018). DCs are also important in the development of adaptive immune responses to biologic drugs such as ADAs.

2.2.1.4 Eosinophils

Eosinophils are granulocytes that differentiate from bone marrow precursors after stimulation by GM-CSF, IL-3, and IL-5. Mature eosinophils circulate in the blood and are also present in the tissues, particularly respiratory, gastrointestinal, and genitourinary mucosas. Eosinophils are important for defense against extracellular helminths when they are triggered, such as via IgE or IgG Fc receptors, to release their granule contents that are toxic to helminths. However, after allergen exposure, eosinophils can participate in allergic inflammation by releasing allergic mediators and inflammatory cytokines and augmenting the function of basophils and mast cells. Eosinophils can also interact with and regulate the functions of many other adaptive immune cells, such as presenting antigens to promote T helper type 2 (T_H2) cell differentiation and supporting the differentiation and survival of plasma cells (Chu et al., 2011; Fulkerson & Rothenberg, 2018).

2.2.1.5 Basophils

Basophils are granulocytes that constitute less than 1% of the circulating leukocytes and are the rarest of these leukocytes. Basophils can develop from bone marrow precursors with the stimulation of IL-3 in humans and mice or with the stimulation of thymic stromal lymphopoietin (TSLP) in mice. They are normally rare in tissues but their numbers can increase by recruitment from the blood to some inflammatory sites. Basophils are capable of synthesizing many of the same mediators as mast cells. Like mast cells, basophils express the high-affinity IgE receptor FcεRI, bind IgE, and can be triggered by antigen binding to the cell surface IgE that is bound to FcεRI. Because basophil numbers are low in tissues, their importance in host defense and allergic reactions is not very clear. Recent studies have suggested that basophils may be important for the development of T_H2 immune responses triggered by allergens by being one of the initial cell types that release IL-4 (Sokol et al., 2009).

2.2.1.6 Mast Cells

Mast cells are bone marrow-derived cells present in the skin and mucosal epithelial tissues, which upon activation release many inflammatory mediators that defend against parasite infections, or cause symptoms of allergic disorders. The cytokine stem cell factor, or c-Kit ligand (KitL), is essential for mast cell development. Mature mast cells are not normally found in the circulation but are present in tissues. Their cytoplasm contains numerous membrane-bound granules filled with preformed inflammatory mediators, such as histamine, and acidic proteoglycans that bind basic dyes, giving a dark blue color to the granules when special stains are used. Various stimuli can activate mast cells to release their cytoplasmic granule contents into the extracellular space, as well as to synthesize and release cytokines and other inflammatory mediators. The histamine and other mediators promote changes in the blood vessels that cause inflammation. Mast cells express the high-affinity IgE receptor FcεRI and are usually coated with IgE antibodies that are bound to the cell surface FcεRI. When the antibodies on the mast cell surface bind antigen, signaling events are induced that lead to mast cell activation. Mast cells are also activated when they recognize microbial products, independently of IgE, and function as tissue sentinels of the innate immune system or immunoregulatory cells by secreting regulatory cytokines.

2.2.1.7 Innate Lymphoid Cells

Innate lymphoid cells (ILCs) are lymphocytes that develop from the common lymphoid progenitor but do not express the diversified and clonotypic antigen receptors through somatic recombination as T cells or B cells do. Instead, they carry germline-encoded receptors, a characteristic feature of the innate immune system, but lack the expression of cell-surface molecules that identify other differentiated immune cell types, and so are defined as cell lineage marker-negative (Lin^-) cells. Under physiological conditions, ILCs are largely tissue-resident cells that are intimately involved in the homeostasis of the tissues (Gasteiger et al., 2015). ILCs comprise the cytotoxic NK cells and three distinct groups of non-cytotoxic cells: group 1 ILCs (ILC1s), group 2 ILCs (ILC2s), and group 3 ILCs (ILC3s) including lymphoid tissue inducer cells. The non-cytotoxic ILC subsets are defined based on their differential requirements for transcription factors during development, their patterns of expression of effector cytokines, and the acquisition of distinct effector functions. Functionally, ILCs are the innate counterparts of T lymphocytes, while NK cells functionally resemble $CD8^+$ cytotoxic T cells, and ILC1s, ILC2s, and ILC3s functionally resemble $CD4^+$ T_H1, T_H2, and T_H17 cells, respectively (Artis & Spits, 2015; Vivier et al., 2018). ILC1s are capable of producing IFN-γ and TNF, and have been implicated in immunity against intracellular bacteria, parasites, and tumors. ILC2s can produce T_H2 cytokines, such as IL-4, IL-5, IL-9, IL-13, and/or amphiregulin, and promote type 2 inflammation for immune defense against helminths, allergic inflammation, and tissue repair. Depending on the stimulus, ILC3s produce IL-2, IL-17A, IL-17F, IL-22, GM-CSF, and/or TNF, and can promote immunoregulation, antibacterial immunity, chronic inflammation, or tissue repair. Although similar to T helper cell subsets in cytokine expression and effector function, various subsets of ILCs perform their functions in the absence of antigenic specificity.

2.2.1.8 Splenic Marginal Zone B Cells

Spleen filters blood to remove old blood cells but it is also an important immune organ and is exposed to all the circulating antigens and pathogens. Splenic marginal zone B cells (MZBs) are cells located as sentinels at the interface between the circulation and lymphoid tissue, which enables them to rapidly respond to blood-borne antigens. To be fit for this function, MZBs are highly mobile and exhibit long membrane extensions similar to those of DCs to facilitate the capturing of blood-borne antigens and then transferring them to the inner splenic follicles (Arnon et al., 2013). In addition, unlike conventional B cells, which are a major component of the adaptive immune system and express specific antigen receptors encoded by highly diversified V(D)J genes, MZBs express less specific B cell receptors (BCRs) encoded by semi-invariant or poorly diversified V(D)J genes that recognize multiple highly conserved microbial determinants (Cerutti et al., 2013). For this reason, MZBs are deemed "innate-like" lymphocytes. In addition, similarly to DCs, macrophages, and granulocytes, MZBs express high levels of TLRs, which enables them to cross over the conventional boundaries between the innate and adaptive immune systems. Interestingly, engagement of BCR and TLR simultaneously by conserved microbial molecules such as LPS stimulates MZB cells to quickly secrete low-affinity but the first wave of Abs to constrain blood-borne antigens and pathogens, which is critical to cover the 2–3 weeks of gap required for the induction of high-affinity antibody production by conventional follicular B cells (Bendelac et al., 2001; Martin et al., 2001).

2.3 Adaptive Immunity to Biologic Drugs in Local Immune Organs

The adaptive immune system consists of B and T cells. Both B cell-dependent humoral immunity and T cell-dependent cell-mediated immunity have been found to develop against various biologic drugs, depending on the drug's formulation, route of administration, dosage, dosing schedule, the patient's genetic and immunological conditions, concomitant medication, and factors that are not well understood. Presence of product-related impurities or contaminants with adjuvant properties, such as host cell proteins, endotoxins, or oligonucleotides, may lead to the activation of otherwise self-tolerant host T and B cells, leading to the differentiation of drug-specific T cells and B cells and the subsequent production of low- or high-affinity ADAs.

Secondary lymphoid organs, including the lymph nodes, spleen, and components of the mucosal immune system, are where the responses of adaptive lymphocytes, including B cells and T cells, to administrated biologic drugs are initiated and developed. These organs are anatomically organized in ways that optimize the cellular interactions necessary for initiation of adaptive immune responses. In these organs, lymphocytes and APCs are localized and concentrated in anatomically defined areas that are also the sites where foreign antigens are transported and concentrated. This ensures that antigens and antigen-specific näive lymphocytes localize to the same regions so that adaptive immune responses can be initiated.

2.3.1 B Cells

B cells express clonotypic BCRs, which are cell surface immunoglobulin (Ig) molecules, with a single specificity that is distinct from the specificities of the receptors in all other clones. Millions of B cell clones exist in the body and they can undergo further diversification upon encountering antigens in secondary lymphoid organs, enabling the host to recognize and respond to a very large number of foreign antigens. B cells develop in the bone marrow from common lymphoid progenitors that also give rise to T cells and ILCs. Bone marrow differentiating pre-B cells first undergo somatic V(D)J recombination to rearrange their Ig variable region genes and express the cell surface IgM receptor. These pre-B cells then undergo a process of central tolerance to remove self-reactive clones by apoptosis or to render the self-reactive clones anergic (actively unresponsive) to antigens, thereby generating a primary repertoire of B cells. The pre-B cells differentiate into immature B cells with the coexpression of IgM and IgD on the cell surface. They then emigrate out of the bone marrow, maturate, and populate the cortex region of peripheral lymph nodes, follicular regions of the white pulp of the spleen, or marginal zone at the border of the white and red pulp of the spleen, to become various subsets of peripheral mature B cells, such as conventional follicular B cells, B-1 cells, and MZB cells. B-1 cells are found mainly in mucosal tissues and the peritoneal and pleural cavities, and MZB cells are present mainly in the spleen, whereas follicular B cells are the most numerous type of B cells in the body which are found in lymphoid tissues and the blood. In secondary lymphoid organs, B cells encounter cognate antigens through their cell surface IgM and/or IgD receptors, and can undergo limited somatic hypermutation (SHM) and Ig heavy chain class-switch recombination (CSR) with the help of innate immune signals, such as those from DCs, macrophages, and epithelial cells engaging PRRs and the BCRs, and differentiate into short-lived IgM-, IgG-, or IgA-plasmablasts or plasma cells, or undergo extensive somatic SHM and CSR in the germinal centers (GCs) with the help of activated T_H cells, including follicular helper T (T_{FH}) cells that specialize in promoting T cell-dependent B cell antibody diversification and production in the GCs, and differentiate into long-lived IgG-, IgA-, or IgE-plasma cells to provide high-affinity, long-lasting humoral immunity or differentiate into memory B cells. Follicular B cells give rise to most of the high-affinity antibodies and memory B cells that confer protection from repeated infections by the same microbes. In contrast, B-1 and MZB cells make up a minority of B cells and produce antibodies with limited diversity.

2.3.1.1 Antibody and Ab Isotypes

While antibody specificity is determined by V(D)J recombination and SHM, antibody function is mainly determined by its heavy chains. There are five classes of antibodies: IgM, IgD, IgG, IgA, and IgE. These classes have different binding preference to variant Fc receptors (FcRs) on different cell types (Table 2.1). Almost all approved therapeutic mAbs are of the IgG class. There are four subclasses (isotypes) of IgG in humans: IgG1, IgG2, IgG3, and IgG4, with different functions (Table 2.2) (French et al., 2013).

Table 2.1 Antibody isotypes and functions.

Receptor/ ligand	IgG1	IgG3	IgG2	IgG4
C1			++	+++
FcγRI			+++	+++
FcγRII	+	?	++	
FcγRIIIa/b			+	+
FcRn		+	+	+
	Ligation of all Fc receptors, including FcγRI in monomeric form (IgG3 > IgG1)		Restricted ligation of Fc receptors and only when complexed, particularly large complexes	Restricted ligation of Fc receptors and only when complexed, particularly large complexes
	Potent complement activation through the classical pathway (IgG3 > IgG1)		Weak complement activation	No complement activation
			Most resistant of all IgG isotypes to proteolytic degradation	Produced after chronic immune stimulation, particularly parasite infections
			Predominant IgG subclass in plasma IgM–IgG complexes	Regulated similarly to IgE
			Only IgG subclass to undergo covalent dimerization	May form bispecific antibodies due to Fab arm (half molecule) exchange with other IgG4 molecules in vivo
			Predominant IgG subclass in phagocytic antibodies to polysaccharide antigens	

Based on the therapeutic purpose for either depleting (antibody–drug conjugate, rituximab), activating (anti-CD27), or inhibitory (Keytruda) the targets/targeting cells, different IgG subclasses with modifications of either the amino acid sequence or the glycosylation are preferred. As shown in Figure 2.1A (adapted from Beers et al., 2016), depleting mAb binds to the target and recruits cytotoxic immune cells such as NK cells, neutrophils, or macrophages expressing activating FcγRs (FcγRIIA or FcγRIIIA) to induce antibody-dependent cellular cytotoxicity to kill the targeted tumor cells. Inhibitory FcγRIIB, on the other hand, prevents this killing process. Hence, Fc

Table 2.2 Ligand binding/activation and different function of an IgG subisotype antibody response (adopted and modified from French et al., 2013).

Function	IgM	IgD	IgA1, IgA2	IgG1, 2, 3, 4	IgE
	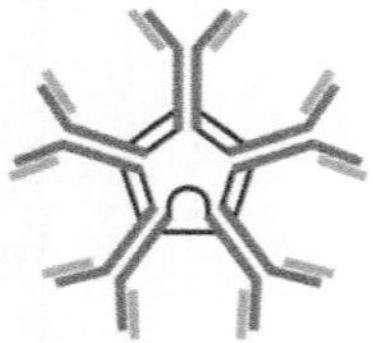	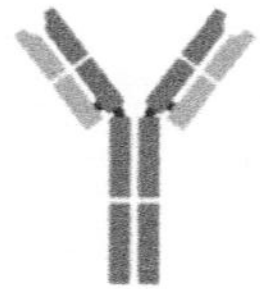	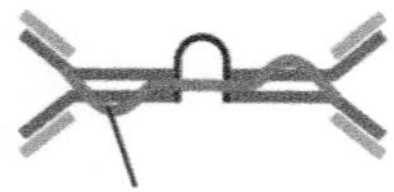	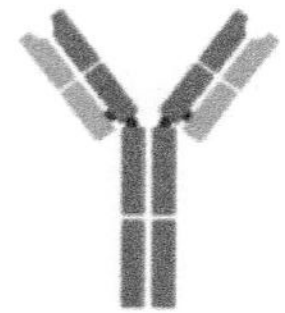	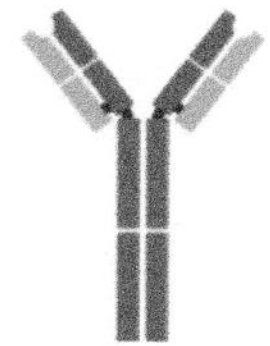
Ag binding sites (n)	10	2	4	2	2
Crosses placenta	No	No	No	Yes	No
Fixes complement	Yes	No	No	Yes	No
FcRs					
Percentage of total Ab in serum (%)	6	1	13	80	0.002
Biological function	B cell receptor; dominant Ab of first wave response; best at complement fixation	B cell receptor; activates basophils and mast cells	Secreted into mucus such as saliva, tears, breast milk, and intestine; protects against mucus pathogens	Major Ab in serum; neutralizes toxin, virus, opsonization	Antiparasitic activity; involved in allergic response

modification G236A, S239D, 330L, and I332E to increase the activating:inhibitory FcγR binding ratio have been identified and some of them are currently in clinical trials (Kellner et al., 2013; Lazar et al., 2006). Similarly, glycol-engineered antibodies lacking core fucose residue from oligosaccharides attached at Asn297 of the Fc have significantly enhanced binding affinity to FcγRIIIA and are more potent and efficacious at mediating antibody-dependent cellular cytotoxicity (Shields et al., 2002; Shinkawa et al., 2003).

Agonistic mAbs are designed to bind to cell surface targets and enhance cell activation. This requires either clustering through crosslinking by FcγRIIB on targeted cells with enhancers such as the SE/LF mutation in IgG1 or by using IgG2 subtypes which uniquely undergo covalent dimerization (bottom) (Figure 2.1B). Lastly, blocking mAbs such as Keytruda to block PD-1/PDL1-mediated immune suppression prefer to have isotypes with minimal FcγR binding, such as hIgG4 or "FcγR null" mutant to prevent FcγR engagement (Figure 2.1C). However, due to the unique ability of human IgG4 to undergo Fab swap (half molecule exchange with other IgG4 molecules) in vivo, which can give rise to bispecific IgG4 Abs, a S228P point mutation at hinge modification has been introduced to both Keytruda and Opdiva to prevent such exchange between Fab half molecules (Yang et al., 2015). All these mutations in IgGs enable enhanced functions; however, they may also lead to increased immunogenicity of these therapeutics.

2.3.2 T Cells

Committed T cell precursors develop in the bone marrow and migrate to the thymic cortex for further development. Within the thymus, T cell receptor (TCR) complexes are expressed during early intrathymic T cell maturation after the formation of functional TCR genes by somatic V(D)J rearrangement at the TCR loci. At this time, the specification of γδ versus αβ T cells occurs. In addition, cell surface expression of a variety of accessory molecules occurs during thymic maturation. Some of these accessory molecules, such as CD4 and CD8, play important roles in T cell maturation, specifying the $CD4^+$ helper versus the $CD8^+$ cytotoxic T cell lineage, and the activation of mature T cells. After different receptors are expressed on the surface of different clones of developing T cells, the primary repertoire of T cells is modified or shaped by two related selection processes. Positive selection is the process by which the T cell repertoire becomes self-MHC-restricted. This ensures that only T cells expressing TCRs capable of recognizing antigens presented by the individual's own APCs will be permitted to survive and mature. A second process, called negative selection, gives rise to central tolerance by eliminating or inactivating potentially autoreactive T cell clones, especially those specific for antigens that are presented at high concentrations in the thymus by self-APCs, ensuring that

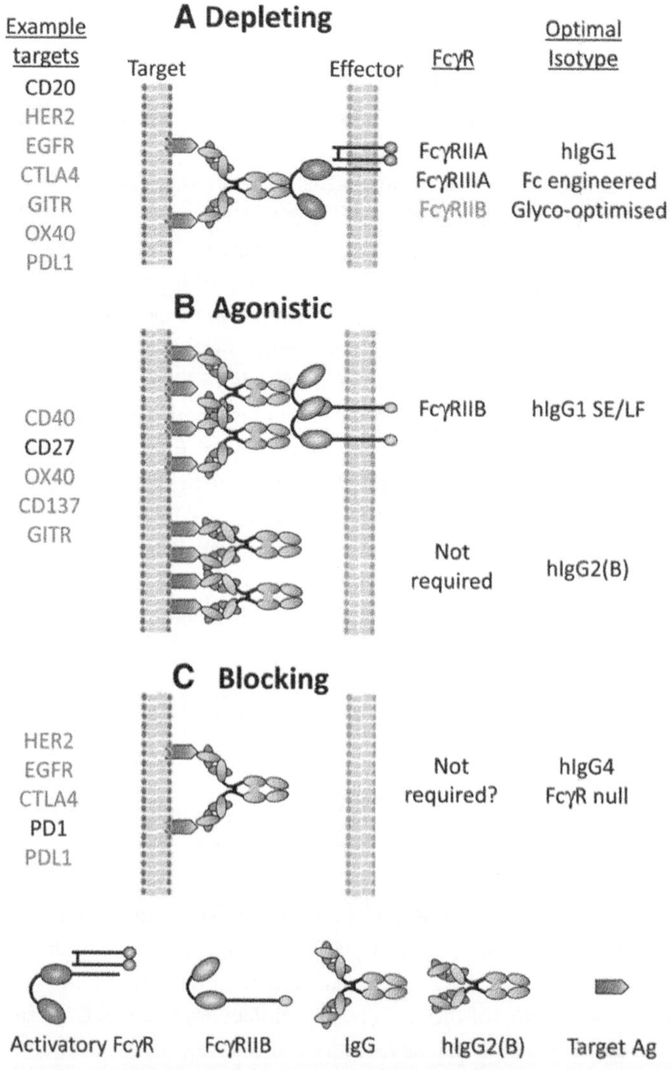

Figure 2.1 Functional therapeutic monoclonal antibody types and role of IgG isotype and FcγR interactions (from Beers SA, Glennie MJ, and White AL. Influence of immunoglobin isotype on therapeutic antibody function. *Blood* 2016; 127: 1097–1101).

an individual is tolerant to these self-antigens. In addition, during this process, some autoreactive T cell clones are converted to regulatory T cells that suppress immune responses to self-antigens. Therefore, the selective growth or death of different cells results in the formation of a foreign-antigen-specific, self-MHC-restricted, self-antigen-tolerant mature T cell repertoire.

2.3.3 Activation of the Adaptive Immune Response to Biologics in Secondary Lymphoid Organs

The development of some ADAs occurs via T cell-independent B cell responses that are directly triggered by innate immune cells, such as DCs, mononuclear phagocytes, and granulocytes. In this case, the administered biologic drug could bind to cognate B cells that exist due to incomplete central tolerance or stimulate the breakdown of B cell peripheral tolerance by inducing the production of various B cell-activating molecules that lead to the activation of anergic B cells specific to the biologic drug. In some instances, such as IFN-β supplementation therapy, the biologic drug itself is a strong B cell activator. This could lead to the direct production of ADAs with limited affinity maturation and class switching.

The great majority of the development of ADAs occurs via T cell-dependent B cell responses that take place in secondary lymphoid organs. Upon migration out of the thymus and into peripheral secondary lymphoid organs, mature näive T cells populate the paracortex region of the lymph node and the white pulp of the spleen, also called the T cell zone. For tissue-delivered antigens, such as those biologic drugs administered by the subcutaneous (SC) or intradermal (ID) route, professional APCs, such as DCs, that have taken up the antigens from the tissues, will migrate to the paracortex region of the tissue-draining lymph nodes and present antigens on the MHC-I complex to cognate $CD8^+$ T cells and/or MHC-II complex to cognate $CD4^+$ T cells if these antigen-specific T cell are present. With the additional help of APC-derived costimulatory molecules, such as CD80 and CD86 that engage CD28 on the T cells, as well as cytokines, the $CD4^+$ T cells will be activated by the APCs, including expression of the molecule IL-2 and CD40 ligand (CD40L), and migrate towards the cortical region where the T cell–B cell follicle border is located. Some activated $CD4^+$ T cells migrate into the B cell follicles and differentiate into T_{FH} cells that specialize in supporting the GC reaction of antibody diversification and affinity maturation by B cells. At the T cell–B cell follicle border, activated T cells will engage the MHC-II complex presenting peptides derived from the biologic drug on the surface of the B cells via the TCR if the T cell is specific to the peptide–MHC-II complex, and additionally through CD40L to CD40 on the surface of B cells. B cell clones with BCRs that are specific to the administered biologic drug and that are presenting the drug-derived peptides to the specific $CD4^+$ T cells can stably interact and be activated by the $CD4^+$ T cells. Some of the activated and proliferating B cells become short-lived plasma cells and form extrafollicular foci, and some other B cells undergo rapid proliferation to form the GC. The GC can further be divided into two zones: the light zone and dark zone. In the dark zone, B cells undergo rapid clonal expansion accompanied with high frequency of somatic hypermutation in BCRs to introduce more antibody diversification. Expanded B cells will relocate to the light zone, where follicular dendritic cells (FDCs) coated with immune complex will select the new clones with the highest BCR binding affinity through binding competition. The clones with the highest binding affinity will then differentiate into either memory B cells or long-lived plasma cells. T_{FH} cells that express B cell-help factors including CD40L, BAFF, and IL-21 are also enriched in the GC light zone to help the selection of high-affinity B cell clones (Kurosaki et al., 2015) (Figure 2.2).

2.4 Administration of Biotherapeutics

There are multiple administration routes for biotherapeutics, such as intravenous (IV), SC, intramuscular (IM), ID, oral (OR), intratumor, and intravitreal. IV delivery has traditionally been the standard route of administration for biologic drugs, such as antibodies or recombinant hormones and cytokines. However, there are major problems associated with IV delivery of

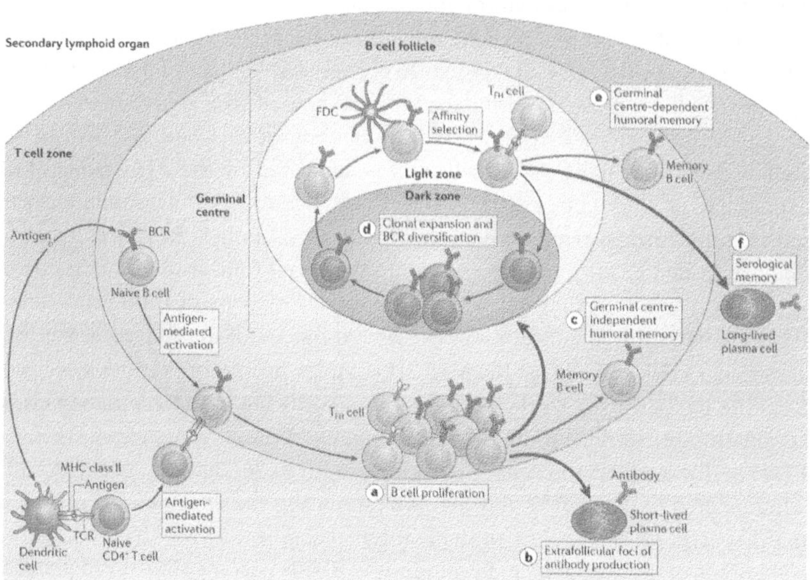

Figure 2.2 B cell and antibody response in germinal center (from Kurosaki et al., 2015).

biologics, such as systemic toxicity, rapid clearance, adverse events due to high peak drug dose, and even catheter-mediated infection. Alternative administration methods such as SC, ID, or OR provide advantages by creating a relative depot effect at the site of injection/administration, prolonging the time for the biologic drug to drain into the circulation, and increasing the half-life of the drug while simultaneously lowering the peak drug level in various body compartments and reducing the frequency of dosing. In particular, there is an increasing tendency to use the SC route due to several compelling advantages, such as patient comfort, potential for self-administration, better therapy adherence, and reduced health care cost. Interestingly, most of the vaccines are historically given via the SC route and it is generally accepted that SC administration leads to a higher immunogenicity compared to IV administration for a given vaccine (Cook, 2008). However, due to the differences between vaccines and biologics, mainly the use of adjuvants in vaccine to deliberately enhance immune response against vaccine, and the self-mimic nature of many of biologics to intentionally reduce the immune response against biologics (Fathallah et al., 2013), the higher immunogenicity of vaccines through the SC route may not be the case for biologics, as shown by recent evidence (Hamuro et al., 2017; Schiff et al., 2014). In the next section we detail the translocation route and immune cell interaction with SC and IV administration.

2.4.1 SC Route of Administration

Skin as the largest organ in the body has a critical role in immune defense in addition to its function as a physical barrier. There are three layers in the skin, each with different structures and critical immune cells. The most superficial layer is the epidermis, a protective layer containing keratinocytes and a special subset of DCs, the LCs. As shown in Figure 2.3 (Jarvi & Balu-Iyer, 2021),

LCs form a dense immunosurveillance network in the epidermis. Due to their frontline location and active surveillance function, any injected foreign antigens including biotherapeutics will be taken up and processed by LCs, which then translocate from the epidermis to regional lymph nodes (LNs) and initiate systemic immune responses (Jakob & Udey, 1999; Jakob et al., 2001). The middle layer is the dermis in which reside dermis DCs (DDCs), which have a distinct function from that of LCs in the epidermis (Nagao et al., 2009). A porous membrane structure separates the dermis and epidermis which supports the vascular network for nutrient supply to the epidermis. The last layer below the dermis is called the hypodermis, also referred to as

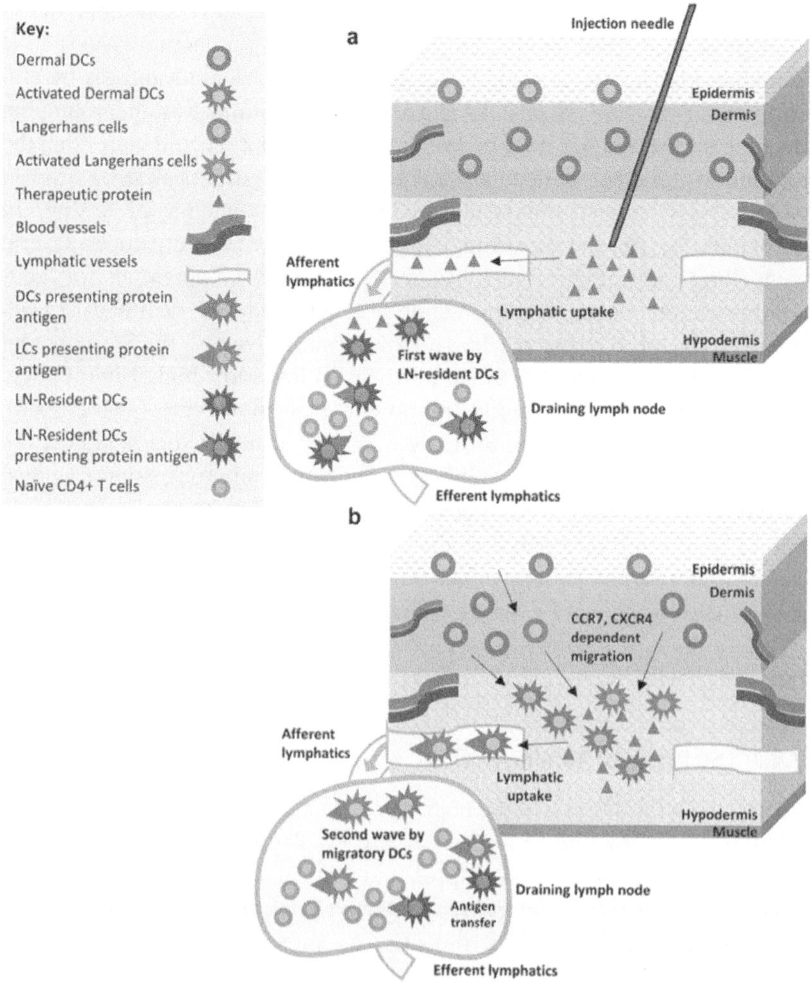

Figure 2.3 Schematic representation of the subcutaneous (SC) space and mechanism involved in immunogenic response against SC-administered therapeutic proteins. DC, dendritic cell; LC, Langerhans cell (Jarvi & Balu-Iyer, 2021).

SC connective tissue. While blood capillaries as well as nerve endings are enriched in the upper dermis layer, larger blood vessels and lymphatic networks facilitating immune cell transportation and fluid homeostasis are mainly located between the dermis and hypodermis. Since the hypodermis has a less compact cell density, the majority of the SC administrated fluids containing biotherapeutics will be received here. From here, biotherapeutics are further taken up and distributed to the systemic circulation through two routes: lower molecular weight drugs (<16–20 kDa), including small molecules and small proteins, can be absorbed directly across the endothelial cells of blood capillaries, whereas larger molecular weight protein drugs (>20 kDa), including mAbs and many recombinant proteins, are expected to traffic primarily through the peripheral lymphatic system before entering the general circulation. Either way, biotherapeutics arrive within hours to LNs where LN-resident DCs process and subsequently present to T cells. This is the first wave of Ag-presentation after SC injection. The second wave of Ag-presentation happens much later, within 1–3 days, involving activation and migration of cutaneous DCs. SC injection devices cause tissue damage, and exposure of danger signals, which activates the innate immune responses and induces inflammatory cytokines such as IL-1β and TNFα (see later), which (1) triggers migration of LCs and DDCs to the hypodermis area through upregulation of chemoattractant receptors and (2) upregulates local vascular endothelial growth factor C (VEGF-C) production and induces expansion of local lymphatic vessels. It is expected that the extent of physical tissue damage is correlated with the parameters of the injection device such as needle gauge, needle length, and injection depth/angle, with bigger gauge and deeper injection causing more tissue damage and exposure of danger signals (Liu et al., 2005), which in turns induces a stronger immune response. However, the link between injection site reactions and impact on incidence of immunogenicity is missing due to lack of studies. There was a single phase III study reporting that anti-trastuzumab ADAs in the Herceptin SC (HannaH study) were not associated with increased administration-related reactions in the IV group (Ismael et al., 2012). A better understanding of the mechanisms that regulate danger signal exposure, proinflammatory cytokine production, as well LC and DDC trafficking might offer new opportunities for therapeutic interventions to suppress cutaneous immune responses against biotherapeutics.

One major challenge for SC administration is minimizing dosing volume and maximizing dosing concentration, which in turn increases the chance of protein aggregation. Aggregated proteins tend to expose neoepitopes and be retained longer at the injection site as well as draining LNs upon SC administration, thus increasing immune responses. Hence many substances referred as "excipients" are introduced to SC formulations to increase their solubility and stability. Among them, polysorbates are used in most of the biotherapeutics to prevent aggregation. However, autoxidization of polysorbates leads to a variety of side products such as aldehyde and peroxides, which in turn modify the side chains of various amino acids in biotherapeutics, which could generate new epitopes/structures to either directly activate DCs (Mueller et al., 2009) or stimulate the T cell/B cell response (Kishore et al., 2011). In addition, zinc ion (Zn^{2+}) is commonly used in SC-administered peptides due to its ability to reduce aggregation in the vial (Yu et al., 2013). However, once injected, neutral pH in the injection site may facilitate peptide precipitation and Zn^{2+} may increase the immune response as an adjuvant (Matsumura et al., 2010). Thus, components of formulation should be tested not just for in-vial stability but also for stability/immune response after SC administration.

To really accommodate large SC dosing volume and reduce its accompanying uncomfortableness such as tightness and pain, recombinant human hyaluronidase (rHuPH20) enzyme is used to temporarily break down hyaluronan, the gel-like, nonstructural component of the SC extracellular matrix, to facilitate dispersion of subcutaneously delivered fluids and drugs (Bookbinder et al., 2006). Although studies have demonstrated overall low immunogenicity to either rHuPH20 or coadministered biotherapeutics such as trastuzumab, and there was no clinical relevance in terms of safety, efficacy, and trough concentration (Ctrough) level (Ismael et al., 2012), there has been no detailed report regarding the impact of rHuPH20 on local proinflammatory cytokine production and cutaneous DC mobilization/activation.

2.4.2 IV Route of Administration

Unlike the SC route, biotherapeutics through the IV route enter the systemic circulation directly, where they first encounter the mononuclear phagocyte system (MPS), previously known as the reticuloendothelial system (Domínguez-Andrés et al., 2019), circulating immune cells and soluble factors, and are quickly enriched in liver and spleen. While the liver is mainly an immune-privileged organ to detoxify, the spleen is a large secondary immune organ. In contrast to the SC route, where DCs are the primary cells for Ag presenting, Zhong and colleagues demonstrated that splenic non-Ag-specific B cells or so-called MZB cells are the first wave (around 4 hours) and the primary cell types that engage in Ag processing and presentation in the IV route (Zhong et al., 1997). This was followed by presenting by DCs after 24 hours. Interestingly, the authors implied that Ag presenting through non-specific B cells in the absence of contaminating endotoxin may contribute to the induction of high-dose T cell tolerance or immune deviation instead of immune activation. This may explain why despite having high Ctrough levels via the IV route, biotherapeutics may not necessarily have higher immunogenicity than the SC route (Hamuro et al., 2017; Schiff et al., 2014).

2.5 Detrimental Consequences of ADAs

The unwanted development of ADAs against biologic drugs is a major factor that influences the pharmacokinetic and pharmacodynamic properties as well as the efficacy and side effects of the drug. ADAs that bind to the biologic drugs (binding ADAs) can, in some instances, cause the formation of local and/or systemic immune complexes, and through interaction with Fc receptors and/or phagocytic cells affect the distribution, effective dose, effective exposure time, and clearance of the drug and can cause hypersensitivity reactions to the drug. Development of the ADAs in this instance results in a measurable decrease in the circulating levels of the drug after the induction of the ADAs and in increased clearance of the drug from the circulation. Such an effect ultimately results in lowered efficacy, as the drug is now not as available as designed at the pharmacologic site for the length of time required to mediate its biologic effect. Among the binding ADAs there are neutralizing ADAs that can directly interfere with the biologic activity of the drug by blocking the interaction of the drug with its receptors or ligands. In addition, when a biologic drug supplements an endogenous and physiologically important but deficient homolog molecule, binding or neutralizing ADAs can target the endogenous homolog, with unforeseeable consequences for physiological function even long after treatment cessation (Sominanda et al., 2010).

2.6 Summary

There are lots of differences depending on the routes of administration: drug concentration, Ctrough drug level, speed of absorption and biodistribution; formulation components which have differing impacts on immune cells and the immune response; immune cells encountered, and immune organ involved, as summarized in Table 2.3 (Hamuro et al., 2017). It is possible that an intervention can be introduced at some of the steps, which in turn will modify the downstream immune response against the biotherapeutics to achieve the best efficacy, for example, by using a less invasive injection device or inhibiting cutaneous DC mobility/activity.

Table 2.3 Comparison of the subcutaneous and intravenous routes (modified from Hamuro et al., 2017).

Consideration	Subcutaneous	Intravenous
Drug absorption	Potential for depot formation	Rapid dilution
Drug concentration:		
(1) Injection site	High	Low
(2) Systemic	Low	High
Formulation impact	Moderate to high	Minimal to low
First pass of lymphatic exposure	<16–20 kDa systemic before lymphatic >16–20 kDa lymphatic before systemic	Systemic before lymphatic
First pass of immune cell interactions	Langerhans cells in epidermis/dermis Dermis dendritic cells B cells in lymph nodes	Kupffer cells in liver marginal zone B cells in spleen
Other interactions	Extracellular matrix, collagen, lymphatic venules, nodes, target binding, FcRn uptake/recycling at injection site	Systemic proteins (soluble target, preexisting monoclonal antibodies, other circulating binding partners), circulating immune cells
First-pass absorption (listed in chronological order)	Injection site Target at injection site Lymphatic Target site Systemic	Systemic reticuloendothelial system in liver and spleen Lymphatic Target site

References

Arnon, T. I., Horton, R. M., Grigorova, I. L., & Cyster, J. G. (2013). Visualization of splenic marginal zone B-cell shuttling and follicular B-cell egress. *Nature*, 493(7434), 684–688. https://doi.org/10.1038/nature11738

Artis, D., & Spits, H. (2015). The biology of innate lymphoid cells. *Nature*, 517(7534), 293–301. https://doi.org/10.1038/nature14189

Barton, G. M., & Kagan, J. C. (2009). A cell biological view of Toll-like receptor function: Regulation through compartmentalization. *Nature Reviews Immunology*, 9(8), 535–542. https://doi.org/10.1038/nri2587

Beers, S. A., Glennie, M. J., & White, A. L. (2016). Influence of immunoglobulin isotype on therapeutic antibody function. *Blood*, 127(9), 1097–1101. https://doi.org/10.1182/blood-2015-09-625343

Bendelac, A., Bonneville, M., & Kearney, J. F. (2001). Autoreactivity by design: Innate B and T lymphocytes. *Nature Reviews Immunology*, 1(3), 177–186. https://doi.org/10.1038/35105052

Bookbinder, L. H., Hofer, A., Haller, M. F., Zepeda, M. L., Keller, G. A., Lim, J. E., Edgington, T. S., Shepard, H. M., Patton, J. S., & Frost, G. I. (2006). A recombinant human enzyme for enhanced interstitial transport of therapeutics. *Journal of Controlled Release*, 114(2), 230–241. https://doi.org/10.1016/j.jconrel.2006.05.027

Boraschi, D., & Italiani, P. (2018). Innate immune memory: Time for adopting a correct terminology. *Frontiers in Immunology*, 9, 799. https://doi.org/10.3389/fimmu.2018.00799

Cerutti, A., Cols, M., & Puga, I. (2013). Marginal zone B cells: Virtues of innate-like antibody-producing lymphocytes. *Nature Reviews Immunology*, 13(2), 118–132. https://doi.org/10.1038/nri3383

Chu, V. T., Frohlich, A., Steinhauser, G., Scheel, T., Roch, T., Fillatreau, S., Lee, J. J., Lohning, M., & Berek, C. (2011). Eosinophils are required for the maintenance of plasma cells in the bone marrow. *Nature Immunology*, 12(2), 151–159. https://doi.org/10.1038/ni.1981

Cook, I. F. (2008). Evidence based route of administration of vaccines. *Human Vaccines*, 4(1), 67–73. https://doi.org/10.4161/hv.4.1.4747

Demaria, O., Cornen, S., Daeron, M., Morel, Y., Medzhitov, R., & Vivier, E. (2019). Harnessing innate immunity in cancer therapy. *Nature*, 574(7776), 45–56. https://doi.org/10.1038/s41586-019-1593-5

Domínguez-Andrés, J., Novakovic, B., Li, Y., Scicluna, B. P., Gresnigt, M. S., Arts, R. J. W., Oosting, M., Moorlag, S., Groh, L. A., Zwaag, J., Koch, R. M., Ter Horst, R., Joosten, L. A. B., Wijmenga, C., Michelucci, A., Van Der Poll, T., Kox, M., Pickkers, P., Kumar, V., Stunnenberg, H., & Netea, M. G. (2019). The itaconate pathway is a central regulatory node linking innate immune tolerance and trained immunity. *Cell Metabolism*, 29(1), 211–220 e5. https://doi.org/10.1016/j.cmet.2018.09.003

Fathallah, A. M., Bankert, R. B., & Balu-Iyer, S. V. (2013). Immunogenicity of subcutaneously administered therapeutic proteins–A mechanistic perspective. *The AAPS Journal*, 15(4), 897–900. https://doi.org/10.1208/s12248-013-9510-6

French, M. A., Abudulai, L. N., & Fernandez, S. (2013). Isotype diversification of IgG antibodies to HIV Gag proteins as a therapeutic vaccination strategy for HIV infection. *Vaccines (Basel)*, 1(3), 328–342. https://doi.org/10.3390/vaccines1030328

Fulkerson, P. C., & Rothenberg, M. E. (2018). Eosinophil development, disease involvement, and therapeutic suppression. *Advances in Immunology*, 138, 1–34. https://doi.org/10.1016/bs.ai.2018.03.001

Gasteiger, G., Fan, X., Dikiy, S., Lee, S. Y., & Rudensky, A. Y. (2015). Tissue residency of innate lymphoid cells in lymphoid and nonlymphoid organs. *Science*, 350(6263), 981–985. https://doi.org/10.1126/science.aac9593

Hamuro, L., Kijanka, G., Kinderman, F., Kropshofer, H., Bu, D. X., Zepeda, M., & Jawa, V. (2017). Perspectives on subcutaneous route of administration as an immunogenicity risk factor for therapeutic proteins. *Journal of Pharmaceutical Sciences*, 106(10), 2946–2954. https://doi.org/10.1016/j.xphs.2017.05.030

Ismael, G., Hegg, R., Muehlbauer, S., Heinzmann, D., Lum, B., Kim, S. B., Pienkowski, T., Lichinitser, M., Semiglazov, V., Melichar, B., & Jackisch, C. (2012). Subcutaneous versus intravenous administration of (neo)adjuvant trastuzumab in patients with HER2-positive, clinical stage I-III breast cancer (HannaH study): A phase 3, open-label, multicentre, randomised trial. *The Lancet Oncology*, 13(9), 869–878. https://doi.org/10.1016/S1470-2045(12)70329-7

Iwasaki, A., & Medzhitov, R. (2015). Control of adaptive immunity by the innate immune system. *Nature Immunology*, 16(4), 343–353. https://doi.org/10.1038/ni.3123

Jakob, T., & Udey, M. C. (1999). Epidermal Langerhans cells: From neurons to nature's adjuvants. *Advances in Dermatology*, 14(5), 209–258. discussion 259. https://doi.org/10.1067/mai.2001.118797

Jakob, T., Ring, J., & Udey, M. C. (2001). Multistep navigation of Langerhans/dendritic cells in and out of the skin. *The Journal of Allergy and Clinical Immunology*, 108, 688–696. https://www.ncbi.nlm.nih.gov/pubmed/10643500

JARVI, N. L. & BALU-IYER, S. V. 2021. Immunogenicity challenges associated with subcutaneous delivery of therapeutic proteins. *BioDrugs*, 35, 2, 125–146. DOI: 10.1007/s40259-020-00465-4

Kawai, T., & Akira, S. (2010). The role of pattern-recognition receptors in innate immunity: Update on Toll-like receptors. *Nature Immunology*, 11(5), 373–384. https://doi.org/10.1038/ni.1863

Kellner, C., Zhukovsky, E. A., Potzke, A., Bruggemann, M., Schrauder, A., Schrappe, M., Kneba, M., Repp, R., Humpe, A., Gramatzki, M., & Peipp, M. (2013). The Fc-engineered CD19 antibody MOR208 (XmAb5574) induces natural killer cell-mediated lysis of acute lymphoblastic leukemia cells from pediatric and adult patients. *Leukemia*, 27(7), 1595–1598. https://doi.org/10.1038/leu.2012.373

Kishore, R. S., Kiese, S., Fischer, S., Pappenberger, A., Grauschopf, U., & Mahler, H. C. (2011). The degradation of polysorbates 20 and 80 and its potential impact on the stability of biotherapeutics. *Pharmaceutical Research*, 28(5), 1194–1210. https://doi.org/10.1007/s11095-011-0385-x

Kurosaki, T., Kometani, K., & Ise, W. (2015). Memory B cells. *Nature Reviews Immunology*, 15(3), 149–159. https://doi.org/10.1038/nri3802

Lanier, L. L. (2013). Shades of grey–The blurring view of innate and adaptive immunity. *Nature Reviews Immunology*, 13(2), 73–74. https://doi.org/10.1038/nri3389

Lazar, G. A., Dang, W., Karki, S., Vafa, O., Peng, J. S., Hyun, L., Chan, C., Chung, H. S., Eivazi, A., Yoder, S. C., Vielmetter, J., Carmichael, D. F., Hayes, R. J., & Dahiyat, B. I. (2006). Engineered antibody Fc variants with enhanced effector function. *Proceedings of the National Academy of Sciences of the United States of America*, 103(11), 4005–4010. https://doi.org/10.1073/pnas.0508123103

Liu, L., Zhou, X., Liu, H., Xiang, L., & Yuan, Z. (2005). CpG motif acts as a "danger signal" and provides a T helper type 1-biased microenvironment for DNA vaccination. *Immunology*, 115(2), 223–230. https://doi.org/10.1111/j.1365-2567.2005.02150.x

Martin, F., Oliver, A. M., & Kearney, J. F. (2001). Marginal zone and B1 B cells unite in the early response against T-independent blood-borne particulate antigens. *Immunity*, 14(5), 617–629. https://doi.org/10.1016/s1074-7613(01)00129-7

Matsumura, M., Nagata, M., Nakamura, K., Kawai, M., Baba, T., Yamaki, K., & Yoshino, S. (2010). Adjuvant effect of zinc oxide on Th2 but not Th1 immune responses in mice. *Immunopharmacology and Immunotoxicology*, 32(1), 56–62. https://doi.org/10.3109/08923970903124627

Mueller, R., Karle, A., Vogt, A., Kropshofer, H., Ross, A., Maeder, K., & Mahler, H. C. (2009). Evaluation of the immuno-stimulatory potential of stopper extractables and leachables by using dendritic cells as readout. *Journal of Pharmaceutical Sciences*, 98(10), 3548–3561. https://doi.org/10.1002/jps.21672

Nagao, K., Ginhoux, F., Leitner, W. W., Motegi, S., Bennett, C. L., Clausen, B. E., Merad, M., & Udey, M. C. (2009). Murine epidermal Langerhans cells and langerin-expressing dermal dendritic cells are unrelated and exhibit distinct functions. *Proceedings of the National Academy of Sciences of the United States of America*, 106(9), 3312–3317. https://doi.org/10.1073/pnas.0807126106

Netea, M. G., Quintin, J., & Van Der Meer, J. W. (2011). Trained immunity: A memory for innate host defense. *Cell Host & Microbe*, 9(5), 355–361. https://doi.org/10.1016/j.chom.2011.04.006

Rathinam, V. A., Vanaja, S. K., & Fitzgerald, K. A. (2012). Regulation of inflammasome signaling. *Nature Immunology*, 13(4), 333–342. https://doi.org/10.1038/ni.2237

Saxena, M., & Bhardwaj, N. (2018). Re-emergence of dendritic cell vaccines for cancer treatment. *Trends Cancer*, 4, (2), 119–137. DOI: 10.1016/j.trecan.2017.12.007.

Schiff, M., Weinblatt, M. E., Valente, R., Van Der Heijde, D., Citera, G., Elegbe, A., Maldonado, M., & Fleischmann, R. (2014). Head-to-head comparison of subcutaneous abatacept versus adalimumab for rheumatoid arthritis: Two-year efficacy and safety findings from AMPLE trial. *Annals of the Rheumatic Diseases*, 73(1), 86–94. https://doi.org/10.1136/annrheumdis-2013-203843

Shi, C., & Pamer, E. G. (2011). Monocyte recruitment during infection and inflammation. *Nature Reviews Immunology*, 11(11), 762–774. https://doi.org/10.1038/nri3070

Shields, R. L., Lai, J., Keck, R., O'Connell, L. Y., Hong, K., Meng, Y. G., Weikert, S. H., & Presta, L. G. (2002). Lack of fucose on human IgG1 N-linked oligosaccharide improves binding to human Fcgamma RIII and antibody-dependent cellular toxicity. *Journal of Biological Chemistry*, 277(30), 26733–26740. https://doi.org/10.1074/jbc.M202069200

Shinkawa, T., Nakamura, K., Yamane, N., Shoji-Hosaka, E., Kanda, Y., Sakurada, M., Uchida, K., Anazawa, H., Satoh, M., Yamasaki, M., Hanai, N., & Shitara, K. (2003). The absence of fucose but not the presence of galactose or bisecting N-acetylglucosamine of human IgG1 complex-type oligosaccharides shows the critical role of enhancing antibody-dependent cellular cytotoxicity. *Journal of Biological Chemistry*, 278(5), 3466–3473. https://doi.org/10.1074/jbc.M210665200

Sokol, C. L., Chu, N. Q., Yu, S., Nish, S. A., Laufer, T. M., & Medzhitov, R. (2009). Basophils function as antigen-presenting cells for an allergen-induced T helper type 2 response. *Nature Immunology*, 10(7), 713–720. https://doi.org/10.1038/ni.1738

Sominanda, A., Lundkvist, M., Fogdell-Hahn, A., Hemmer, B., Hartung, H. P., Hillert, J., Menge, T., & Kieseier, B. C. (2010). Inhibition of endogenous interferon beta by neutralizing antibodies against recombinant interferon beta. *Archives of Neurology*, 67(9), 1095–1101. https://doi.org/10.1001/archneurol.2010.218

Song, W. M., & Colonna, M. (2018). Immune training unlocks innate potential. *Cell*, 172(1–2), 3–5. https://doi.org/10.1016/j.cell.2017.12.034

Steinhagen, F., Kinjo, T., Bode, C., & Klinman, D. M. (2011). TLR-based immune adjuvants. *Vaccine*, 29(17), 3341–3355. https://doi.org/10.1016/j.vaccine.2010.08.002

Takeda, K., Kaisho, T., & Akira, S. (2003). Toll-like receptors. *Annual Review of Immunology*, 21, 335–376.https://doi.org/10.1146/annurev.immunol.21.120601.141126

Vivier, E., Artis, D., Colonna, M., Diefenbach, A., Di Santo, J. P., Eberl, G., Koyasu, S., Locksley, R. M., McKenzie, A. N. J., Mebius, R. E., Powrie, F., & Spits, H. (2018). Innate lymphoid cells: 10 years on. *Cell*, 174(5), 1054–1066. https://doi.org/10.1016/j.cell.2018.07.017

Wu, J., & Chen, Z. J. (2014). Innate immune sensing and signaling of cytosolic nucleic acids. *Annual Review of Immunology*, 32, 461–488. https://doi.org/10.1146/annurev-immunol-032713-120156

Yang, X., Wang, F., Zhang, Y., Wang, L., Antonenko, S., Zhang, S., Zhang, Y. W., Tabrizifard, M., Ermakov, G., Wiswell, D., Beaumont, M., Liu, L., Richardson, D., Shameem, M., & Ambrogelly, A. (2015). Comprehensive analysis of the therapeutic IgG4 antibody pembrolizumab: hinge modification blocks half molecule exchange in vitro and in vivo. *Journal of Pharmaceutical Sciences*, 104(12), 4002–4014. https://doi.org/10.1002/jps.24620

Yu, L., Li, K., Liu, X., Chen, C., Bao, Y., Ci, T., Chen, Q., & Ding, J. (2013). In vitro and in vivo evaluation of a once-weekly formulation of an antidiabetic peptide drug exenatide in an injectable thermogel. *Journal of Pharmaceutical Sciences*, 102(11), 4140–4149. https://doi.org/10.1002/jps.23735

Zhong, G., Reis, E., Sousa, C., & Germain, R. N. (1997). Antigen-unspecific B cells and lymphoid dendritic cells both show extensive surface expression of processed antigen-major histocompatibility complex class II complexes after soluble protein exposure in vivo or in vitro. *Journal of Experimental Medicine*, 186(5), 673–682. https://doi.org/10.1084/jem.186.5.673

3

Platform and Instrument Considerations in Bioanalytical Testing

Dominic Warrino[a] *and Franklin Spriggs*[b]

[a]*KCAS Bioanalytical and Biomarker Services, Biopharma and Biomarker Services (corresponding author)*
[b]*Bioanalytical Consulting LLC*

3.1 Introduction

The goal of this chapter is to introduce the reader to the wide range of platforms available for the quantitation of biotherapeutics and assist in the selection of the right platform depending on the bioanalytical requirement(s). This chapter focuses on pharmacokinetic (PK) and pharmacodynamic/biomarkers (protein quantitation) and not immunogenicity (anti-drug antibodies and neutralizing antibodies) due to some of the analytical differences of the analyte of interest.

For the purposes of this chapter, a platform is defined as the technology used in an analytical method to translate a biochemical event into a measurable output or signal. This signal allows the scientist to accurately and reproducibly make measurements of a drug or analyte of interest. An instrument is defined as the tool utilized to measure a platform's output and convert the resultant signal into a quantifiable measurement. Many platforms employ optical signals including the absorbance of light through a medium or the emission of fluorescence, or luminescence (Spriggs et al., 2012). A variety of light detectors are used to measure these optical signals including photo diodes, charge-coupled device cameras, and photomultiplier tubes.

Bioanalytical Aspects in Biological Therapeutics, First Edition. Edited by Xiaohui Xu and Weifeng Xu.
© 2022 John Wiley & Sons, Inc. Published 2022 by John Wiley & Sons, Inc.

3.2 Common Large Molecule Platforms for PK Analysis

Platforms for support of large molecule quantitation can be further categorized into five areas as follows:

- Plate-based platforms
- Bead-based platforms
- Label-free platforms
- Liquid chromatography-mass spectrometry (LC-MS)/MS platforms
- Other (radioimmunoassay [RIA] and emerging).

While there may be additional categories and platforms available, the common platforms will be discussed in detail with a brief overview of a few emerging platforms, including RIA and Gyrolab.

3.3 Considerations to be Taken Into Account When Choosing a Platform

The choice of a platform for bioanalytical method development is multifaceted; however, as with many choices, Occam's razor should remain the guiding principle. One should start the decision process before ever running an experiment. To choose the right platform(s) to explore, one must understand the need of the method being developed. What do we mean by this? Questions need to be asked such as: what is the anticipated sample volume, what sensitivity is needed, what sensitivity is wanted, what is the availability and quality of reagents, and so on. The answers to these questions can quickly narrow the scope of the method development while ensuring that the method to be developed will answer the bioanalytical questions that need to be answered in order to best support the drug development program (Figure 3.1).

3.4 Plate-Based Platforms

A plate-based platform refers to a method in which the final assay is contained in a plate, typically a 96-well or 384-well plate. The first critical step in a plate-based platform is the selection of the plate type. A plate typically comprises 96 wells with a well working volume of 25–200 μL. There are numerous types of plates that have a major impact on the method, including the material/binding properties of the plate, the design of the bottom of the plate,

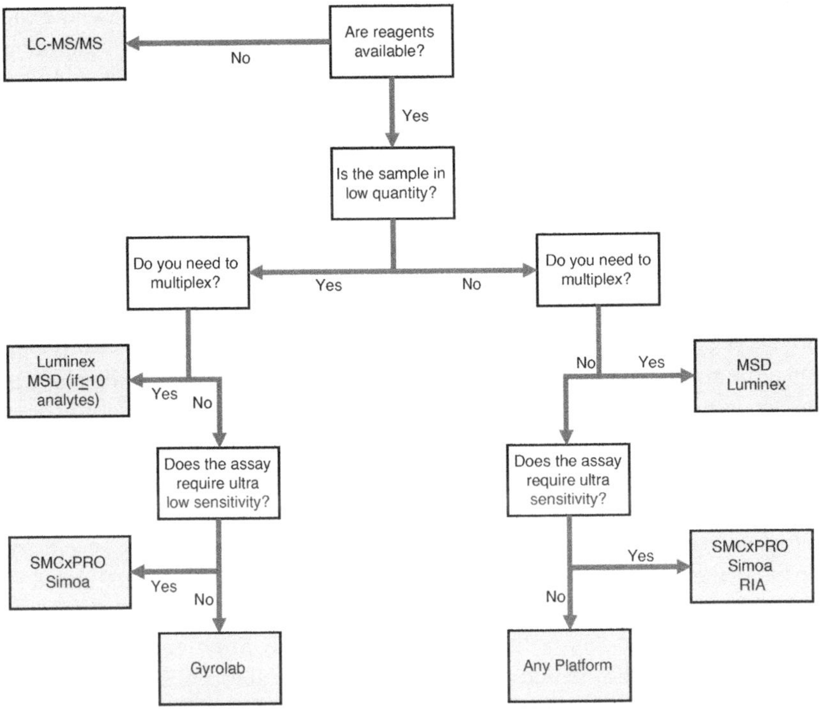

Figure 3.1 Flow chart to aid in the platform decision-making process. LC, liquid chromatography; MS, mass spectrometry; MSD, Meso Scale Diagnostics; RIA, radioimmunoassay.

the color of the plate, and for specialty plates the expiration date and procurement of sufficient plates to complete method development, validation, and sample analysis.

3.4.1 Descriptions

Colorimetry: Colorimetric assays use reagents that undergo a measurable color change in the presence of the analyte (Wikipedia, 2020). The most common form of colorimetric assay is the enzyme-linked immunosorbent assay (ELISA). This technique utilizes an enzymatic label that in the presence of a substrate undergoes a color change proportional to the enzyme present in the well. The enzyme is present in the well as it is conjugated to a detection molecule that is specific to the analyte of interest and therefore indicates the amount of analyte present in the well as interpolated from a standard curve.

The most common form of enzyme used is horseradish peroxidase and the substrate tetramethylbenzidine, resulting in a blue end product. The addition of an acid to the well changes the color to yellow and drastically slows/stops the enzymatic reaction. The colorimetric end product can be read by an absorbance reader. Other color-producing enzymes and substrates exist, but follow the same principle.

Colorimetric assays are able to detect and quantify a variety of analytes such as peptides, small proteins like hormones, and large proteins like antibodies. ELISA is a sensitive, specific, cost-effective, and common platform for PK analysis. There are four kinds of ELISA tests (direct, indirect, sandwich, and competitive). The sandwich, which utilizes a capture antibody and detection antibody with the analyte of interest forming "the sandwich," is the most common and is in the authors' opinion the most suitable/robust for PK analysis.

There are other colorimetric assays that the bioanalytical laboratory can use, and one that is particularly common is the bicinchoninic acid assay for measuring total protein or copper, iron, and other ion/metal assays. These types of assays are typically utilized as a tool in the laboratory, but lack the specificity to be used to measure anything more specific then the moieties stated above.

Finally, colorimetric assays can be used for enzyme activity assays but utilization of a luminescent or fluorometric method may improve the robustness and sensitivity of the method (see next section on Luminescence/Fluorometry).

Luminescence/Fluorometry: Luminescence is any process whereby photons are emitted without heat being the cause. Fluorescence is a specific type of luminescence whereby a photon is absorbed at one wavelength, causing an atom to be in an excited state. The atom returning to a ground state then releases energy in the form of light at a lower wavelength (Nipun, 2015). For the purposes of this PK chapter the specific focus will be on fluorometry. Why is this relevant for bioanalytical assays? It is an excellent alternative method for PK analysis should the two more popular methods (ELISA and electrochemiluminescence [ECL]) not be able to provide a reliable and defendable bioanalytical method. Fluorometric methods can routinely be optimized to obtain sensitive, specific, accurate, precise, and robust PK methods. Fluorometric methods are commonly used for cell-based, tissue-based, and enzymatic activity methods in the bioanalytical laboratory.

Electrochemiluminescence (ECL): ECL is another type of luminescence produced during electrochemical reactions in solutions. The Meso Scale Diagnostics platform (MSD-ECL) uses electrochemiluminescent labels to generate light when stimulated by electricity in the appropriate chemical environment. This reaction is incorporated into immunoassays to provide the light signal used to measure analytes of interest. How does electrochemiluminescence work, specifically the technology utilized by MSD?

- High-binding carbon electrodes in the bottom of a MULTI-ARRAY microplate allow for easy attachment of biological reagents (10× greater binding capacity than polystyrene).
- MSD assays use electrochemiluminescent labels that are conjugated to detection antibodies.

A voltage is applied to the plate electrodes by an instrument leading to an electronically excited state. The relaxation of the intermediates to a lower state then generates light. Light intensity is then measured in photons to quantify the amount of label found on the plate, and therefore is directly proportional to the amount of analyte in the sample.

3.4.2 Ideal Usage (Pros)

The assay will depend on the investment in the capture antibody, detection antibody (or the detection reagent), preparation process of standards and quality controls, screening of patient matrix, and the optimization of the method. Assuming we have these building blocks to enable us to select a platform we prefer to do a feasibility study of the capture and detection using multiple platforms (ELISA and MSD) as a way of evaluating the reagent. If it is not possible to employ multiple platforms we share our thoughts as to the ideal usage of the platform.

Colorimetric/ELISA assays: The ELISA platform is a great starting point for most large molecule therapeutics including monoclonal antibodies, fusion proteins, and recombinant proteins. This does not mean to imply it cannot be used for other laboratory methods such as antibody–drug conjugates, bispecifics, multivalent proteins, and oligonucleotides, but it might not be ideal for these types of drugs. This platform is still the most common method and has the most widely available reagents. Finally, the cost of the equipment as well as the consumables tend to be lowest for this type of analytical method.

Luminescence/Fluorometry: If the available instrument has the ability to measure luminescence and fluorescence and low sensitivity is needed this is an excellent alternative to the traditional ELISA. It is also a preferred choice for enzyme activity assays and cell-based proliferation assays.

MSD-ECL: This platform is widely used and has fairly straightforward reagent generation. It started as the primary method to detect antibodies against large molecule therapeutics, but has grown significantly over the last decade. The availability of kits for single- and multiplex kits is limited, but continues to expand annually. Finally, as safety profiles have continued to improve in the biotherapeutics arena, dose ranges have become larger, making the

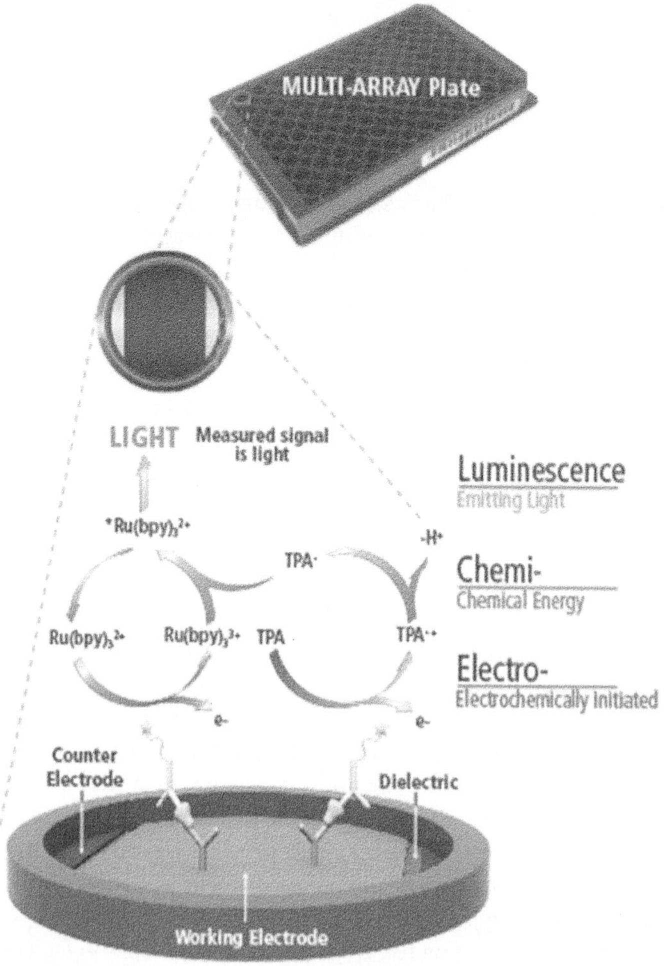

Figure 3.2 Schematic of how the Meso Scale Diagnostics (MSD) platform produces a signal in an assay.

MSD-ECL's larger potential dynamic range a significant benefit. One major differentiator of the MSD platform is the ability to multiplex up to 10 analytes in a single well if needed.

3.4.3 Challenges with the Platforms (Cons)

There is no "one size fits all" platform and, as with all choices, one must critically evaluate the positives and negatives of each option and choose the pathway that provides the most benefit while minimizing the risks associated with the negatives. This section will outline what the authors see as the critical points to consider when deciding the analytical path to pursue for the long-term success of the associated methods essential for a drug development program.

Colorimetric/ELISA assays: The major con that scientists associate with ELISA is the lack of sensitivity to detect at the picogram level. It can also generate a higher than desired background that results in a loss of sensitivity. One cannot multiplex in the platform either.

Luminescence/Fluorometry: Both luminescent and fluorescent methods have consumable costs similar to colorimetric methods. In the case of luminescence, the availability of reagents is also widespread. This may seem to be the best of both worlds; however, there are some drawbacks. These drawbacks are: sensitivity gains are not always guaranteed, requirement for either new or a more complex plate reader, potential challenges with cross-talk, and, finally, specific to fluorometry, the reagent pool can be lower (but not always).

MSD-ECL: MSD-ECL is a very useful technology, and MSD provides a great number of well-characterized kits; however, there are two drawbacks to using this technology. First, the instrument, consumables, and the label for conjugation/detection of analyte are generated from a single vendor. This is a particular concern when one is looking at long-term support of a clinical phase program. Articles have been written about the risks associated with single suppliers (Yohrling, 2009; King et al., 2014) with some discussion on how to mitigate this risk.

The second drawback to the MSD-ECL platform is much more tangible and that is cost. Compared to the standard colorimetric or even luminescent method, the consumables as well as the instrumentation cost are much higher than the other technologies discussed here. We do not mean to imply that cost should be the only consideration when developing an analytical method, but we also understand that it is a factor that needs to be considered, especially for long-term support of a program.

3.5 Bead-Based Platforms

Until recently, the majority of bead-based platform usage has tended toward the measurement of biomarkers. More recently, these platforms have continued to evolve and be used for other areas of drug development such as PK analysis and immunogenicity. Bead-based platforms have only started to gain traction in the non-biomarker arena as they have been on the market long enough now to have been used in support of regulatory submissions, essentially getting vetted by regulators such as the US Food and Drug Administration (FDA) and the European Medicines Agency (EMA).

3.5.1 Descriptions

Simoa: The Single Molecule Arrays (SiMoA or Simoa) utilizes paramagnetic particles coupled with capture antibodies. After the target is bound, similar to other ligand binding assay (LBA) platforms, a detection antibody with a fluorescent tag binds forming an immunocomplex. This complex is loaded into the Simoa disc containing 200,000 microwells. In theory each microwell is large enough to hold only one bead. A camera can then quantify each immunocomplex.

Luminex: Luminex is based on polystyrene or paramagnetic microspheres, or beads, that are internally dyed with red and infrared fluorophores of differing intensities. Each dyed bead is given a unique number, or bead region, allowing the differentiation of one bead from another which can be utilized for multiplexing.

Each dyed bead (over 500 different beads are available) can be coated with a capture antibody. Similar to other LBA methods, the analyte of interest is captured by the antibody-coated bead and detected with a reporter dye-labeled antibody. The instrument uses a light source to excite the internal dyes that identify each microsphere particle, and also any reporter dye captured during the assay.

Sigma Millipore SMCxPro: The SMCxPro is the second generation of the single molecule counting (SMC) technology originally developed by Singulex in the Erenna instrument. The SMCxPro utilizes a workflow similar to a traditional ELISA assay and is depicted schematically in Figure 3.3. The assay diverges from traditional quantitative assays by a proprietary elution step being performed that allows the disassociation of the fluorescently labeled antibody to be detected by the instrument. This unique aspect of the assay and instrument enables the user to obtain high levels of sensitivity (femtograms/milliliter) by reducing signal-to-noise ratios. Note that disassociation is a technique frequently used independent of platform to increase the sensitivity. Briefly, disassociating the analyte of interest from a complex increases the sensitivity.

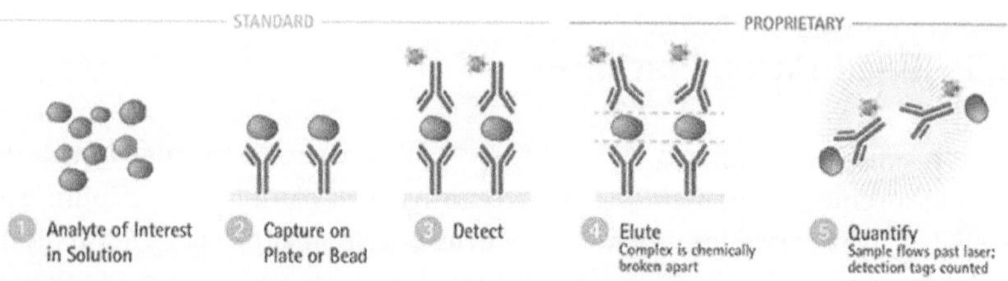

Figure 3.3 Illustration of the typical enzyme-linked immunosorbent assay (ELISA) workflow (Steps 1–3) and the additional steps (Steps 4 and 5) performed that make the SMCxPRO unique and able to detect low quantities of analyte in the sample.

3.5.2 Ideal Usage (Pros)

Simoa: The Simoa platform is a powerful tool that is useful in the measurement of several established kits produced by Simoa. The biggest advantage of the Simoa is the fact that the system is a fully automated system. This reduces the human factor and allows for an analyst to work on other activities that require specific expertise.

Luminex: The Luminex system is the only platform discussed here that allows for multiplexing. The multiplexing is done by labeling different colors of beads with specific capture reagents This method of differentiating analytes by bead color has the advantage that detection reagents can use the same conjugated fluorescent label with a reduced level of complexity as the complication of "cross-talk" between different fluorophores is eliminated. The ability to multiplex on the Luminex comes with another important aspect, namely sample volume. The Luminex platform allows for the quantitation of a large number of analytes (up to 500 by some counts) (Bio-Rad, 2020) with a single sample, making the Luminex the ideal bead-based platform for screening biomarkers at an early stage to identify predictive markers to pursue further.

SMCxPro: The SMCxPro platform is the ideal high-sensitivity, bead-based platform when the method being developed is a de novo assay (Fischer et al., 2015). This platform is also amenable to a higher throughput (ability to run 384-well plates). Additionally, as the instrument itself is a plate reader, the read time for a single plate is approximately 1 hour, making it possible to have a single instrument with multiple users in a single day.

3.5.3 Challenges with the Platforms (Cons)

Simoa: The Simoa's advantage is also one of the platform's challenges, namely that the fully automated aspect of the instrument reduces the throughput to allow for only one run in a standard 8-hour work day. This challenge can be overcome through the purchase of multiple instruments, but for laboratories without a high capital equipment budget this can be a hurdle that is too high to overcome. Additionally, while it is possible to develop de novo assays on the Simoa, this can be cumbersome and a long-term project. With today's rapid drug development timelines, this is a significant challenge with this platform for developing assays that are not already produced as a kit. Finally, as with other platforms, the limited availabity of kits due to there being a single vendor generating kits is not without risk of stunted supply chain.

Luminex: The Luminex's challenge is also associated with the advantage it has over the other platforms. The Luminex bead-based system is highly dynamic and customizable due to the fact that the system is similar to a flow cytometer. The system's dynamic and customizable nature and ability to multiplex can add a high level of complexity to utilization of the platform. With training and time working with the platform this can be overcome, but additional time will be needed to ensure the expertise is developed to generate reliable, reproducible results for any number of analytes. An additional challenge with the Luminex platform is that maintenance of the instrument is more involved due to the fluidics of the system. This is certainly not an insurmountable challenge, but should not be dismissed when weighing which platform is best for an individual laboratory.

SMCxPro: The SMCxPro, similar to the Luminex, requires time using the system to acquire the requisite expertise. The familiarity of the first part of the assay can lead to overconfidence in the method. The ultrasensitive aspect of this platform will amplify errors not seen until the plate has been read. This platform requires users to maintain a high level of precision and accuracy in their techniques to ensure that the results are comparable between plates and across studies.

3.6 Other and Emerging Technologies

There are platforms that do not fit neatly into any of the above categories and therefore these are discussed in this section. These platforms include RIA, Gyrolab, and other emerging technologies.

3.6.1 Descriptions

Radioimmunoassay (RIA): RIAs have long been used in the bioanalytical field, although they have fallen out of favor in recent years due to safety and regulatory (Nuclear Regulatory Commission [NRC]) challenges. An RIA is typically done in larger borosilicate tubes and the detection reagent has a radioactive isotope that is used to measure the relative quantity of the analyte of interest.

Gyrolab: The Gyrolab platform uses a streptavidin bead column that has been manufactured into a compact disc with microstructures that control the volume of solution that is flowed over the column. The samples are drawn over the column through centripetal force by spinning the compact disc. The detection reagent is conjugated to a fluorophore that is excited via laser and the resulting signal measured on the instrument (Figure 3.4). The advantages of this platform are low volume, fast assay development, and semi-automation.

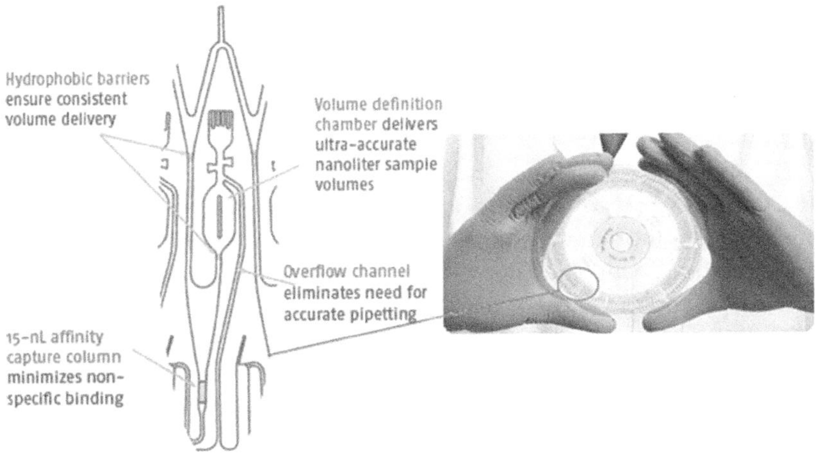

Figure 3.4 The Gyrolab uses a nanoscale column as the solid phase of a traditional ligand binding assay method. The different channels control the volume to ensure the same volume is drawn over the bead columns through the centripetal force generated by spinning the compact disc (reproduced with permission of Gyros Protein Technologies, Sweden https://www.gyrosproteintechnologies.com/gyrolab-technology).

Emerging Technologies: The diversity of large molecule modalities continues to grow, and in response so do the platforms used to quantitate those modalities. These platforms can be surface plasmon resonance systems, quantitative polymerase chain reaction, Western blots, and so on. None of these are new technologies, but rather old technology that has been brought up to date with modern instruments. An attempt to outline these technologies in this chapter is simply not practical because as soon as this book is published the latest technology will have changed, rendering this section outdated. Suffice to say that all emerging technologies should be assessed by the individual laboratory to ensure they are useful in answering the critical question that needs to be answered. There are always going to be certain advantages to a new technology, and conversely there will be times when a new technology may not be the optimal choice. Each platform and each situation must be evaluated as a blanket "this is the best platform" is neither practical nor possible.

3.6.2 Ideal Usage (Pros)

Radioimmunoassay (RIA): Despite the environmental risks, the use of RIA for small peptides and hormones should still be considered due to the sensitivity, cost-effectiveness, and the challenges that intrinsically exist when working with antibodies and ligands of highly conserved sequences across species and small molecular weights. The RIA field is safer than ever, but it is generally agreed that an alternative to radioactive material is preferred.

Gyrolab: The Gyrolab system has two distinct advantages that may lead one to pursue the use of this platform. The first of these is the sample size, and the second is the automated throughput (Dai et al., 2020). Due to the low volume requirement of the microstructures, the sample plates can have very low sample volumes (25 μL total after dilution). This low volume makes serial PK sampling from rodents including mice feasible. This can reduce the overall cost of the PK portion of a study as each PK time point does not require multiple animals. Additionally, by performing serial PK sampling of the animals, subject-to-subject variability can be observed.

The other advantage to the Gyrolab is the fact that the system is automated once samples have been diluted and placed in the instrument. The top end instrument allows for up to five discs to be run at once, making this a higher throughput platform compared to some of the other platforms discussed in this chapter.

3.6.3 Challenges with the Platforms (Cons)

Radioimmunoassay (RIA): By far the greatest challenge with utilizing RIA in a laboratory is the infrastructure required. Not only are licenses with the NRC required, but monitoring of total amounts of radiolabeled material, specialized equipment, processes for ensuring proper and sufficient clean-up, and regular training from a radioactive safety officer are required to maintain use of this technology. As with the other technologies discussed previously, these are not insurmountable challenges, especially if the benefits of the platform outweigh the challenges.

Gyrolab: With the multiple advantages that Gyrolab offers, there are also some challenges with using the instrument. One consideration is the fact that the instrument does not use disposable tips when transferring samples or reagents to the discs. This has resulted in some carry over when high concentration samples are transferred. Gyros has worked with their customers to develop wash protocols to minimize this occurrence, but it should still be assessed when developing and utilizing this platform for unknown sample analysis.

Another consideration is related to the reagents (specifically capture and detection antibodies) to be used in the assay. Due to the short assay times (approximately 5 hours to run five discs), the reagents are not in contact with the samples for a long period of time. This necessitates all the reagents to be of high quality and to possess strong binding ability.

3.7 Questions to Consider When Choosing Between Platforms

3.7.1 Key Drivers

A key consideration when choosing a platform for supporting bioanalysis are desirable analytical performance characteristics. LBA platforms for support of protein quantitation require consideration of the following characteristics: sensitivity, precision, dynamic range, matrix interference, robustness, ruggedness, total assay time, and suitability for measurement of the analyte in the matrix of interest (DeSilva et al., 2003; Lee et al., 2006; US Department of Health, 2018).

The key drivers for choosing a platform are the biology of the laboratory method, the history of the laboratory method, what reagents/platforms are readily available, and reliability and scalability over time.

The biology of the laboratory method includes items such as the dosage, potency, the type and volume of matrix available for analysis, and any other related items that impact the design of the PK method.

Additional considerations that are more specific to biomarkers and a pharmacodynamics platform are should one multiplex, does the method require functional testing, is it an activity assay, and does it require live cells for quantification of the marker of interest or proliferation of cells.

3.7.2 When to Choose an Alternative to a Plate-Based Platform

The primary alternative to plate-based platforms are bead-based. There are a few additional items to consider once a decision has been made to evaluate a bead-based platform over a plate-based platform. The first question to consider is whether one is evaluating a single analyte or if the intention is to examine multiple analytes in the same sample? Only one platform is amenable to multiplexing within the same sample (Luminex). If multiple analytes are of interest, however, and sensitivity of a specific marker is of the utmost importance, then a combination of platforms may be the best approach as the Simoa and SMCxPro are ultrasensitive platforms that can provide quality data in the subpicomolar and femtomolar concentrations. Finally, as with all LBAs, availability of reagents and the ability to modify them in such a way as to allow their use in the different platforms may also influence the decision on the platform to select.

3.8 Conclusions

Drug development has continued to evolve at an ever-increasing pace, generating drug modalities that were unheard of only a couple of decades ago. With these innovative modalities has also come an increase in potency, therefore reducing the dose provided in both toxicology and pharmacokinetic studies. These factors have impacted the bioanalytical world – whether with improved LC-MS/MS systems or new LBA platforms being developed, bioanalysis has kept pace with the constantly evolving therapeutic landscape.

3.8.1 Strategy for Implementation of New and Emerging Technologies

Many considerations need to be taken into account when evaluating a new and emerging technology. The obvious considerations are the cost and complexity of the platform (is there a need to hire a specific technician or scientist to run the instrument?).

3.8.2 Recap on Decision for Platform Choice

As stated earlier in this chapter, the authors' objective was to provide information about the different kinds of platforms that are available to use when developing LBAs. Hopefully this primary objective has been accomplished, but the authors also hope that they have successfully conveyed another important aspect, namely that a single platform cannot be relied upon to fulfil every bioanalytical requirement that a laboratoy might have.

No single platform will be able to meet all the needs of today's or tomorrow's bioanalytical laboratory. While a colorimetric-based platform will suit many laboratories' initial needs, as the therapeutic progresses down the drug development path to a Biologics License Application submission it is almost certain that additional tools will be needed to answer all the critical questions required to ensure a successful submission and eventual approval.

This chapter cannot provide all the information needed to make the best decision for an individual program, but it is meant to initiate a discussion to determine the optimum course of action. The ultimate question is: Can data be generated that are of use on this platform? An additional question that should also be contemplated is: Can this platform also be utilized for the next program? Keeping these questions at the forefront of the decision-making process will help ensure success in the bioanalytical arena.

References

Bio-Rad Laboratories. (2020). *Bio-Plex Multiplex Immunoassay System*. https://www.bio-rad.com/en-us/category/bio-plex-multiplex-immunoassay-system?ID=85824182-db2a-4c6e-ac82-4e07dd9ab904 (accessed 06 August 2020)

Dai, Z., Juneja, J., Schneeweis, L. Cohen, D., Marsilio,F., Morin,P., & DasGupta, R. (2020). Application of the Gyrolab microfluidic platform to measure picomolar affinity of a PD-L1-binding Adnectin™ radioligand for positron emission tomography. *BioTechniques*, 69(3), 201–206. https://doi.org/10.2144/btn-2020-0080

DeSilva, B., Smith, W., Weiner, R., Kelley, M., Smolec, J., Lee, B., Khan, M., Tacey, R., Hill, H., & Celniker, A. (2003). Recommendations for the bioanalytical method validation of ligand-binding assays to support pharmacokinetic assessments of macromolecules. *Pharmaceutical Research*, 20(11), 1885–1900. https://doi.org/10.1023/b:pham.0000003390.51761.3d

Fischer, S., Joyce, A., Spengler, M.Yang, T., Zhuang, Y., Fjording, M., Mikulskis, A. (2015). Emerging technologies to increase ligand binding assay sensitivity. *American Association of Pharmaceutical Scientists Journal*, 17(1), 93–101. https://doi.org/10.1208/s12248-014-9682-8

Hwang, J., Banerjee, M., Venable, A., Walden, Z., Jolly, J., Zimmerman, C., Adkisson, E., Xiao, Q. (2019). Quantitation of low abundant soluble biomarkers using high sensitivity Single Molecule Counting technology. *Methods*, 158, 69–76. https://doi.org/10.1016/j.ymeth.2018.10.018

King, L. E., Farley, E., Imazato, M., Keefe, J., Khan, M., Ma, M., Pihl, K. S., & Sriraman, P. (2014). Ligand binding assay critical reagents and their stability: recommendations and best practices from the Global Bioanalysis Consortium Harmonization Team. *The AAPS Journal*, 16(3), 504–515. https://doi.org/10.1208/s12248-014-9583-x

Lee, J. W., Devanarayan, V., Barrett, Y. C., Weiner, R., Allinson, J., Fountain, S., Keller, S., Weinryb, I., Green, M., Duan, L., Rogers, J. A., Millham, R., O'Brien, P. J., Sailstad, J., Khan, M., Ray, C., & Wagner, J. A. (2006). Fit-for-purpose method development and validation for successful biomarker measurement. *Pharmaceutical Research*, 23(2), 312–328. https://doi.org/10.1007/s11095-005-9045-3

Nipun. (2015). *Difference Between Florescence and Luminescence*. https://pediaa.com/difference-between-fluorescence-and-luminescence (accessed 17 July 20).

Spriggs, F. P., Zhong, Z. D., Safavi, A., Jani, D., Dontha, N., Kant, A., Ly, J., Brilando, L., Österlund, K., Rouleau, N., Fischer, S. K., Boissonneault, M., & Ray, C. (2012). Ligand binding assays in the 21st century laboratory: platforms. *The AAPS Journal*, 14(1), 113–118. https://doi.org/10.1208/s12248-012-9321-1

U.S. Department of Health and Human Services Food and Drug Administration Center for Drug Evaluation and Research (CDER) Center for Veterinary Medicine (CVM). Bioanalytical Method Validation Guidance for Industry. May 2018 Biopharmaceutics. https://www.fda.gov/regulatory-information/search-fda-guidance-documents/bioanalytical-method-validation-guidance-industry

Wikipedia. (2020). *Colorimetric analysis*. https://en.wikipedia.org/wiki/Colorimetric_analysis (accessed 17 July 2020)

Yohrling, J. (2009) Ligand-binding assays: risk of using aplatform supported by a single vendor. *Bioanalysis*, 1(3). https://doi.org/10.4155/bio.09.46

4

Pharmacokinetic Assays

Tong-Yuan Yang[a] *and Eric Wakshull*[b]

[a]*Janssen BioTherapeutics, Janssen R&D, LLC (corresponding author)*
[b]*Consulting*

4.1 Introduction

A *protein therapeutic* or *biologic* is a unique type of biopharmaceutical whose primary component is manufactured in a living system such as a microorganism, plant, or animal cells using recombinant DNA technology. Most biologics are very large, complex molecules with heterogeneity due to posttranslational modifications (PTMs) (e.g., glycosylation, deamidation, isomerization, etc.). One common class of biologic is monoclonal antibodies (mAbs) which on average have a molecular weight (MW) of ~ 150,000 Da, compared to small chemical drug entities, for example, aspirin with a MW of ~ 180 Da. Other examples, such as cytokines, growth factors, and replacement enzymes, can range in size from 6000 Da (insulin) to almost 300,000 Da (full-length factor VIII). The size and complexity of these molecules can be further increased through chemical modifications such as pegylation (to increase half-life) or covalent addition of toxins (e.g., antibody–drug conjugates [ADCs]). The first biologic recombinant protein, human insulin (Humulin), was approved for the US market in 1982. Since then more than 200 biologics have entered the market, which include cytokines, enzymes, and mAbs, for the treatment of cardiovascular, autoimmune, infection, and oncology indications.

Bioanalytical Aspects in Biological Therapeutics, First Edition. Edited by Xiaohui Xu and Weifeng Xu.
© 2022 John Wiley & Sons, Inc. Published 2022 by John Wiley & Sons, Inc.

The measurement of drug concentrations from early discovery/proof-of-concept studies done in the research setting and throughout the drug development process and beyond drug approval is an essential component of the drug discovery and development process. In the early discovery stages, in vivo disease models provide some assurance of relevant biological activity on hypothesized mechanism of actions (MOAs). Often the pharmacokinetic (PK) and efficacy (pharmacodynamic [PD]) data thus derived form the basis of the desired target clinically efficacious drug concentration. During nonclinical development, the value of PK – sometimes termed toxicokinetic data when obtained from toxicology studies – is to assess the exposure–safety relationships and allow for the determination of the projected safe dose in "first in human" studies. In clinical studies, PK data are critical for determining both exposure–efficacy and exposure–safety relationships which then allow determination of the dosing window (also known as the "therapeutic index") for achieving optimal efficacy with minimal safety risks. The therapeutic index can be expressed as either a delta or ratio between efficacious and tolerable (acceptably safe) drug concentrations, either in terms of amount of drug dosed ("mg/kg" or simply total "mg" for flat dosing) or various PK parameters such C_{max}, C_{min}, or exposure (area under the curve). The choice of which parameter is used is based on PK/PD data modeling, understanding of the putative MOA, combined with the more nuanced risk/benefit assessment, which are topics beyond the scope of this chapter.

The in vivo bioactivity or function of a biologic depends on its higher-order structure. Quite often during nonclinical and clinical studies, "bioactive" drug concentration measurement is required to establish PK/PD/safety relationships for biologics. Since their emergence as biopharmaceuticals, the standard methodology for measuring the "active" drug concentration of biologics has been immunoassays or ligand binding assays (LBAs) which typically use very minimal and mild sample preparation, primarily dilution into a buffer system that causes no disruption of the intact higher-order structure of the biologic, and often uses reagents that are specific to the functional active site. This is in contrast to the bioanalytical process for small chemical entity drugs as well as some protein therapeutics in which liquid chromatography-mass spectrometry (LC-MS) is the primary analytical platform after sample extraction, usually under harsh biophysical and biochemical conditions with organic solvents and precipitations. In the latter case, quantitation is achieved by measuring a signature peptide (a proteolytic fragment unique to the biologic) in serum samples resulting in a so-called "total" drug concentration measurement.

Since the first development of an immunoassay as an analytical method in the 1950s by Berson and Yalow (Yalow, 1973; Yalow & Berson, 1960), a variety of LBA formats, including multiple detection modalities (such as colorimetric, fluorescent, chemiluminescent, or electrochemiluminescent, etc.) with instrumentation to match, have been developed over the years that provide equivalent or better sensitivity compared to the original radioimmunoassays thus eliminating the use of radioisotopes along with the issues of exposure to and waste management of these radioisotopes. The advent of these immunoassay technologies has provided greater flexibility for bioanalytical scientists when developing "fit for purpose" PK methodologies and the robustness required for scaled-up batch size runs using automation systems for better thoroughput during bioanalysis. The most common LBA format is a "sandwich" assay employing a drug-specific capture and detection reagent. The capture and detection reagents can be anti-idiotypic (anti-ID) antibodies generated against a given biologic drug, or can be target proteins the drug binds to if available. Capture reagents bound to a solid phase support are used to isolate drug from a given biological matrix while detection reagents are used to detect the drug after isolation by the capture reagents. In general, the most specific reagent is used as capture, as matrix is then removed in a washing step prior to addition of the detection reagent, which removes most interfering molecules. In this case, the need for a specific detection reagent (e.g., anti-ID) is reduced and can be replaced by a more generic reagent (e.g., an anti-human IgG coupled to a detection moiety). This can be an advantage if obtaining a second specific reagent is problematic. However, the use of two specific assay reagents if available can reduce background signals, improving assay sensitivity and overall performance. These types of LBAs have been used in the last 20 years for supporting traditional biologics, such as mAb therapeutics, cytokines, enzyme replacement therapy, and vaccines. There are various publications that describe how to develop immunoassays with more technical details (Khan & Findlay, 2009; Wild, 2013).

With the advancement of more novel and complex modalities of biologics, such as antibody-conjugated cytotoxins, peptide conjugated to half-life extension frame work, chimeric antigen receptor T cell (CAR-T) therapy, cancer vaccines, and gene therapy programs, different methodologies such as LC-MS/MS, real-time quantatitive polymerase chain reaction (qPCR), and flow cytometry are being more and more frequently used for PK method development (Park et al., 2016; Ramagiri & Moore, 2016). The selection of each methodology depends on the bioanalytical strategies required to support a particular molecule modality through nonclinical and clinical development. In most instances, more than one methodology will be required for PK characterization of these more complex modalities.

The application of each methodology in PK assay development for in vivo characterization is discussed in the following paragraphs.

4.2 Critical Reagents

Critical reagents, such as recombinant target proteins, anti-ID antibodies, aptamers, affimers, and so on, used in the development of LBAs, immunoaffinity pulldown in LC-MS assays, or antigen enumeration in flow cytometry, are macromolecules with complex structure also. The importance of the quality of critical reagents in maintaining the assay performance during its life cycle of a given LBA method has been discussed extensively within the bioanalytical community (Geist et al., 2013; O'Hara & Theobald, 2013). The same principles described in these publications during LBA method development will apply to other bioanalytical methods if a similar class of critical reagents is being used. Generation of these reagents is an essential part of bioanalytical method development to support biologics and usually will take a longer time (usually 3–6 months) from immunization to purification of anti-IDs leads in the case of anti-ID antibodies. These critical reagents should be generated as soon as a lead drug candidate is selected in order to meet the timelines for bioanalytical method development.

During LBA method development it is common for standard amine conjugation chemistry to be used to label these reagents for the purpose of signal amplification and robustness of assay performance. The cloning of recombinant reagents, particularly antibody-based reagents, into expression constructs and transfection into stable production cell lines such as CHO or 293 cells often improves the yield and reproducibility of reagent quality. Care should be taken to evaluate the performance of cloned reagents as sometimes changes in binding properties can occur. Case studies have shown that poor quality or lot-to-lot variability of these critical reagents can have a dramatic impact on the performance of LBA assays (Geist et al., 2013). In addition, conjugation processes can change the physical behavior of conjugated reagents in solution, such as isoelectric point (pI), affinity, and specificity, as well as stability, for instance, inducing reagent aggregation, which may lead to aberrant assay results. Therefore, a number of biochemical, biophysical, and/or immunological methods should be developed to characterize and ensure the quality, specificity, selectivity, and the batch-to-batch consistency of the source materials of assay critical reagents and their conjugates. For those drugs that successfully navigate the research and development process to become approved therapeutics, a consistently reliable PK assay will be needed for many years. Consistent production of source proteins and their conjugated versions is the foundation for the robust performance of a given LBA method over the life cycle management of a biologic.

The standard calibrators and quality control (QC) samples used in any PK method are prepared from manufacturing lots of a given biologic. During the early discovery phase sometimes the biophysical properties or stability of early lead candidates may not be well characterized. Experimental investigation is necessary to define in vitro and ex vivo stability of the drug in order to guide the proper formulation and storage conditions of the drug and understand potential aberrant bioanalytical method performance during method development. Once a drug candidate is selected, the manufacturing process is usually readily defined according to good manufacturing practices regulation and the storage conditions and stability properties will be quickly established by the Chemistry, Manufacturing, and Controls (CMC) organization. Obaining this information from CMC colleagues is a prerequisite for initiation of any bioanalytical method development to guide further stability assessments in biological matrices.

If intact proteins or peptide fragments are being used as internal standards for biologic quantitation with an LC-MS/MS method, the stability of these internal standards also has to be established in stock solution and in biological matrices which have been extensively reviewed in the literature and defined by regulatory documents (Faria & Halquist, 2018).

Real-time qPCR assays are the standard methodology for in vivo transgene measurement to support CAR-T and gene therapy programs, although there are currently no regulatory guidances for assay validation required before implementation for regulated bioanalysis. The critical reagents in qPCR method development are DNA plasmids encoding transgenes for standard calibrators and QC sample preparation and oligonucleotides for sequence-specific primers and probes. It has been well established that DNA molecules have demonstrated long-term storage stability in appropriate buffers when stored at any temperature below −20°C.

4.3 Various Methodologies for PK Characterization

4.3.1 LBA Methods

There are a variety of technology platforms that are commonly used for developing "fit for purpose" LBA PK methods depending on the stage of drug development. These platforms differ primarily in their detection system: colorimetric, fluorescent, chemiluminescent, or electrochemiliuminescent readouts. All these detection systems use tags that are chemically conjugated (primarily by amine coupling) to detection reagents, for example, an antibody detection reagent, that binds directly or indirectly (via a secondary bridging antibody) to the analyte. Variability in conjugating the tags to the detection reagent will impact assay reproducibility and robustness when different lots of reagents are incorporated into existing assays, so carefully defined and controlled coupling chemistry as well as establishing prespecified performance metrics are important to ensure consistent batch production resulting in reliable assay performance during the lifespan of the PK assay. Each detection system has its own strengths and weaknesses. Colorimetric assays generally use horseradish peroxidase/tetramethylbenzidine reactions to create a color for which absorption is proportional to the drug concentration in a sample. It is the most common platform used in discovery studies; however, the dynamic range is often < 100-fold which becomes a less favorable choice for supporting nonclinical studies with a wide range of drug concentrations. Chemiluminescent assays typically either use acridinium esters coupled to the detection reagent + H_2O_2 (usually injected for a "flash" light output requiring a specialized injector system) or horseradish peroxidase/luminol or alkaline phosphatase/adamantyl 1,2-dioxetane arylphosphate (AMPPD) (producing a sustained "glow" output). Chemiluminescent assays can be more sensitive than colorimetric or fluorescent assays with a wider dynamic range. However, one issue with this detection system is that the high sensitivity can lead to significant background signals if there is nonspecific binding (Cinquanta et al., 2017). Instrumentation for these three readouts are available from multiple vendors, and multimodal readers that support all three readouts in a single instrument are also available. Lastly, the electrochemiliuminescent detection system which utilizes a ruthenium-based conjugate has become the primary workhorse in recent years in many bioanalytical laboratories due to its low background/high sensitivity, wide dynamic range capabilities, and scalability on automation. These features are favored over other detection systems particularly in the situation when the PK methods need to go through an extensive validation exercise under regulatory requirement for the purpose of supporting good laboratory practice (GLP) or clinical studies ("regulated bioanalysis"). However, there is currently only one instrument vendor (Meso Scale Discovery, MSD) that provides the consumables (e.g., assay plates and labeling kits) and instrumentation, although other varieties appear to be available (Gross et al., 2016).

It should be noted that new bioanalytical technologies are constantly being developed and emerging, although reaching the stage of being a reliable and robust platform takes time and needs to be evaluated by bioanalytical laboratories which demand faster assay turnaround, scalability with automation, and low cost in consumables and instrumentation. Additionally, new systems may need to be validated (both hardware and software) and CFR21 Part 11 compliant for use in support of regulated bioanalysis. The adoption of such new technologies can be beneficial, such as platforms capable of achieving ultrasensitivity (e.g., ~ pg/mL in 100% serum matrix for a mAb therapeutic with a

MW of ~ 150 kDa) (Fischer et al., 2015), but when intended to support regulated assay work extensive evaluation with method development and validation will be needed.

Regardless of the detection system used, the suitable performance of a given LBA method for PK measurement depends on the evaluation of multiple assay parameters (DeSilva et al., 2003). The accuracy and precision of a LBA PK method depends on the recovery of standard calibrators and quality control samples (QCs). In general, standard calibrators and QCs are prepared by spiking drug solution into close to 100% biological matrix, such as serum or plasma. For more nonroutine or rare matrices (cerebral spinal fluid, aqueous humor, nasal or bronchial lavage, tissue extracts), surrogate matrix, such as buffer, can be used as a substitute provided an appropriate minimal sample dilution is established during assay development to show analyte recovery is equivalent between buffer and the diluted matrix. Generally a tiered approach will be used for developing and validating a "fit for purpose" LBA PK method depending on its usage at different stages of drug development (Watson et al., 2017). Spike/recovery experiments during assay validation for supporting regulated bioanalysis (such as GLP and clinical studies) (see later) should be done in a limited number of real spiked matrices obtained from representative individual subjects in the intended study (e.g., healthy vs. diseased individuals) for "selectivity" assessment. The standard calibrators are required to have at least six non-zero concentrations while QCs will have five designated concentrations spanning the quantifiable range of the assay, for example, LLOQ (lower limit of quantitation), LQC, MQC, HQC, (low, mid, high QCs), and ULOQ (upper limit of quantitation) (at least during validation; three QC levels are adequate during production runs) (DeSilva et al., 2003; Findlay et al., 2000). The LLOQ and ULOQ should be represented in the calibrator set as per regulatory guidances (EMA, 2012; FDA, 2018). The QC samples are usually prepared in 100% pooled matrix then aliquoted into single-use volumes and stored in the same storage conditions as study samples, such as −80°C, while standard calibrators are often prepared fresh on the day the assay is performed. Alternatively, pre-made frozen standard curves in neat matrix can be stored frozen in aliquots for single use if storage stability can be demonstrated. For LBAs, the dose–response relationship of the standard curve (e.g., instrument signal vs. drug concentration) is often nonlinear which will require a nonlinear logistic curve fitting model for back calculation of sample concentration. Selection of the right curve fitting model for standard curves should be based on the recovery of standard calibrators and QCs (Findlay et al., 2000; Gottschalk & Dunn, 2005; Liu & Wu, 1998). Due to the relative narrow quantitation range of standard curves, in particular for colorimetric assays, and potential hook effects (i.e., prozone), dilutional linearity needs to be confirmed by spiking drug concentrations in matrices higher than the top standard calibrator (preferably at the predicted maximum drug concentration based on PK modeling; C_{max}) and serially diluting the sample into the quantitative range of the assay. All dilutions within the quantitative range should recover close (see section on Validation, later) to the nominal value of the neat (undiluted) sample after dilution factors are incorporated. Assay robustness (e.g., performance tolerance), which in a large part is dependent on the reagents used and the assay design, is a critical aspect of PK assays that often can be used over the course of many years in support of clinical development and post-approval indication expansion studies. Furthermore, assay robustness can dictate how well it can be further scaled up in batch runs on a robotic-assisted automated liquid-handling system.

Typical mAb therapeutics are dosed in the single or low double-digit milligrams/kilogram range, where obtaining sufficient PK data (optimally providing about five half-lives of concentration-time data) requires assay sensitivity ~ 100 ng/mL in 100% serum matrix. With the development of multispecific biologics which are designed to target multiple targets or engage with different effector cells for enhanced efficacy such as T cell engaging bispecific mAb therapeutics, dosing levels for human studies can be reduced significantly, particularly in the early dose-escalation studies which might utilize "minimum anticipated biological effect level" (MABEL) dosing schemes due to potential safety concerns. Such dosing can result in very low C_{max} (peak) and C_{min} (trough) concentrations, perhaps as low as single-digit nanograms/milliliter or less. Another example where very sensitive assays are required is when direct injection into a specific tissue (e.g., intravitreal injections for retinal diseases) may have very low serum concentrations as drug eliminated from the injection site is diluted significantly upon entering the blood circulation (the central compartment). Such assay sensitivity levels will usually require an ultrasensitive LBA method (~100 pg/mL in 100% matrix) which is often beyond the detection limit of the conventional platforms discussed earlier. Novel LBA platforms are available to achieve ultrasensitivity in biomarker research which can be also used for developing an LBA-based PK method (Fischer et al., 2015). Generally these novel LBA platforms achieve ultrasensitivity by using one or both of the following techniques. (1) Sample enrichment. Magnetic beads conjugated with analyte specific reagents (e.g., target or anti-ID antibodies) will be used to isolate and enrich analyte from matrices. The enrichment can be further enhanced using large volume of samples (e.g., 300–500 μL) which can be procured from clinical samples. (2) Signal amplification with digital readout by laser detection system or PCR amplification. Laser detection systems can detect a single positive fluorescent event when coupled with a fluidic sampling system or other separation means. Single positive event detection significantly increases the signal-to-noise ratio which translates into improved assay resolution with better sensitivity as applied in the Singulex Erenna and Quanterix Simoa platforms. The immunoPCR platform represented by Imperacer from Chimera Biotech uses proprietary oligo-conjugated detection reagent followed by PCR amplification, and can in some cases improve the assay sensitivity 2 to 3 logs over conventional immunoassays.

Multiplexing bioanalysis has been implemented for small molecule LC-MS/MS PK analysis and LBA methods for biomarker detections. The same principle can also be applied to detection of coadministered multiple biologics in combination, an increasingly common therapeutic approach. With more combination studies using multiple biologics for improved efficacy, development of multiplexing LBA PK methods for the measurement of different drugs in the same sample has the potential to offer significant bioanalytical efficiencies resulting in significant time, cost, and sample acquisition savings. In addition, with molecular design of peptides conjugated to half-life extension platforms, such as Fc, polyethylene glycol (PEG), or albumin conjugates, different drug variants can exist in a study sample due to in vivo biotransformation. Provided the appropriate reagents are available, multiplexing LBA methods for measuring different drug variants, such as "intact/or active" drug versus "total" framework from the same study samples, can not only improve bioanalytical efficiencies but also reduce interassay variabilities resulting in less variable bioanalytical data for more accurate PK/PD modeling and prediction.

Some chemically synthesized peptide drugs and antisense oligonucleotide (ASO) therapeutics are not considered as biologics but still use LBAs for PK characterization due to required assay sensitivity, when critical antibody reagents can be generated against these chemically synthesized drugs. In the case of ASOs, hybridization immunoassays are a reasonably straightforward assay format that uses a sequence complementary oligonucleotide fragment conjugated with a tag, such as digitoxin, to hybridize and pull down the ASO drug followed by detection, for example, with a fluorescence-conjugated anti-digitoxin antibody. Generally the cost for establishing LBAs in a bioanalytical laboratory is very low regarding instruments, consumable goods, and desired sample throughput with the additional benefit of being readily adaptable to robotic-assisted automation when compared to other methodologies such as LC-MS/MS.

4.3.2 Cell-Based PK Assay

Cell-based assays are routinely used during the discovery phase to support intended MOAs and selection of lead candidates with superior in vitro potency which generally correlates with in vivo efficacy in animal disease models. Due to their intrinsic variabilities, cell-based assays are very challenging to be validated for supporting GLP or clinical studies according to current bioanalytical guidance (see later). As robust recombinant cell lines are being generated with improved dose responses, cell-based assays can be validated to support bioanalytical testing besides potency release testing. Blincyto (blinatumomab) has used a flow cytometry-based T cell activation bioassay for monitoring its drug concentration in serum samples from clinical studies (Trivedi et al., 2017). Cell-based bioactivity assay can truly measure the "bioactivity" from all drug variants when there is existing soluble target proteins, potential anti-drug antibodies, or lack of adequate critical reagents which limit LBA ability to measure true "active" form of drug variants.

4.3.3 LC-MS/MS PK Methods for Biologics

LC-MS/MS can complement and in some cases replace LBA methods in providing bioanalytical characterization of biologics with its unique capabilities depending on the needs of the project and the structure of the drug (Becker & Hoofnagle, 2012). This methodology was applied early in the development of ADCs due to their large molecule/small molecule hybrid structure and the need to monitor the payload of small molecule toxins (Shi et al., 2018a, 2018b). Also, during early discovery and development phases when critical reagents sometimes are not available for developing an LBA-based PK method, an LC-MS/MS method which can require less stringent or no drug-specific capture reagent may be appropriate. With a "top down" approach, a "total" drug concentration measurement can be achieved with LC-MS/MS once a signature peptide is identified based on the primary amino acid sequences of the biologic (Qiu et al., 2018). With the increasing appreciation for in vitro and in vivo biotransformation of therapeutic proteins (most commonly deaminidation, isomerization, oxidation of various amino acids), LC-MS/MS can also be used to elucidate the structure/function relationship at the molecular level by identifying PTMs and molecular stability in vitro and ex vivo which can impact critical quality attributes in vivo.

With the increasing introduction of these new and complex modalities into drug development, such as peptide agonist/or antagonist conjugated to half-life extension moieties and ADCs, the bioanalytical strategy around ADME (absorption, distribution, metabolism, and excretion) or biotransformation becomes increasingly complex which LBA-based PK methods alone are sometimes insufficient to address. Therefore, when there is evidence of or a reasonable concern that biotransformation is occurring that might alter drug activity and confound interpretation of PK data, methods suitable for elucidating biotransformations should be employed. For example, PTMs (the most common being deamindation, oxidation, isomerization) (Mo et al., 2012) or proteolytic degradation (Frohman et al., 1986) at key amino acids can cause loss of drug activity and efficacy. In the case of ADCs, monitoring the payload on the intact ADC as well as the free payload is essential to the understanding of exposure–efficacy and toxicity relationships.

Additionally, measuring the concentration of biologics with half-life extenders (PEG; hyaluronic acid; peptides made from random amino acid sequences, etc.) can be problematic with LBAs due to interference. In these cases, LC-MS/MS becomes a complementary, orthogonal, and sometimes the only method to support PK characterization of these compounds. When coupled with sample isolation by immunoaffinity (IA) pulldown (also known as hybrid IA-LC-MS/MS methods), LC-MS/MS assay sensitivity can be improved significantly. This latter approach can also be used to distinguish "free" from "total" drug (although measuring "free" drug accurately is difficult) (Lee et al., 2011). In addition to ever-improving instrumentation, techniques such as nano LC and new data analysis software for MS are providing improved detection sensitivity and quantitation. The assay sensitivity sometimes can be comparable to LBAs, but as yet is unable to match the best that LBAs can do. Furthermore the use of LC-MS/MS to structurally characterize the relationship and correlation between in vitro stability samples as well as in vivo samples for PTMs can also drive the decision on modality design, and lead candidate selection during the early stage of drug discovery by improving critical quality attributes, such as replacing labile amino acids to improve in vivo stability without significantly compromising in vivo bioactivity. LC-MS/MS can also be used to measure "free" drug concentration from nonclinical and clinical study samples by using a so-called "bottom up" approach. In this approach, drug-specific reagent (e.g., either an anti-ID antibody or its target protein) can be conjugated to magnetic beads and used for isolating intended biologics from study samples, a process termed "immunoaffinity pulldown," before they are analyzed on LC-MS/MS using a signature peptide as described earlier or intact analysis without the signature peptide (Qiu et al., 2018). The in vivo stability and detailed molecular structure can be elucidated with intact analysis on LC-MS/MS with resolution of a single amino acid difference, which can further guide the molecule design for a better candidate by eliminating structure liabilities.

The disadvantages of LC-MS/MS methodology when compared to LBA methods are more expensive capital investment, lower sample throughput, and poorer assay sensitivity (although this is improving), which may not be desirable during implementation for late stages of clinical development. The extensive review on applying LC-MS for biologics characterization are discussed in Chapter 5 "Recent Progress in Biomarker Bioanalysis and Target Engagement Assessment" and Chapter 11 "Bioanalytical Aspects in Biological Therapeutics: Biotransformation" of this book.

4.3.4 Real-Time qPCR and Flow Cytometry-Based PK Methods

Real-time qPCR-based assays are the primary methods for supporting CAR-T and gene therapy programs. With the approval of multiple CAR-T and gene therapy products, these new modalities show great promise and are another alternative for improving our ability to therapeutically intervene in previously difficult-to-treat disease settings.

CAR-T therapy is a novel adoptive cell therapy in which an antitumor antigen receptor (chimeric antigen receptor [CAR]) is transduced by a viral expression vector into a patient's T cells ex vivo, and the CAR-expressing T cells are reintroduced into the same patient (autologous CAR-T therapy). In vivo efficacy of CAR-T therapy correlates with persistence of CAR-T cells in circulation, which can be considered the cellular equivalent of drug exposure. To monitor the number of CAR-T cells in circulation, the methodology currently in use is qPCR assay to measure the copy numbers of the transgene in a whole blood sample. Briefly, genomic DNA (gDNA) isolated from a patient's whole blood sample is probed with primer sets specific for the transgene as well as a housekeeping (reference) gene, such as albumin or Apo B genes. The concentration of transgene copy numbers from gDNA calculated against a standard curve prepared with plasmid DNA containing a single copy of the transgene will be normalized against the copy number of the reference gene and total genomic DNA (μg), which serves as a "surrogate" concentration of CAR-T cells in circulation expressed as "copy number per μg gDNA." The same principle is also adopted for monitoring gene therapy products in vivo such as tissue distribution and PK/PD studies. In addition, LBA or LC-MS methodology can be used for monitoring the tissue expression of the transgenes encoded by gene therapy product. Sometime reverse transcription coupled qPCR (RT-qPCR) is used to monitor any low abundance transgene expression within target tissues.

In order to confirm the correlation of copy number of transgene to expression of CAR protein on the surface of CAR-T cells, an orthogonal flow cytometry-based method is often used to measure surface expression of CAR protein on CAR-T cells. A similar flow cytometry-based method can also be applied to enumeration of CAR-T cells in circulation. The biggest limitation of the flow cytometry-based PK method is its poor sensitivity and general requirement for freshly prepared cells which prevent its use as the primary PK method other than complementing the qPCR method when CAR-T cell concentration is high, such as during the expansion phase after administration.

The validation of qPCR and flow cytometry methods can be very challenging due to inherent assay variabilities and there being no regulatory guidelines to follow. Despite this lack of health authority guidance on the appropriate assay validation process for qPCR and flow cytometric methods, these topics are frequently discussed at scientific conferences and there are white papers which provide the consensus on how to validate these methods in order to support GLP or clinical studies (Der Strate et al., 2017).

4.4 Automation and Method Validation

Automation has been widely adopted into the daily activities of bioanalytical laboratories (Ware et al., 2018) providing substantial benefits including the replacement of manual operation with automated liquid handling leading to improved accuracy and precision of the assay (Joelsson et al., 2008), reduced repetitive motion injuries, improved sample throughput without commensurate need for more personnel, and greater reproducibility within and across studies (eliminating analyst-to-analyst variability) over time. In addition, the programing capabilities of automation software systems can facilitate a more complex design-of-experiment (DOE) approach for more efficient and robust assay development. However, implementation of automation is not an easy process due to multiple factors including the lack of familiarity with these systems by laboratory personnel, concern over job replacement by automated systems, small assay changes beyond DOE can be cumbersome when run on automated systems, less user-friendly interfacing with relatively complex hardware and software systems, plus the fact that manually developed assays may require modification to perform as required on an automated system. In this regard, it is advantageous to know in advance if an assay will be transferred to an automated platform to account for this translation. Additionally, automated liquid handlers can improve the reproducibility of standard calibrators and QC preparation, in turn improving assay performance and reproducibility.

However, automation is not a panacea for bioanalytical laboratories. The "total laboratory automation (TLA)" proposed for routine clinical chemistry laboratories (Armbruster et al., 2014) is not always the best strategy for the application of automation to the development and implementation of LBA methods in bioanalytical laboratories. Routine PK assay methods can be developed and optimized very efficiently by an experienced analyst without the need for automation. Method development and bioanalysis groups can often share the same instruments/equipment for the reasons of reduced capital cost, maximal efficiency of equipment use, and laboratory space. With multiple different methods running daily, relying solely upon TLAs will not provide operational flexibility due to the limited access to the instrument when TLAs are in operation for one or multiple methods, or require redundant systems for purposes of development and production. However, if a templated method is programed and applied routinely within a laboratory then a TLA system can readily provide assay development runs that can be performed. It has been shown that head-to-head a human analyst can equal or outperform an automated system, though over the long haul repetition fatigue can lead to errors that generally will not occur on a (regularly maintained) automation system when continuous large batch runs are being performed.

PK method validation is performed according to industry white papers and regulatory guidelines when the use of a given method is intended to support safety/efficacy/bioequivalence studies (DeSilva et al., 2003; EMA, 2012; FDA, 2018). Details of method validation parameters and processes will not be presented here, and the interested reader should consult the references cited. The PK method validation report is part of the submission documents when an Investigational New Drug (IND) (although the format for INDs is changing to be closer to that of a Clinical Trial Application, which contains fewer reports such as method validations) and Biologic License Application (BLA) are filed. The validation process and data from a sponsor, whether the work was done directly by the sponsor or at a contract research organization, will be subjected to inspections from regulatory authorities prior to the market approval of BLAs/or Marketing Authorization Applications (MAAs). The validation process consists of two phases: pre-study versus in-study validation. During pre-study validation, key assay parameters, accuracy, precision, selectivity, specificity, dilutional linearity, spike/recovery, and stability, will be evaluated against prespecified assay acceptance criteria (see guidance documents). The specific requirements for method validation differ between LC-MS/MS and LBA methods, as detailed in the various published regulatory guidance documents. In-study validation will include testing of incurred sample reanalysis and parallelism when study samples are available for the assessment. Success in completing validation is the foundation for producing high-quality and reliable bioanalytical data.

For the more nontraditional, less-established PK methods described (i.e., hybrid IA LC-MS/MS, qPCR, flow cytometry, cell-based methods) where no guidance currently exists, appropriate application of the parameters listed (or others as deemed necessary for the specific method) and assay acceptance criteria with due regard to good scientific practices, as well as direct discussions with health authorities, will be needed.

4.5 Conclusions

PK methods are critical in the development of biologics for elucidation of in vitro and in vivo structure/function, PK/PD, and exposure–toxicity/efficacy relationships evaluated in nonclinical and clinical studies. PK methods are required to be validated according to regulatory guidance in order to support GLP, pivotal clinical, and bioequivalence studies.

During the last 20 years of development of biologics, a single LBA-based PK method is usually sufficient for supporting drug concentration measurement during nonclinical and clinicial development. With recent advancements in complex molecular design and modalities of biologics, such as multispecifics and ADCs, no single PK method is sufficient to provide answers to critical bioanalytical questions related to characterization of ADME and PK. Multiple PK methods to measure different drug variants, such as "intact/or active," "total" drug forms, or ratio of drug to payload, will need to be developed across different analytical platforms, usually in combination with LBA, LC-MS/MS, cell-based assays, and so on.

Reagent generation is very important in the development of PK methods for biologics regardless of the analytical platform used. This is due to the fact that the in vivo bioactivity of a biologic depends on its tertiary or higher-order structure and the desire to measure structurally intact drug forms to establish in vivo PK/PD/toxicity relationships. Critical reagents specific to biologics, typically anti-ID antibodies or recombinant target proteins, are essential for isolating a biologic drug from complex biological matrices for further detection or molecular characterization/measurement. The life cycle management of critical reagents plays an important role in maintaining consistent and robust performance of PK methods during the "end-to-end" development of biologics. An arsenal of biochemical and biophysical methods should be implemented for characterizing critical reagents, source proteins, and their conjugates.

Recent breakthroughs in gene therapy and CAR-T therapies have provided novel therapeutic opportunities for significantly impacting diseases with high unmet medical needs. For example, Luxturna is the first gene therapy approved by the US Food and Drug Administration (FDA) to treat a rare, inherited retinal disease due to mutations in the RPE65 gene that can lead to blindness; two CAR-T therapies, Kymriah and Yescarta, both targeting B cell lymphomas, have been recently approved by the FDA. PK characterization of these types of therapies present unique challenges that will require novel bioanalytical approaches, such as real-time qPCR methods for transgene detection or flow cytometry-based methods for enumeration of circulating CAR-T cells, differing significantly from the more conventional LBAs or LC-MS/MS for traditional biologics. It is expected that as more novel therapeutic modalities are designed and implemented to treat diseases currently underserved by conventional approaches, yet more innovative methods for assessing PK will need to be developed.

Automation not only can improve method precision but significantly enhance throughput for study sample analysis, which contributes to reduction of cost in bioanalytical laboratories. Artificial intelligence (AI)-assisted automation systems can provide a true "walk away" system which can fully automate method development, validation, and bioanalysis from sample storage, movement, testing, and data reporting in a highly compliant manner assisted by bar coding systems. The fully automated method development/validation/bioanalysis driven by AI will be the future of bioanalytical laboratories.

PK method validation is essential not only to assure satisfactory method performance resulting in high-quality and reliable PK data but also for the successful approval of BLAs or MAAs. The validation process should follow the recommendations and priniciples from regulatory guidelines for LBA and LC-MS/MS issued by health authorities, while exceptions in meeting prior critieria during method validation can be made by scientific justifications ("fit for purpose"). But the complexity of immunoaffinity pulldown-coupled LC-MS/MS, qPCR, cell-based assays, and flow cytometry-based methods brings technical and regulatory challenges for their method validation, and for these novel methods both industry and health authorities are still in the process of discussing what constitutes best practice. With experience, regulatory expectations will crystallize into informed guidance documents.

References

Armbruster, D. A., Overcash, D. R., & Reyes, J. (2014). Clinical chemistry laboratory automation in the 21st Century – Amat Victoria curam (Victory loves careful preparation). *The Clinical Biochemist. Reviews/Australian Association of Clinical Biochemists*, 35(3), 143–153. https://www.ncbi.nlm.nih.gov/pmc/articles/PMC4204236

Becker, J. O., & Hoofnagle, A. N. (2012). Replacing immunoassays with tryptic digestion-peptide immunoaffinity enrichment and LC-MS/MS. *Bioanalysis*, 4(3), 281–290. DOI: 10.4155/bio.11.319

Cinquanta, L., Fontana, D. E., & Bizzaro, N. (2017). Chemiluminescent immunoassay technology: What does it change in autoantibody detection? *Autoimmunity Highlights*, 8(1), 9. DOI: 10.1007/s13317-017-0097-2

Der Strate, B. V., Longdin, R., Geerlings, M., Bachmayer, N., Cavallin, M., Litwin, V., Patel, M., Passe-Coutrin, W., Schoelch, C., Companjen, A., & Fjording, M. S. (2017). Best practices in performing flow cytometry in a regulated environment: Feedback from experience within the European Bioanalysis Forum. *Bioanalysis*, 9(16), 1253–1264. DOI: 10.4155/bio-2017-0093

DeSilva, B., Smith, W., Weiner, R., Kelley, M., Smolec, J., Lee, B., Khan, M., Tacey, R., Hill, H., & Celniker, A. (2003). Recommendations for the bioanalytical method validation of ligand-binding assays to support pharmacokinetic assessments of macromolecules. *Pharmaceutical Research*, 20(11), 1885–1900. DOI: 10.1023/b:pham.0000003390.51761.3d

European Medicines Agency (EMA). (2012). Guideline on Bioanalytical Method Validation. https://www.ema.europa.eu/en/documents/scientific-guideline/guideline-bioanalytical-method-validation_en.pdf

Faria, M., & Halquist, M. S. (2018). Internal standards for absolute quantification of large molecules (proteins) from biological matrices by LC-MS/MS. In ed Mark, T. *Calibration and Validation of Analytical Methods – A Sampling of Current Approaches*. INTECHOPEN LIMITED. https://www.intechopen.com/chapters/60447

Findlay, J. W., Smith, W. C., Lee, J. W., Nordblom, G. D., Das, I., DeSilva, B. S., Khan, M. N., & Bowsher, R. R. (2000). Validation of immunoassays for bioanalysis: A pharmaceutical industry perspective. *Journal of Pharmaceutical and Biomedical Analysis*, 21(6), 1249–1273. DOI: 10.1016/s0731-7085(99)00244-7

Fischer, S. K., Joyce, A., Spengler, M., Yang, T. Y., Zhuang, Y., Fjording, M. S., & Mikulskis, A. (2015). Emerging technologies to increase ligand binding assay sensitivity. *The AAPS Journal*, 17(1), 93–101. DOI: 10.1208/s12248-014-9682-8

Food and Drug Administration (FDA). (2018). Bioanalytical Method Validation: Guidance for Industry. https://www.fda.gov/files/drugs/published/Bioanalytical-Method-Validation-Guidance-for-Industry.pdf

Frohman, L. A., Downs, T. R., Williams, T. C., Heimer, E. P., Pan, Y. C., & Felix, A. M. (1986). Rapid enzymatic degradation of growth hormone-releasing hormone by plasma in vitro and in vivo to a biologically inactive product cleaved at the NH2 terminus. *Journal of Clinical Investigation*, 78(4), 906–913. DOI: 10.1172/JCI112679

Geist, B. J., Egan, A. C., Yang, T. Y., Dong, Y., & Shankar, G. (2013). Characterization of critical reagents in ligand-binding assays: Enabling robust bioanalytical methods and lifecycle management. *Bioanalysis*, 5(2), 227–244. DOI: 10.4155/bio.12.304

Gottschalk, P. G., & Dunn, J. R. (2005). The five-parameter logistic: A characterization and comparison with the four-parameter logistic. *Analytical Biochemistry*, 343(1), 54–65. DOI: 10.1016/j.ab.2005.04.035

Gross, E. M., Maddipati, S. S., & Snyder, S. M. (2016). A review of electrogenerated chemiluminescent biosensors for assays in biological matrices. *Bioanalysis*, 8(19), 2071–2089. DOI: 10.4155/bio-2016-0178

Joelsson, D., Moravec, P., Troutman, M., Pigeon, J., & DePhillips, P. (2008). Optimizing ELISAs for precision and robustness using laboratory automation and statistical design of experiments. *Journal of Immunological Methods*, 337(1), 35–41. DOI: 10.1016/j.jim.2008.05.012

Khan, M. N., & Findlay, J. W. (2009). *Ligand-Binding Assays: Development, Validation, and Implementation in the Drug Development Arena*.

Lee, J. W., Kelley, M., King, L. E., Yang, J., Salimi-Moosavi, H., Tang, M. T., Lu, J. F., Kamerud, J., Ahene, A., Myler, H., & Rogers, C. (2011). Bioanalytical approaches to quantify "total" and "free" therapeutic antibodies and their targets: Technical challenges and PK/PD applications over the course of drug development. *The AAPS Journal*, 13(1), 99–110. DOI: 10.1208/s12248-011-9251-3

Liu, G. H., & Wu, J. T. (1998). Impact of assay parameters on the accuracy of free PSA test: Source and stability of calibrator, calibration curve fitting, and level of total PSA in the serum. *Journal of Clinical Laboratory Analysis*, 12(5), 304–309. DOI: 10.1002/(SICI)1098-2825(1998)12:5<304::AID-JCLA10>3.0.CO;2-Z

Mo, J., Tymiak, A. A., & Chen, G. (2012). Structural mass spectrometry in biologics discovery: Advances and future trends. *Drug Discovery Today*, 17(23–24X), 1323–1330. DOI: 10.1016/j.drudis.2012.07.006

O'Hara, D. M., & Theobald, V. (2013). Life cycle management of critical ligand-binding reagents. *Bioanalysis*, 5(21), 2679–2696. DOI: 10.4155/bio.13.241

Park, J. H., Geyer, M. B., & Brentjens, R. J. (2016). CD19-targeted CAR T-cell therapeutics for hematologic malignancies: Interpreting clinical outcomes to date. *Blood*, 127(26), 3312–3320. DOI: 10.1182/blood-2016-02-629063

Qiu, X., Kang, L., Case, M., Weng, N., & Jian, W. (2018). Quantitation of intact monoclonal antibody in biological samples: Comparison of different data processing strategies. *Bioanalysis*, 10(13), 1055–1067. DOI: 10.4155/bio-2018-0016

Ramagiri, S., & Moore, I. (2016). Hybridizing LBA with LC-MS/MS: The new norm for biologics quantification. *Bioanalysis*, 8(6), 483–486. DOI: 10.4155/bio.16.9

Shi, C., Goldberg, S., Lin, T., Dudkin, V., Widdison, W., Harris, L., Wilhelm, S., Jmeian, Y., Davis, D., O'Neil, K., Weng, N., & Jian, W. (2018a). Bioanalytical workflow for novel scaffold protein-drug conjugates: Quantitation of total Centyrin protein, conjugated Centyrin and free payload for Centyrin-drug conjugate in plasma and tissue samples using liquid chromatography-tandem mass spectrometry. *Bioanalysis*, 10(20), 1651–1665. DOI: 10.4155/bio-2018-0201

Shi, C., Goldberg, S., Lin, T., Dudkin, V., Widdison, W., Harris, L., Wilhelm, S., Jmeian, Y., Davis, D., O'Neil, K., Weng, N., & Jian, W. (2018b). LC/MS/MS bioanalysis of protein-drug conjugates-the importance of incorporating succinimide hydrolysis products. *Analytical Chemistry*, 90(8), 5314–5321. DOI: 10.1021/acs.analchem.8b00411

Trivedi, A., Stienen, S., Zhu, M., Li, H., Yuraszeck, T., Gibbs, J., Heath, T., Loberg, R., & Kasichayanula, S. (2017). Clinical pharmacology and translational aspects of bispecific antibodies. *Clinical and Translational Science*, 10(3), 147–162. DOI: 10.1111/cts.12459

Ware, M., Shankar, G., Breslin, P., & Yang, T. Y. (2018). Application of automated liquid handling in ligand-binding assay-based bioanalytical method development: The practical considerations. *Bioanalysis*, 10(7), 425–430. DOI: 10.4155/bio-2017-0212

Watson, R. G., Clements-Egan, A., Schantz, A., Ware, M., Wu, B., Yang, T. Y., Shankar, G., & Marini, J. C. (2017). Implementing a tiered approach to bioanalytical method validation for large-molecule ligand-binding assay methods in pharmacokinetic assessments. *Bioanalysis*, 9(18), 1407–1422. DOI: 10.4155/bio-2017-0044

Wild, D. (2013). *Theory and applications of ligand binding, ELISA and related techniques*. (4th ed.).

Yalow, R. S. (1973). Radioimmunoassay. Practices and pitfalls. *Circulation Research*, 32(Suppl 1), 116–128. https://pubmed.ncbi.nlm.nih.gov/4351355

Yalow, R. S., & Berson, S. A. (1960). Immunoassay of endogenous plasma insulin in man. *Journal of Clinical Investigation*, 39, 1157–1175. DOI: 10.1172/JCI104130

5

Recent Progress in Biomarker Bioanalysis and Target Engagement Assessment

Yan G. Ni,[a] *Lindsay E. King*[b] *and Carmen Fernández-Metzler*[c]

[a]*Passage Bio (corresponding author)*
[b]*Pfizer Inc.*
[c]*PharmaCadence Analytical Services*

5.1 Section 1: Introduction

Box 1: Biomarkers categories
Susceptibility/risk biomarker Diagnostic biomarker Monitoring biomarker Prognostic biomarker Predictive biomarker Pharmacodynamic/response biomarker Safety biomarker See FDA-NIH Biomarker Working Group BEST document (https://www.ncbi.nlm.nih.gov/books/NBK338448)

Biomarkers include a diverse array of modalities, from blood chemicals, nucleotides, and proteins to functional imaging. As such, biomarker bioanalysis leverages a variety of technical platforms to meet the analytical needs in drug development. The "fit-for-purpose" (FFP) assay validation principle (Lee et al., 2006) has provided a unifying approach for validating biomarker assays on different technical platforms and for different context of use (COU). FFP validation is sometimes mistaken as less than full validation. On the contrary, an FFP-validated assay is fully validated based on COU (Piccoli, 2019). COU

Bioanalytical Aspects in Biological Therapeutics, First Edition. Edited by Xiaohui Xu and Weifeng Xu.
© 2022 John Wiley & Sons, Inc. Published 2022 by John Wiley & Sons, Inc.

refers to a concise description of a biomarker's specific intended use. The COU for a given biomarker assay in the drug development process may evolve with changes in drug candidate, patient population, and other factors, and consequently should define the extent of analytical validation.

There are many challenges intrinsic to biomarker assay validation due to their diverse technical platforms and unique COU. In recent years, industry experts and working groups, many led by the American Association of Pharmaceutical Scientists (AAPS), have proposed solutions to address the unique issues in biomarker bioanalysis such as reference calibrator characterization, parallelism assessment, biomarker specimen preservation, and stability assessment, and validation approaches for flow cytometric testing, commercial kits, and multiplexing assays (Cowan et al., 2017; Dakappagari et al., 2017; Jani et al., 2016; Khan et al., 2015; Lee et al., 2011; O'Hara et al., 2012; Stevenson & Purushothama, 2014).

Many targets for biotherapeutics are soluble proteins, peptides (e.g., cytokines), or cell surface molecules (e.g., CD20). Target levels and target engagement assessment provide important information for understanding the pharmacokinetics-pharmacodynamics (PK-PD) relationship and dose selection. Protein target levels can be measured using multiple technical platforms, which can bring another set of unique challenges in biomarker bioanalysis. Expression levels of membrane proteins are most commonly assessed using immunohistochemistry and flow cytometry; while soluble protein target levels in circulation are often measured using ligand binding assays (LBA) and liquid chromatography-mass spectrometry (LC-MS). In recent years, LC-MS and LBA have also been used for the measurement of target levels in the tissues. A combination of these approaches can provide information on both anatomical distribution and cell type information of the target, as well as quantitative information on levels of expression.

This chapter will review recent progress in biomarker bioanalysis with a focus on emerging consensus addressing unique challenges in biomarker assay validation, and their application in the development of quantitative target biomarker assays using LBA, LC-MS, and flow cytometry. We will also provide examples on how knowledge of target levels can inform decisions during clinical drug development.

5.2 Section 2: The "Fit-for-Purpose" Approach and Considerations on Biomarker Method Validation

The diversity in biomarker COU, modality (e.g., DNA, RNA, protein, cells), and technical platforms for biomarker quantification (e.g., next-generation sequencing, LBA, LC-MS, enzyme activity assays, functional imaging) has made it challenging to establish a single guidance document for all biomarkers.

Unlike PK assays, the majority of biomarker quantification occurs in clinical practice and outside pharmaceutical research and is therefore subject to Clinical Laboratory Improvement Amendments (CLIA). Furthermore, the experimental practices used for biomarker validation are varied. For example, LBA has been the main technology used for protein biomarkers and, therefore, many protein biomarker assay validation practices reflect the conventions used in LBAs. The increase in biotherapeutic programs has led to the use of LBA for PK measurement which, along with recent use of LC-MS for biomarker quantitation, has brought a perception that the same extent of experimental validation applies to both biomarker and PK methods. This perception is misleading and neglects the fact that biomarker and PK measurement have entirely different COU, and often employ different technical platforms. The need to identify a unifying scheme for biomarker method validation was first addressed by the FFP approach recommended in the 2006 landmark white paper by Lee et al. (2006) and endorsed by the bioanalytical field (DeSilva, 2012; Houghton, 2009, 2012) and the FDA (Booth et al., 2015). The key principle of the FFP approach is that the objective of a biomarker method validation is to ensure the analytical performance characteristics of a method are sufficient to support its COU. This requires active communication and collaboration between assay developers and assay end users, and can be an iterative process especially when there is change in the COU.

Regulatory guidance on the use of LBA or LC-MS for PK and toxicokinetics (TK) analysis to support pharmaceutical research and development has been available for the past two decades beginning with the FDA Guidance on Bioanalytical Methods Validation in 2001 (FDA BMV, 2001). Currently, there is still no regulatory guidance document geared specifically for biomarker method validation; although in 2013 the FDA announced in a draft update to their 2001 BMV Guidance that included biomarker methods within scope (FDA BMV, 2013). In the finalized BMV guidance (FDA, 2018), biomarker method validation is included in a very small section, with the recommendation that (1) for assays intended to support early drug development (e.g., candidate selection, go-no-go decisions, proof-of-concept), the sponsor should incorporate the extent of method validation they deem appropriate and (2) when biomarker data will be used to support a regulatory action, such as the pivotal determination of safety and/or effectiveness or to support labeled dosing instructions, the assay should be fully validated. Here the FDA recommends using drug assay validation elements as a starting point in biomarker assay validation and recognizes that not all the characteristics of a PK/TK validation fit the biomarker assay validation.

The inclusion of biomarker methods in the updated FDA BMV (FDA-Biopharmaceutics-2001, 2013; FDA, 2018) resulted in active discussions on biomarker method validation within the AAPS community at the Crystal City V meeting (2013) (Booth et al., 2015), and organization of the Crystal City VI meeting (2015) (Arnold et al., 2016). It is now well recognized that biomarker and PK method validation face very different challenges. For example, for many biomarker assays, including the target assays, the calibrator proteins may not be identical to the endogenous analyte, which leads to the need to demonstrate similarity between the calibrator and endogenous analyte and a different perspective of assay accuracy (Cowan et al., 2017). A draft AAPS white paper representing industry consensus and written in conjunction with the Critical Path Institute (Piccoli, 2019) describes more detailed current thinking on biomarker assay validation, and scientific expectations for biomarker methods validation in support of Drug Development Tools (DDT), with the recognition that the method employed in a biomarker study can influence what is being measured and the definition of a biomarker (e.g., early assays for prostate-specific antigen). In addition, although LC-MS-related considerations were included in this C-Path paper (Piccoli, 2019), an AAPS Biomarker subteam, now the AAPS Biomarker and Precision Medicine Community, is working on an in-depth review/white paper on this subject that includes a short primer to introduce relevant concepts and topics germane to biomarker assay validation (Fernández-Metzler and colleagues, personal communication, 2020).

The 2018 FDA guidance document was specific for LBA or LC-MS platforms. Clinical flow cytometry applications have been defined by key specific publications from flow cytometry organizations such as the International Clinical Cytometry Society. For clinical laboratories operating under College of American Pathologists (CAP) and/or CLIA, specific cytometry assay guidance has been provided through the Clinical & Laboratory Standards Institute (CLSI; see https://clsi.org) for immunophenotypic analysis of lymphocyte subsets and hematopoietic stem cells (H42), neoplastic hematolymphoid cell analysis (H43), and red blood cell diagnosis tests (H52). The Standards for Validation of Methods Performed by Flow Cytometry (H62) was released in 2021 (https://clsi.org/standards-development/documents-for-public-review). It should be noted that CLSI documents are scientific recommendations and not regulatory guidance.

5.3 Section 3: Recent Advances in Addressing Specific Challenges Intrinsic to Biomarker Assay Validation

Biomarker quantification in biological matrices by LBA or LC-MS assays shares similarities with bioanalysis of therapeutic agents, including the need for sensitive, specific, and reproducible detection of the analyte and the

Box 2: Key Assay Characterization Parameters

Assay sensitivity can be defined by Limit of Quantitation (lowest and highest analyte concentrations that can be measured with acceptable precision) or Limit of Detection (lowest analyte concentration that can be detected, typically 3 standard deviations, above background).

Assay specificity determines the ability of an assay to distinguish between the analyte of interest and other structurally similar analytes within the endogenous sample matrix.

Assay precision: "The closeness of agreement among a series of measurements obtained from multiple sampling of the same homogenous sample under the prescribed conditions" (FDA BMV, 2018).

Accuracy: "The degree of closeness of the determined value to the nominal or known true value under prescribed conditions." (FDA BMV, 2018).
Analytical quantification allows the determination of whether two measurements are the same or different. Relative quantification – good precision (e.g., <20%) but accuracy is unknown; absolute quantification – both precision and accuracy are good (e.g., <20%).
Absolute quantification requires a consensus (e.g., WHO or NIST) calibrator that is exactly the same as the endogenous analyte. Very few biomarker assays, including small molecule assays, can meet these requirements

Parallelism: An experimental parameter determined by measurement of serially diluted samples with endogenous analytes in comparison to serial dilutions of the assay calibrator which is typically a recombinant protein. Used to demonstrate the endogenous analyte in matrix is recognized similarly to the calibrators by the assay reagents.

Dilution linearity: For LC-MS assays, "dilutional linearity demonstrates the accurate measurement of concentrations of samples exceeding the quantitation range when serially diluted to within the quantitative assay range." (FDA BMV, 2018)

Stability: A measure of the lack of loss or degradation in a given matrix under specific storage and use conditions relative to the starting material for given time intervals. (FDA BMV, 2018).

concomitant quality-defining attributes. However, biomarker method validation poses very specific challenges not present in PK/TK assays: the lack of clear definition of what needs to be measured, the difficulty in obtaining reference material which is identical to the endogenous molecule, particularly for protein biomarkers, and the lack of an analyte-free matrix for the preparation of calibrators and quality control (QC) samples. The diverse nature of biomarker modalities also leads to challenges in controlling preanalytical variables, including sample stability, to ensure that the study results reflect the endogenous analyte concentration at the time of sample collection. Additional issues include the challenges associated with free target assessment post-therapeutic treatment, and the use of multiplexing assays and commercial kits. Emerging consensus in addressing these unique challenges in biomarker bioanalysis will be discussed further in this and subsequent sections.

5.3.1 Biomarker Assay Calibrator

Calibrator material is of paramount importance for assay quality. Often, an assay calibrator is not only used to generate the assay standard curve, and derive the quantity of analyte in a sample; it is also the starting point for capture reagent screen when developing an immune-based assay. Therefore, it is critical that the bioanalytical scientist has a clear understanding of the assay attributes from the assay calibrator perspective. A key challenge in building a biomarker assay is to identify a calibrator material that is structurally and functionally representative of the endogenous analyte of interests (Cowan et al., 2017). This includes determining the correct protein isoform/isoforms to be measured, the physicochemical properties (i.e., whether the analyte is a monomer or multimer), posttranslational modifications, and higher-order structures of the analyte. This is fundamentally different from PK method development, in which the therapeutic agent itself can be spiked in a sample matrix to generate an assay calibrator, and thus provides more confidence that the calibrator is identical to the analyte of interests. Biomarker assay calibrators are generally recombinant or synthetic materials and are spiked into a buffer matrix. Consequently, in many biomarker assays, calibrator materials are not identical to the endogenous analyte of interest. Therefore, most biomarker assays, including target assays, provide relative quantitative measurements. Biomarker assay validation should include a plan to characterize the assay calibrator and ascertain that it is suitable for the intended use.

For antibody drug development, total and free target assays are mainstays for assessing baseline target levels and degree of target engagement post-treatment. For total and free target assays, the same unbound recombinant target proteins are commonly used as calibrator material. Finally, it is also possible to develop assays to measure the concentration of target in complex with a drug or endogenous binding partner. The calibration curve for these assays may be generated by incubating the recombinant target protein with drug to form target–drug complex just before use in the assay (Harth et al., 2019).

There are several international reference standard sources, such as the National Institute for Biological Standards and Control (NIBSC: www.nibsc.org/products/brm_product_catalogue/ rs), the European Directorate of the Quality of Medicines (EDQM: www.edqm.eu/en/ph-eur-reference-standards-627.html), the US Pharmacopeial Convention (USP: www.usp.org/reference-standards), the National Institute for Standards and Technology (NIST: www.nist.gov/srm/index.cfm), and the World Health Organization (WHO: http://www.who.int/bloodproducts/ref_materials/en) (Cowan et al., 2017). In general, these organizations serve the needs of clinical diagnostic assays. Although some of the clinical diagnostic assays are used as drug development tools, the majority of biomarker assays used in drug discovery and development lack well-defined and characterized reference material. For protein biomarkers with high molecular complexity, obtaining appropriate reference materials and fully characterizing these materials is challenging and requires a systematic approach.

5.3.1.1 Protein Biomarker Reference Standard Characterization

Due to the lack of knowledge of the endogenous analyte and the heterogeneity which often exists, finding calibrator material which is identical to the endogenous analyte of interest is a major challenge in protein biomarker assay development. In some organizations, assay developers may collaborate with protein engineers to obtain well-characterized target material, as well as drug–target-specific data that would support biomarker assay development and serve as calibration material. However, this is not common.

To address this issue, an AAPS Biomarker Focus Group has published a white paper (Cowan et al., 2017) and recommended that, to the best extent possible, protein assay reference materials should be characterized to ascertain the identity of the calibrator matches that of the analytes of interest, and the concentration of the calibrator is accurate. The assay developers/end users should consider including the following key aspects during the selection and characterization of assay reference materials:

- Ascertain the correct form of endogenous analyte (e.g., monomer or polymer, isoforms, and posttranslational modifications) when possible.
- If a recombinant protein is used as calibrator, acquire information on the protein expression systems (e.g., *Escherichia coli*, insect cells, or eukaryotic expression system), the degree of purity/impurities, lot-to-lot differences, and potential issues related to misfolded protein and protein stabilities.

Table 5.1 Recommendations for protein biomarker reference material characterization (adapted from Cowan et al., 2017).

Characterization element	Proteins
Identity	Review product data sheet, peptide mapping, and sequence confirmation if needed
Quantification	If the material is protein carrier-free, confirm the information provided on the product datasheet using measurement of A280 with the proper extinction coefficient, bicinchoninic acid (BCA) or Bradford protein assay
Impurity and physicochemical properties	Understand whether the calibrator material is a monomer or polymer, and the appropriate protein isoform based on the study needs. Review product data sheet, conduct confirmation studies using techniques such as electrophoresis, size exclusion column, mass spectometry, dynamic light scattering, etc.
Stability	Both short-term (i.e., freeze–thaw, 4°C, room temperature, etc.) and long-term (multiyear −70°C) should be assessed based on the study needs
Parallelism	Calibrator should represent endogenous analyte, disease-state samples should be included as appropriate
Lot-to-lot variability	Test multiple lots from the same vendor if possible to determine variability. Can also substitute the kit calibrator with a well-characterized material of the same protein from an independent source

- Conduct a parallelism assessment between the calibrator and endogenous material during method development to ensure that calibrator material and the endogenous analyte are detected similarly by the assay.

In addition, the authors also recommended a tiered approach for biomarker calibrator material characterization, so that the level of characterization on the key properties, such as identity, quantity or concentration, and impurity, is based on the intended use of the data (Table 5.1), with more rigor being required for assays intended for use in making patient treatment decisions.

5.3.1.2 Small Molecule Biomarker Reference Standard Characterization

Unlike protein reference materials, calibrator material for small molecule biomarkers may be more accessible because they can be chemically synthesized in both unlabeled and labeled form. For simple small molecules, purity and identity can be assessed and confirmed using physicochemical approaches. For example, it may be sufficient to verify identity via standard chromatographic separations and nuclear magnetic resonance (see Table 5.2). For complex

Table 5.2 Recommendations for small molecule biomarker reference material characterization.

Characterization element	Small molecules and peptides
Description/structure/composition (molecular information)	Structure and empirical formula; salt form
Lot number	√
Preparation date	√
Test/retest date	√
Storage condition/stability	√
Appearance	√
Identity (IR, NMR, etc.)	LC/UV, LC/MS, NMR
Purity (method)	LC/UV, NMR

IR, infrared; LC, liquid chromatograpy; MS, mass spectrometry; NMR, nuclear magnetic resonance; UV, ultraviolet.

small molecules such as lipids, however, reference materials may be mixtures of several components which are difficult to separate as an individual standard.

A certificate of analysis which includes structural information, lot number, preparation/test/retest dates, storage condition, identity, purity, and appearance must be obtained for each reference material. Since characterization methodologies can vary from laboratory to laboratory, each type of characterization should be carefully selected based on the reliability of the platform for the measurement. It may also be necessary to provide supplemental information. For example, if purity is low, potential impurities which may impact assay reliability should be identified. If a compound is recognized as hydrophilic, details on water content and reagent storage information to prevent hydration may be required. Many ancillary techniques can be applied to further characterize and ensure reference material quality. These techniques may be added over time as a biomarker assay progresses through its assay life cycle. Examples of ancillary characteristics used for small molecule reference standard characterization include loss on drying, water content (Karl Fischer titrations), residual solvents (gas chromatography), light sensitivity, hygroscopicity characteristics, pH stability, density, inorganic matter/residue on ignition (ash), heavy metals, comparison to compendial standards, solubility, mixture percent if there are multiple components, and biological activity.

5.3.2 Critical Reagents

Critical reagents are defined most broadly as materials without which the assay would no longer meet acceptance criteria. While there may be occasions when uncoated enzyme-linked immunosorbent (ELISA) plates or buffer additives fall into this category, critical reagents mostly refer to assay calibrators or capture and detection reagents which are typically recombinant proteins and antibodies. These protein reagents are usually conjugated to other molecules such as biotin, enzymes (in ELISA), ruthenium (in electrochemiluminescence [ECL]), and fluorescent molecules. Thus, both the parent and the conjugate must be considered from the perspective of the supply and reagent stability. For example, for LBA or hybrid LBA-LC-MS/MS-based assays, the capture antibody used is a critical reagent and, as such, should be procured and characterized for the appropriate duration of use (Santockyte et al., 2019). If needed, a reagent bridging strategy should also be put in place.

During the past 10 years, several white papers have been published to address the challenges of managing and characterizing critical reagents (King et al., 2014; O'Hara et al., 2012; Staack et al., 2011). These papers should be consulted when needed. However, a number of issues remain to be resolved, including the degree of reagent characterization needed to match the intended use of the assay. For example, for diagnostic assays with Premarket Approval (PMA) or 510 K, complete control of the reagents is critical. At the other end of the spectrum, for so-called "home brew" assays, a single lot of reagents can be used over a very short time frame with good success and minimum reagent management. In addition, commercial reagents and complete assay kits (including multiplex formats) are widely used in biomarker studies. The convenience of these tools provides significant benefits but there are also caveats as described by Khan and others (Khan et al., 2015; Nowatzke et al., 2010; Rifai et al., 2013). Reagent specificity and affinity are of particular importance for assay development. For antibodies, specificity is often selected for by immunogen design and screening strategy, and optimal affinity clones can also be selected during this time. However, for commercial products these data may not be fully available to the user and may require additional experimental confirmation.

5.3.3 Parallelism

For LBA-based biomarker assays, parallelism assessment refers to a dilution linearity experiment using a sample matrix containing the *endogenous* analyte. The goal for this study is to demonstrate that the assay calibrator is immunologically similar to the endogenous analyte; in other words, the reagents see assay calibrator and endogenous analyte as the same. This is a critical experiment that needs to be conducted early in assay development, and again during assay validation (King, 2016; Lee et al., 2005; Stevenson & Purushothama, 2014) for each COU. For example, parallelism should be reassessed when there are changes in sample matrix, assay range, or analyte forms. Parallelism is demonstrated when the response from the diluted sample with endogenous analyte is proportional to the dilution factor, and the dilution curve is parallel to the calibration curve. When this occurs, the assay provides relative quantitative data. If there are issues with parallelism assessment, a decision should be made to either communicate with the end user and discuss the potential impact on data or to continue optimizing the assay (e.g., reconsider calibrator selection).

There are a number of challenges associated with this assessment including defining acceptance criteria, determining sample size, and obtaining relevant disease samples with levels of analyte sufficient for the study. Parallelism assessment can be used to select the minimal required dilution (MRD) and enable the use of a surrogate matrix. In fact, it is important to define these before final assessment of parallelism/similarity, because matrix effects that are managed by MRD and MRD buffer formulation can be confounding. In addition, the precision of the assay needs to be established so that analytical error is not mistaken for lack of parallelism. Finally, if all samples dilute linearly and similarly, but the response is not parallel to the standard curve, the assay may still provide useful data depending on the COU.

Parallelism is used in LC-MS biomarker assay development as a way to determine matrix equivalence. For LC-MS biomarker assays, the use of surrogate matrices is common. LC-MS biomarker methods use extraction and chromatographic separation during sample preparation and introduction to the mass spectrometer. These separation methods routinely disrupt binding partner interactions. Consequently, synthetically prepared surrogate matrices can be used. A parallelism study is used to assess whether the analyte concentration–response relation is equivalent for the surrogate and authentic biological matrix across the range of analysis. Jones et al. (2012) recommend three methods for parallelism assessment to qualify a surrogate matrix for subsequent use in validation: (1) spike-recovery, (2) dilutional linearity, and (3) standard addition. The back-calculated error of each point on an authentic matrix curve provides an indication of spike-recovery when measured relative to the surrogate matrix calibration curve. In the case of small molecule biomarkers, stable isotope-labeled internal standards may be used to compensate for sample-to-sample variation in matrix composition, as well as differences between the control and surrogate matrix.

For protein biomarker assays using hybrid LBA-LC-MS/MS approaches, the objective of the parallelism determination is two-fold: (1) to determine matrix equivalence between surrogate and native matrices and (2) to confirm that the assay calibrator is immunologically similar to the endogenous analyte. One added advantage of using an LC-MS method for protein biomarkers is that a matrix from a different but closely related species may be used as a surrogate due to differences in the peptide sequence of the analyte to be measured.

5.3.4 QC Samples

There are a wide range of practices for biomarker assay QCs. It is important to differentiate between spiked QCs from endogenous QCs, as well as QCs used to estimate precision and accuracy during analytical validation, from those used when conducting longitudinal studies to control assay performance and accept runs (Lee et al., 2006).

Spiked QCs can be used to assess relative accuracy and precision. The well-defined principles and approaches for generating QCs in PK methods are applicable here (Azadeh et al., 2019; Jenkins, 2016). A suitably sized matrix pool with no or low levels of endogenous analyte can be used for generating spiked QCs. A World Health Organization (WHO) reference standard, or the equivalent, is often not available. Therefore, for protein biomarker assays, a recombinant calibrator is used as the spiking material. In addition, because a matrix with undetectable levels of endogenous analyte may not be available, the recovery is the sum of the spiked and endogenous analyte levels. The ability to recover after addition of calibrator is a useful confirmation of the assay's ability to measure the endogenous analyte (e.g., confidence in assigned endogenous concentration). However, subtraction of endogenous level from total material levels after spiking has been recommended for use in determining relative accuracy (e.g., if 500 ng/mL spiking material is added to 100 ng/mL endogenous, 600 ng/mL measured concentration would be equivalent = 100% recovery) (Marcelletti et al., 2015).

Endogenous analyte sample QCs are needed for assessing assay precision and for monitoring biomarker assay performance over time to prevent assay drift (Jenkins, 2016). The concentration range of the endogenous analyte in these samples can be assigned by repeat analysis during assay validation. Ideally the concentrations of endogenous analyte in these QCs should cover the expected concentration range in samples to be analyzed. Pools of multiple donors can be used to produce the amount of QCs needed for assay validation and sample analysis. Admixing of pools of various concentrations may be used to generate different levels of endogenous analyte (Grant & Hoofnagle, 2014).

Assay monitoring by charting QC performance, widely practiced in clinical chemistry, is recommended (Azadeh et al., 2019).

Obtaining or defining flow cytometry QC can be very challenging and many research-grade assays do not include assay-specific QCs. However, all longitudinal or multicenter studies should have QC samples. These can be developed in house using cell lines, target transfected cells, or commercial cellular controls. The latter can include stabilized human leukocytes and erythrocytes, for example, BD Multi-Check controls (https://www.bdbiosciences.com/en-us/reagents/clinical-reagents/quality-and-process-control), R&D Systems Status Flow controls (http://www.rndheme.com/ProductDetail.aspx?prodCode=WBFCP&InstrumentCode=CXL), or Streck Labs controls (https://www.streck.com/controls). Many commercial controls are intended for specific clinical applications such as hematopathology. Some of these can be used to control for antibody staining, red blood cell lysis, instrument setup and performance, and data analysis. Appropriate controls should be selected based on specific goals (e.g., to monitor sample processing) or to detect target levels or cell type of interest. Each laboratory should also establish its own QC acceptance criteria (see der Strate et al., 2017).

5.3.5 Biomarker Specimen Preservation and Stability Assessment

Biomarker specimen preservation and stability assessment have been hot topics in recent years. A key purpose of biomarker studies is to assess the physiological or pathological states of a function of interests at baseline or after therapeutic intervention. Therefore, it is important that the measured analyte concentration in the sample specimen represents, as close as possible, the level of analyte in the body or correlates with other biological observations. Many factors may affect the level of an analyte, and some are beyond the control of the analyst. A recent AAPS white paper (Dakappagari et al., 2017) divided factors contributing to biomarker variabilities into three categories: (1) biological variabilities which stem from inter-/intra-individual difference in genetic makeup and physical factors such as circadian rhythm and dietary intake; (2) preanalytical factors which are related to sample collection and handling, including shipment and storage conditions; and (3) analytical variability which is due to assay limitations and execution (see Figure 5.1). Depending on the stage of the study, there are varying degrees of control of these factors. For example, if desired, the biological variabilities may be reduced at the stage of clinical study design through patient selection and sample collection requirement (e.g., time of day for sample collection, fasting vs. nonfasting, etc.). Preanalytical variabilities may be reduced through the design and execution of the sample collection and shipping manuals. Sample

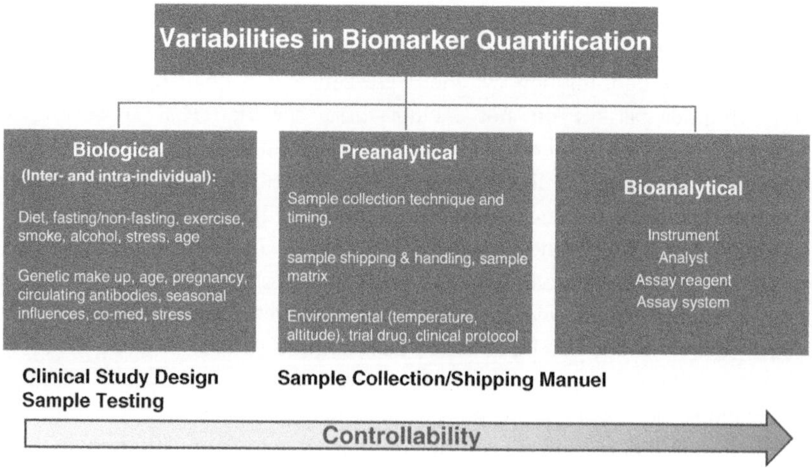

Figure 5.1 Factors contributing to biomarker variability (adapted from Dakappagari et al., 2017).

shipping conditions and testing schedules should take sample stability into consideration to avoid compromising sample quality. Dakappargari and colleagues provided a number of recommendations to mitigate preanalytical variability-related issues (Dakappagari et al., 2017). These include:

- Creating a study-specific biomarker sample collection manual with detailed instructions on preparation of sample labels, sample collection, and processing procedure.
- Standardizing sample collection and processing procedures across investigation sites (including tube type and blood draw order, etc.).
- Conducting rigorous in-person or online site training on the use of a sample collection kit, and on sample storage, packing, and shipping to the testing laboratories.
- Best practices for sample stability testing include setting up a test plan based on the approved clinical study protocol to inform sample shipping and storage conditions in the clinical trial. Stability testing samples should mimic the clinical samples to be collected (with the same sample matrix and endogenous analytes). Freshly prepared (or stability demonstrated) assay calibration standards and QC samples should be used to conduct stability testing.

Sample stability is a significant challenge for classical blood-based cytometry assays (Brown et al., 2015) since typically cells are measured and sample analysis may not be conducted at the same site as the sample collection, and cell viability and marker stability must be evaluated during assay validation. Users need to assess sample stability for each study. General recommendations for this are described by Brown et al. (2015) and Diks et al. (2019). There are commercial products that may extend whole blood marker stability for several days. For example, Davis et al. (2011) assessed the use of CytoChex tubes in comparison to K_3EDTA plasma tubes for storing blood at room temperature. They found that while most markers had similar percent positive levels between the two tube types, T activation markers were preserved better in K_3EDTA plasma.

5.3.6 Multiplex Assay Validation

While many assay validation parameters are shared between multiplex and singleplex assays, the complexity of multiplex biomarker assay poses unique challenges for bioanalytical scientists during both method development and assay validation. Multiplex assays can be developed in many assay platforms (e.g., LBA, LC-MS, flow cytometry, immunohistochemistry, next-generation sequencing). Some technical hurdles are common among different assay platforms and some are platform-specific. Common technical hurdles can include optimizing assay sensitivity for each analyte, assessment of endogenous sample stability, establishment of assay acceptance criteria, and data interpretation. Platform-specific hurdles in multiplex LBA biomarker assays, for example, can include determination of cross-reactivity among reagents, optimization of MRD for each biomarker, parallelism, lot-to-lot reagent changes. A recent white paper on FFP validation of multiplex LBAs detailed these challenges, including those associated with the use of commercial multiplex kits, and recommended solutions (Jani et al., 2016). In the case of LC-MS biomarker assays, the challenges include selection of a suitable stable isotope-labeled (SIL) internal standard for each analyte, and a sample preparation method appropriate for all analytes, optimal chromatographic separation, and scheduling of MS ion detection (scheduled SRM). Increased complexity of validating a multiplex assay, whether using LBA or LC-MS, may require a compromise between choosing the final assay conditions and the appropriate acceptance criteria to meet study needs. Having a Biomarker Assay Work Plan or Assay Validation Plan is a good practice as it can help navigate the validation process. The Biomarker Work Plan outlines the general expectations for assay performance including target acceptance criteria and rigor of validation, and would address considerations for a successful outcome as regards the COU of the assay. In this Plan, each assay in the method should have separate acceptance criteria (Richards et al., 2016). If an analyte does not meet validation acceptance criteria, then it should not be included in the assay. However, if the analyte is to be excluded from the assay, the impact on assay performance needs to be evaluated (Jani et al., 2016).

5.3.7 Use of Commercial Kits (Diagnostics and Research Use Only) in Biomarker Studies

Many academic and clinical biomarker studies take advantage of the large number of research use only (RUO) commercial kits. Some biomarkers, such as inflammatory cytokines, may have kits from multiple vendors; while others only have a single source (which may be distributed by multiple vendors). These RUO kits have the benefit of being readily accessible, though their quality and analytical performance may range considerably. The performance of a RUO kit for the intended use must be carefully assessed experimentally. Khan et al. (2015) recommended that commercial kit assessment should start with setting up a Biomarker Work Plan which describes the project needs and intended use of the data. This would then define assay validation requirements and set the stage for kit selection. Published data from a commercial kit should be replicated, if possible, after an appropriate level of analytical validation before finalizing the kit selection. Kit/vendor selection should also take into consideration the lot change process and access to bulk calibrators and QCs, as both are key challenges associated with the use of commercial kits.

5.3.7.1 (FDA) Regulatory Guidance for Use of Commercial In Vitro Diagnostic Kits in Biomarker Measures

Within the 2018 FDA BMV Guidance (FDA, 2018) there are recommendations regarding the use of commercial in vitro diagnostic (IVD) assays repurposed for biomarker and PK measurement in drug development (which do not apply to the Point of Care Patient Diagnostic Kits and In Vitro Companion Diagnostic Devices). For kits developed for use as clinical diagnostic tools, their suitability for use in PK and PD studies should be demonstrated. Assay performance of these kits should be assessed in the facility conducting the sample analysis. Specifically:

- Site-specific validation: accuracy, precision, and any modification of the kit should be conducted at the site prior to sample analysis.
- Matrix effects: standards and QCs should be prepared in the same matrix as the subject samples. Kits with standards and QCs prepared in a matrix different from the subject samples should be justified.
- Calibration curve integrity: kits that use sparse calibration standards (e.g., one- or two-point calibration curves) should include in-house validation experiments to establish the calibration curve with a sufficient number of standards.
- QCs: QC concentrations should be known. If kit QCs are expressed as ranges, QCs with known concentrations should be prepared and used, independent

of the kit-supplied QCs. Endogenous QCs should be used; they can be prepared and measured repeatedly to assign concentrations.
- Lot-to-lot variability: if multiple kit lots are used within a study, lot-to-lot variability and comparability should be addressed for critical reagents.

Since knowledge of target levels is critical for understanding the PK/PD relationship and first-in-human dose selection, target concentrations are often measured with a RUO commercial kit during preclinical development in animal models. As discussed previously, it is important, in relying on these kits, to recognize that the understanding of endogenous targets may be incomplete in terms of quaternary structure, isoforms, binding partners, catabolism, and turnover rates. Further considerations are needed for factors such as form of analyte (e.g., monomer vs. dimer) and relevance of recombinant calibrator when these kits are used for different preclinical species and novel matrices (Khan et al., 2015). In addition, regulation of the target may not be well understood and there may be up- or down-regulation specific to the disease of interest and/or therapeutic. Moreover, the published literature on target levels may be directionally useful (e.g., higher in disease patients than in healthy individuals), but absolute levels may not be reliable as these values are very dependent on the assigned concentration of the calibrator, the calibrator's source and form, as well as the reagents used in the assay.

5.4 Section 4: Technical Platforms for Assessing Target Biomarkers

Box 3: Common Biomarker Assay Platforms

Ligand binding assay (LBA): An analytic procedure which measures a macromolecule analyte based on its high affinity interaction with a capture- and a detection-reagent (e.g, an ELISA assay).

Liquid chromatography-mass spectrometry (LC-MS): An analytical chemistry technique that combines the physical separation capabilities of liquid chromatography (LC) with the mass analysis capabilities of mass spectrometry (MS). Liquid chromatography separates mixtures with multiple components based on the mass transfer between the liquid stream and the solid sorbent, and mass spectrometry provides structural identity of the individual components with high molecular specificity and detection sensitivity by measuring the mass-to-charge ratio of ions in the gas phase.

(Continued)

Box 2.3 (Continued)

Flow cytometry: Flow cytometry is an analytical technology that allows for single cell analysis using coherent (e.g., laser) and scattered light to measure physical and biochemical properties. Most technologies rely on a focused small diameter fluid stream to rapidly deliver cells though light interrogation point(s). Cells may also be imaged or sorted for collection. Fluorescent activated cell sorting (FACS) is often used to describe this technology because fluorescently labeled reagents specific to cellular molecules are used to analyze cellular phenotype and other characteristics. Mass cytometry uses a similar fluidic flow approach to delivering single cells into the instrument but detection is based on the ability to differentiate very small differences in the mass of rare earth metals used to label the reagents.

5.4.1 COU as a Driver

The concept of COU described by the FDA-NIH Biomarker Working Group in the BEST (Biomarkers, EndpointS, and other Tools) Resource (FDA-NIH Biomarker Working Group, 2018) is intended to be applied to the biomarker qualification process. It also provides a broad framework for biomarker analytical validation. A biomarker's COU should be defined early during the assay development and validation processes, to provide a concise statement on the biomarker category and proposed use. For COU to have practical value there must be an outline of assays and changes needed to move from one context to the next. This requires active communication and collaboration between assay developers and assay end users, and can be an iterative process especially when there is change in the COU.

The BEST biomarker categories in drug development (https://www.ncbi.nlm.nih.gov/books/NBK326791/pdf/Bookshelf_NBK326791.pdf) include susceptibility/risk, diagnostic, disease monitoring, prognostic, predictive, pharmacodynamic/response, and safety. The proposed use should be specific for each study and disease population, and may evolve with the progression of the study. Ideally, during the assay development/validation process, a "minimum detectable change" should be defined for each COU. Minimum detectable change is an important concept in laboratory medicine, which includes consideration of both analytical and biological (intra-individual and inter-individual) variability to help determine if an observed change is clinically meaningful (Bernardia et al., 2004; Lacher et al., 2005; Lorentzon et al., 2019; Shepherd & Lu, 2007). The biological variability data may not be available initially, but should be obtained through phase 0 studies or the iterative FFP validation process.

Biomarkers are often present at very low levels and, in many cases, a small change in biomarker levels needs to be measured. This is a critical design parameter that influences assay platform and reagent selection when developing an assay or assessing a commercial kit. Analyte concentration ranges in healthy and disease populations may not be available to provide guidance. Even when this information is available, there may be a lack of concordance between studies possibly due to assay reagents and preanalytical variables (Cowan et al., 2017) (Mathews et al., 2020). The assay COU may provide whether baseline data are required and the direction and amount of change expected after treatment, which then defines both assay sensitivity and precision goals. Both lower limit of detection (LOD) and lower limit of quantification (LLOQ) can be used to define assay sensitivity. The LOD is usually defined as a concentration with a signal three standard deviations above the signal level of a blank sample. The LLOQ should be defined as the lowest back-calculated concentration of a QC sample which meets the assay acceptance criteria for accuracy and precision.

Depending on the COU, a total target assay which is unable to measure the target concentration at baseline but is able to detect increases post-treatment can be used to support a drug development effort (e.g., IL-1b; Lachmann et al., 2009). Conversely, a free assay capable of measuring target concentration at baseline but unable to detect a decrease after dosing may not be suitable for the intended use. Defining the expected change and how the data will be used is critical for assessing whether an assay is fit for purpose.

5.4.2 Considerations for Assay Design and Assay Platform Selection

LBA, LC-MS, and flow cytometry are the most commonly used analytical platforms for assessing target biomarkers. Choice of the platform depends on the nature of the target (a membrane protein or a soluble protein), reagent availability, project timeline, and sample analysis needs. COU and assay quality attributes are two key drivers guiding analytical platform selection for a target biomarker assay. The COU of a target assay defines the minimally detectable changes, assay precision, and other quality attributes required for a particular testing, which in turn guide the assay design and assay platform selection. For example, complexity of the sample matrix and similarity of the biomarker of interest to other sample components dictate the level of assay selectivity; and the expected concentration range of the biomarker defines the assay sensitivity and analytical range requirement.

The broad range of biologically active molecules, and their dynamic regulation in disease states or post-therapeutic treatment, create both analytical challenges and opportunities for many existing and developing analytical techniques. Numerous analytical techniques have been applied to measuring endogenous compounds, among which LBA and LC-MS technologies are most often used for developing soluble target assays. With the recent advances in LBA (e.g., Gyros technology, Singulex, Simoa Quanterix) and LC-MS (e.g., high-end triple quadrupoles, Q-TOF, Orbitrap mass spectrometers, and the use of immunocapture reagents), high-sensitivity assays can be developed on both assay platforms. While LBA assays are high throughput and are suited to analyzing large numbers of samples, they require two binding reagents with different binding epitopes on the target protein. Conversely, LC-MS assays are comparatively lower throughput and may not require specialized binding reagents, depending on the relative abundance of the biomarker in the biological matrix.

It is important to keep in mind during the design of a LC-MS or LBA assay, that the results are based on what the reagents "see" during a binding event and/or what m/z transitions are monitored. Changes to the analyte or analyte population can not be detected if they do not affect reagent binding to an analyte, and/or occur in parts of the molecule that are not being monitored. Careful experimental demonstration that the assay measures the intended form of the analyte in samples when drug/endogenous binding partner(s) (e.g., soluble receptor) are present over the full range of both target and binding partner(s) is needed. For additional assay design considerations for total and free target assays, see Section 5: Examples of Target Assay Development and Data Application.

5.4.3 Assay Life Cycle

For most new drug candidates entering development, the duration of the program is unknown at the start. It is wise to invest only to the level of resource needed to advance the asset to the next level. The same is true for bioanalytical assay development. Therefore, a life cycle approach to the assay's development and validation which matches the assay quality attributes to the intended data use should be established.

A biomarker assay can evolve with the change of measurement needs. During early drug development (Category 1, Lee et al., 2006), a biomarker assay is often used to answer the following questions: (1) is the biomarker related to the disease, (2) is the biomarker's response to therapy a reflection of the disease modification, or (3) can the assay be used to select patients most likely to respond to therapy? During the course of the study, an existing assay must be assessed continuously as to whether it fits the intended purposes. With each change in its COU, the assay may need to be altered or optimized, or a new assay may be needed.

A sound life cycle approach is a conversation between the bioanalytical scientist and the end user of the data. The assay life cycle considers the biomarker, its biology, and the assay performance characteristics required at each stage, and assay reagents/kit supply, to ensure the generated data are reliable and are created with an appropriate expenditure of resources.

5.4.4 Shared Quality Attributes Between Mass Spectrometry and LBA-Based Assays

Both LC-MS and LBA share many core assay quality attributes and many defining elements of these attributes (see Table 5.3), including:

- Precision determinations (within run, between runs, between days, between operators, and between lots).
- Relative accuracy (assessed by spike and recovery experiment).

Table 5.3 Shared quality attributes between liquid chromatography-mass spectrometry (LC-MS) and ligand binding assay (LBA).

Quality attributes	Elements to define quality attributes	LC-MS	LBA
Accuracy (relative)		√	√
Precision	Within run	√	√
	Between runs	√	√
	Between days	√	√
	Between operators	√	√
	Between lots	√	√
Sensitivity	Analytical measurement range	√	√
	Lower and upper limit of quantitation	√	√
Dilutional linearity		√	√
Selectivity	Blanks, interference testing, matrix effects	√	√
	Parallelism	√	√
	Minimum required dilution	–	√
	Recovery	√	√
	Extraction, ionization, overall method	√	–
Stability	Reference material, reagents, stock solution,	√	√
	samples	√	√
	Bench top, freeze–thaw, short-term, long-term	√	–
	Autosampler, extract		
Robustness		√	√
Ruggedness		√	√

- Selectivity determinations (such as blanks, interference testing).
- Matrix effects, parallelism (different experimental design between LBA and LC-MS), and recovery.
- Stability determinations (for reference material: reagents, stock solution, short-term and long-term stability; for samples: freeze–thaw, bench top, short-term and long-term stability; for LC-MS: extract and autosampler stability).
- QC samples.

The approach and experimental design to characterize these attributes, however, may differ between these two technical platforms. As an example, LBAs are typically conducted directly with matrix samples without the extraction step performed in LC-MS assays.

The assay validation acceptance criteria are also very similar between the two technical platforms (see FDA BMV, 2018). For biomarker assays, the BMV 2018 validation criteria should serve as the starting point and be adjusted based on the COU.

5.4.5 LBAs

LBAs rely on reagents that bind noncovalently and specifically to the analyte of interest. Typically, two critical reagents (e.g., antibodies, aptamers) are used together: one to capture from the matrix and the second to detect after washing away unbound material. Detection of binding events can be amplified in various ways by the use of, for example, conjugation enzyme conjugates, and luminescent and fluorescent molecules. These detection reagents are often highly specific, a quality that is sometimes underappreciated, and can recognize changes in three-dimensional conformations and even loss of a hydroxyl group (Brade et al., 1997; White et al., 1985). Moreover, the binding affinity of these reagents can be very high (picomolar to nanomolar values). These attributes enable the development of highly specific and sensitive biomarker assays. In addition, LBAs offer low cost, high throughput and, on some platforms, very low sample volume requirements.

5.4.5.1 High-Sensitivity Platforms

In recent years, with the introduction of bead-based technology and single molecule detection, the sensitivity of many immune-based assays has greatly improved. Single molecule assay (Simoa) from Quanterix, one of the front runners in this effort, uses a microscopic bead-based immunoassay system. A high bead-to-analyte ratio is used to ensure the capture of low abundance analytes

(Fischer et al., 2015; Rissin et al., 2010). This is a fully automated system and is capable of multiplexing through the use of dye linkers for each individual capture antibody. Interference or "cross talk" among analytes should be evaluated as required for other multiplexing immunoassay platforms. Meso Scale Discovery (MSD) has developed a next-generation multi-array electrochemiluminescent (ECL) technology S-plex with greater than femtogram/milliliter detection limit. This new platform enables the detection of extremely low abundance circulating cytokines and other biomarkers in apparently normal healthy donors. Using this technology, MSD reported that some cytokines (e.g., IL-6, IL-10, IL-17A, TNFα) are present at much lower concentrations than previously reported, possibly due to nonspecific binding interactions (Oberoi et al., 2016). While the Simoa system is laboratory information management system (LIMS) compatible and can be used for regulated bioanalysis studies, S-plex assays are for RUO. Both platforms can achieve femtogram to picogram/milliliter assay limit of detection, with greater than 4 log assay dynamic range, but require specialized training and instrumentation.

5.4.5.2 Specific Challenges and Unique Characteristics Intrinsic to LBA

In a LBA assay, what is being measured may not be identical to the analyte of interest. LBA requires reagents with high specificity for the target analyte, and such reagents may not always be available. Confidence in what is measured by an LBA relies on the specificity of the reagents and is less direct compared to LC-MS. This is because for LBA, colorimetric absorbance, fluorescence, or chemiluminescence/electrochemiluminescence are measured to derive analyte concentration; whereas for LC-MS, the mass of the analyte itself is measured. The depth and breadth of reagent specificity testing may be limited and/or not available to the user if the reagents are from a commercial source, or if sequence heterogeneity and metabolism of the target analyte are not fully defined.

For free- or total-target measurement, complex dissociation is another concern. LBAs are based on noncovalent binding and thus sample dilution and washing steps may result in analyte or analyte complex dissociation. This is of particular concern when there are low-affinity binding interactions that need to be preserved.

5.4.6 LC-MS

MS is well known for trace analysis of endogenous analyte in complex samples due to its direct and broad detection of many chemical species, and high selectivity. In recent years, LC-MS technology has been used to measure soluble and membrane targets. The main drivers for LC-MS assay design are the selectivity and the sensitivity required for the study, which can be achieved by the use of high-sensitivity LC-MS instruments and various sample preparation approaches.

5.4.6.1 Sample Preparation in MS

Sample processing using chemical or biochemical extraction procedures adds both selectivity and the opportunity for enriching the analyte of interest, thus enabling the detection of low-level target proteins and other signaling molecules. The choice of sample preparation strategy will depend on the COU and the level of detection required (see Table 5.4).

For small molecule biomarkers, peptides, and some small proteins, traditional physical chemical separations such as solid phase extraction and two-dimensional LC can be an effective means of isolation and enrichment of the biomarker. For protein biomarkers, there are two common workflows to prepare the sample and quantify the analyte: the surrogate peptide approach and the intact protein approach. With the surrogate peptide approach, the target protein is digested into peptides which are used as surrogate analytes for the protein (Barnidge et al., 2004a; Barr et al., 1996; Kuhn et al., 2004; Oda

Table 5.4 Biomarker liquid chromatograpy-mass spectrometry/mass spectrometry assay sensitivity and sample preparation approach.

Parameter	Small molecules and peptides	High-abundance proteins	Mid- to low-abundance proteins		Low-abundance proteins
Sample preparation	No digestion	Direct digestion	Protein IAC	Peptide IA	Sequential IAC
Sample volume	10's to 100's μL	10's to 100's μL	μL to mL	μL to mL	μL to mL
Process	Protein precipitation/ SPE/LC fractionation	Digestion	Protein IAC followed by digestion	Digestion followed by peptide IAC	Protein IAC followed by digestion then peptide IAC
Sensitivity	Low ng/mL to low pg/mL	10's ng/mL	Low to mid pg/mL	Mid to high pg/mL	Low pg/mL
Notes	May require large volume of sample			Requires large enzyme amount	Cleanest sample introduced to MS

IAC, immunoaffinity capture; LC, liquid chromatography; MS, mass spectrometry; SPE, solid phase extraction.

et al., 1999). With the intact protein approach, the full target protein is analyzed using high-resolution MS, providing a direct measurement of the intact protein. The intact protein approach may also provide structural information on post-translational modifications, but does not often achieve the same level of sensitivity as the surrogate peptide approach (Jian et al., 2013, 2016; Lanshoeft et al., 2017; Ruan et al., 2011).

For protein biomarkers, the initial and simplest LC-MS methods often follow protocols derived from the proteomic experiments in which the candidate biomarkers were first identified (Wiener et al., 2004). These assays typically use direct sample digestion with proteolytic enzyme followed by LC-MS/MS analysis of the resulting peptide mixture. In this workflow, samples may be directly digested. It is also possible to simplify the sample prior to digestion by molecular weight fractionation (filter-aided sample preparation [FASP]; Wisniewski, 2016; Wisniewski et al., 2009); removal of high-abundance proteins; or by sequential precipitation of protein followed by pellet digestion (Neubert et al., 2016). Each of these sample preparation approaches has its advantages but also the potential risk of losing target analyte. In the case of sequential protein precipitation, it is possible to analyze the fractions to measure target small proteins that are still in solution.

An optimized LC-MS method often uses a combination of sample preparation approaches. As summarized in Table 5.4, the most sensitive LC-MS assays incorporate multiple steps of isolation, digestion, and enrichment from potentially large amounts of sample, and are capable of measuring very low abundance proteins. In contrast, direct digestion approaches for low sample volumes, with much fewer isolation steps, are applicable to higher abundance proteins.

5.4.6.2 Hybrid Ligand Binding–LC-MS/MS

To further enhance assay sensitivity, selective enrichment of the target analyte in the sample, a hybrid ligand binding–LC-MS/MS approach may be used. Since analyte detection in LC-MS/MS is mass-based and highly selective, the capture antibody used in a hybrid ligand binding–LC-MS/MS approach needs not be as selective as the one typically used in LBAs. Therefore, many protein biomarker assays now use an immunoaffinity capture (IAC) step prior to enzymatic digestion and surrogate peptide quantification (Neubert et al., 2020). Alternatively, the IAC may be carried out after digestion of the proteins to selectively isolate and enrich peptides of analytical interest, termed SISCAPA (Anderson et al., 2004). The additional selective enrichment provided by the IAC frequently improves the detection and quantification limits by an order of magnitude. Hybrid LBA–LC-MS/MS assays may be sufficiently sensitive to effectively measure relatively high abundance signaling proteins.

An additional gain in detection limit can be reached using sequential (two-step) IAC, a selective enrichment at the intact protein level followed by a second IAC step using an antibody against the surrogate peptide (Neubert et al., 2020; Palandra et al., 2013; Schultz et al., 2016). The approach requires two different antibodies: one against the intact protein and another against the surrogate peptide. Whereas the protein level capture is performed offline in a 96-well plate format, the surrogate peptide capture is most often done online, where the digested protein from the first capture is captured by the second antibody immobilized in an LC column, followed by direct elution and detection with LC-MS/MS. This two-step IAC approach can achieve the highest assay sensitivity and has more relaxed requirements for antibody selectivity during the first capture step (although the anti-peptide antibody does require good specificity). However, this approach is also expensive, and uses a comparatively complicated online analytical system.

5.4.6.3 Chromatography Formats and MS Instrumentation

While sample preparation is arguably the most effective means of improving detection and quantification limits, additional gains can be realized by the use of optimal LC technologies (see Table 5.5). Lower-flow LC separations are most often used when sensitivity and throughput are needed (Lassman & Fernandez-Metzler, 2014). Nanoflow chromatography has the best sensitivity in terms of moles of analyte needed to make the measurement (Wiener et al., 2004), where very small amounts of samples are sufficient for quantification analytes at low levels. The tradeoff of using such instrumentation is throughput, ease of use, and reliability. Nanospray separations often require 30–120 minutes per sample. However, with a clever combination of IAC sample

Table 5.5 Target assay sensitivity and liquid chromatography (LC) format.

Chromatography	Analytical LC	Micro LC	Capillary LC	Nano LC
Column diameter (mm)	2.1–4.6	0.5–1	0.1–0.3	<0.1
Flow rate (μL/min)	500–1000	15–200	1–10	<1
LC cycle time (min)	7–18	7–18	10	10
Sensitivity	Low ng/mL to mid pg/mL	10's to 100's ng/mL for monoclonal antibodies, low to mid pg/mL for peptides	Low to mid pg/mL	Low pg/mL

preparation, nanospray can be effectively used with a sample cycle time as low as 10 minutes (Schultz et al., 2016).

If the amount of sample extract is not limited, loading more analyte on the chromatography column is another way to improve detection. In most LC-MS experiments, separations are run in 0.3–1 mm internal diameter columns with flow rates ranging from a few microliters to 100 μL per minute, or large column diameters (2.1 mm) and high flow rates which allow large amounts of sample extract in the LC-MS system. These high flow rate systems are more robust and easier to use. The tradeoff between different column diameters is the loading capacity, the separating power, and the sample cycle time. The ionization efficiency of the instrument also differs at different liquid flow rates. With careful control, these sample introduction combinations are capable of achieving the analytical resolution required for many biomarker studies.

Like any tool, there are many variations in mass spectrometers designed to perform different tasks. The initial biomarker discovery work is almost exclusively done with the use of high-resolution, accurate mass Orbitrap, and time-of-flight mass spectrometers. While targeted analysis can be effectively accomplished on these LC-MS instruments, they do not match the level of performance of the highest-performing triple quadrupole instruments which are the gold standard for quantitative analysis by MS. Consequently, rigorous biomarker assays use high-performing triple quadrupole LC-MS systems.

5.4.6.4 Assay Calibrators in MS Assays

For small molecule biomarker assays, the chemically synthesized form of the analyte is often readily available. In some cases, a stable isotope-labeled version of the target analyte may be used as the calibrator, termed surrogate analyte. This method is unique to MS and takes advantage of the mass shift of the labeled compound. An advantage of this approach is that standard control matrices may be used to prepare calibrator and QC samples (Jones et al., 2012).

For protein assays using the surrogate peptide approach, the assay calibrator may be the whole protein, or a synthetic tryptic peptide or stable isotope-labeled tryptic peptide (Barnidge et al., 2004a, 2004b, Barr et al., 1996, Oda et al., 1999). The latter two peptides serve as a surrogate analyte for the protein. The preferred assay calibrator is the full protein such that the recovery from the immunocapture step, in the case of hybrid LBA–LC-MS/MS assays, and the digestion efficiency is as similar as possible to the endogenous biomarker protein.

5.4.6.5 Specific Challenges Intrinsic to the Technology

LC-MS has several limitations. Compared to LBA platforms, the equipment is expensive, complicated, and difficult to maintain; and when the sample amount is limited, the limits of detection may be difficult to achieve. However, when good reagents are not available for LBA, LC-MS/MS assays and hybrid LBA–LC-MS/MS are a good starting point in a biomarker assay's life cycle. LC-MS assays can be developed and made ready for use quickly. In the case of hybrid LBA–LC-MS/MS assays, a lower-affinity antibody may be sufficient to enable analyte isolation and analysis. As the LBA assay reagents and assay are developed, an LC-MS/MS assay may be useful for their selection, for assay characterization, and for troubleshooting.

5.4.7 Cytometry

Flow cytometry, while based on fluid-focused single cell analysis and laser light scatter, is another form of LBA which uses fluorescently labeled antibodies that bind specifically and noncovalently to cell surfaces and/or intracellular targets (Shapiro, 2005). There are a tremendous range of applications for this technology; however, the best known is its use in characterizing the phenotype of different cells in whole blood with a combination of antibodies against well-defined cell surface markers. A key strength of this technology lies in its ability to simultaneously detect and quantify multiple cellular targets. In recent years, standardized panels to phenotype cells have been developed, including currently 49 optimized multicolor immunofluorescence panels (OMIPs; see Mahnke et al., 2010 and https://med.virginia.edu/flow-cytometry-facility/resources/optimized-multicolor-immunophenotyping-panels-omips).

Changes in the relative number (frequency) of a specific cell population can be an important biomarker of response. Flow cytometry has been used to measure cellular response to target engagement such as receptor signaling, a variety of intracellular downstream effects, and changes in cell frequency/cell number. In addition, in drug development, flow cytometry can be used in combination with phenotyping to measure cell or target numbers, to help understand target coverage and target occupancy, and to define the relationship between dose and target engagement (Liang et al., 2016). Given this large range of applications, readers are referred to the relevant literature in the field (e.g., Shapiro, 2005). Here, we focus on the measurement of target expression and receptor occupancy using cytometry.

The validation of cytometry-based assays is not covered in the FDA BMVs (2001, 2018). Key validation parameters are described in the literature (Davis et al., 2013; Du et al., 2015; O'Hara et al., 2011; Selliah et al., 2019; Wu et al., 2010) and in CLSI guidance for clinical diagnostics and laboratory-developed tests (LDTs). Because the majority of the CLSI guidance for clinical tests were based on traditional clinical chemistry assays, there is uncertainty regarding their application to flow cytometry as a drug development tool.

5.4.7.1 Target Expression and Calibration Beads

Target expression levels on the surface of cells can be an important biomarker in drug development. Qualitative measurement of target expression among different cell types can be determined relatively easily by flow cytometry. Access to cells from dissociated tissues (e.g., tumor-infiltrating lymphocytes from mouse models) is readily available preclinically for oncology programs – for an in-depth description see Reichard & Asosingh (2019) and Lizotte et al. (2016).

A quantitative approach can also be used to assess antibodies bound per cell and measure target/cell number using calibration beads (Davis et al., 1998; Gratama et al., 1998). There are a number of commercial calibration beads that allow conversion of immunofluorescence intensity per cell to the number of molecules of equivalent intensity (MESF: molecules of equivalent soluble fluorochrome). Assays calibrated with these beads are relatively quantitative as they may underestimate or overestimate the absolute receptor number/cell. The refinement of this approach, using monoclonal antibodies (mAbs) conjugated 1:1 with phycoerythrin, brings this method closer to absolute quantification, albeit with a number of caveats (Davis et al., 1998). These target expression data enable clinical decision-making whether it is to aid in diagnosis (Wang et al., 2006) or as a parameter in a mechanistic PK/PD model (Jager et al., 2011). One important consideration, which is often overlooked, is the need to generate a full binding curve to confirm that the binding is specific and saturable under conditions approaching equilibrium. The use of a single "saturating" concentration of reagent to measure receptor numbers can be very misleading if there is significant nonspecific binding.

Cytometric bead array (CBA) assays are flow-based assays that use microbeads that can be distinguished from each other by their unique fluorescence characteristics (Varro et al., 2007). Typically, they are multiplex assays designed to measure multiple soluble biomarkers. These assays are essentially LBAs using beads rather than plates as the solid phase. Each bead population is precoated with a specific capture antibody, incubated with samples, washed, and then mixed with a labeled detection antibody. Multiplex cytokine CBAs are available commercially from multiple vendors (e.g., BD Biosciences, eBiosciences, Luminex) and can be used on most flow cytometers.

5.4.7.2 Receptor Occupancy

Engagement of membrane targets can be evaluated by receptor occupancy (RO) studies. An important approach to define a PK/PD relationship is to assess cell surface target engagement after drug dosing. This can be achieved through a direct measurement of receptor–drug binding, also known as RO; or through the monitoring of a signaling cascade subsequent to the RO. The validation of RO assays and different assay formats have been thoroughly reviewed by Liang et al. (2016) and Green et al. (2016). Typically, RO assays are performed on phenotyped cells using a near-saturating concentration of a labeled probe that competes with the drug. Therefore, the first requirement of a quantitative RO assay is to demonstrate saturable specific binding to the cell populations of interest. The second requirement is that the labeled probe does not displace bound drugs to reduce the risk of underestimating occupancy. The labeled probe can be a labeled drug with the advantage of having the same binding kinetics as the drug, or a competing agent which binds the target at the same site with the drug. In either case, the probe binding kinetics should be well defined during method development; and probe incubation duration must be managed to permit binding to only free target sites. This is critical for agonist drugs where low receptor occupancy levels may need to be monitored. Finally, a bound drug or labeled probe may internalize after target binding which should be minimized from the time of sample collection. This is typically done by including inhibitors for internalization during sample collection and keeping the sample on wet ice at all times.

RO can be expressed as a percent of baseline (pre-dose) levels. However, this would not be accurate if the receptor expression level changes over the course of the study due to treatment. Therefore, the development and use of a total receptor assay in parallel should be considered. This requires a noncompeting antibody and supporting data at high drug concentrations. It is also important to note that for studies with long duration, antidrug antibodies (ADAs) may develop. The use of labeled drug could lead to an underestimation of free receptor due to interference by neutralizing ADAs; or overestimation due to ADAs that link labeled bound drug to labeled unbound drug creating a “daisy chain” of complex molecules which links drug bound to a receptor to the labeled free drug (see Figure 5.2).

Directly labeled probes are preferred. However, it is important that the degree of labeling of these probes are controlled. Ideally, the labeling should be low to minimize nonspecific binding and the risk of altering specific target binding by the fluorescent labeling. Unstained samples and isotype controls are often used to set gating thresholds for cells that exhibit target expression. While isotype controls are critical when secondary reagents (e.g., goat anti-mouse IgG) are used for detection, they are not appropriate controls for directly labeled probes. This is because small differences in the antibody framework due to different complementarity-determining regions (which may be different among antibodies of the same isotype) (Honegger et al., 2009) can contribute to nonspecific binding. Fluorescence minus one (FMO) controls

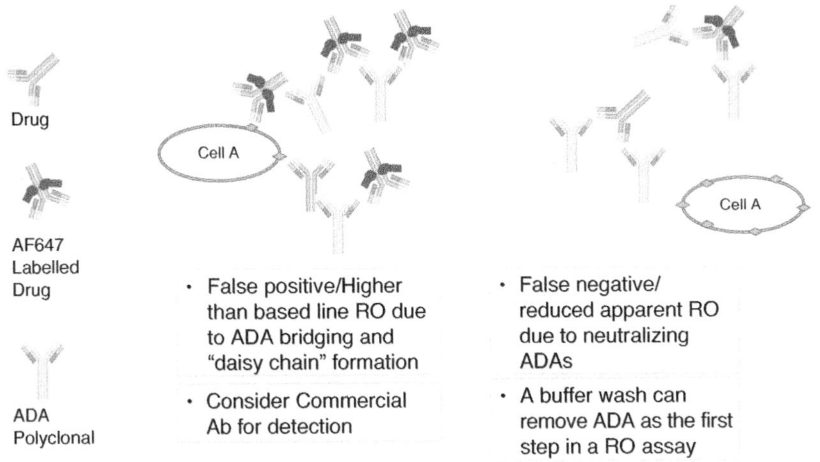

Figure 5.2 Impact of antidrug antibodies (ADA) on receptor occupancy (RO) assay when the labelled drug is used for detection.

should be used as an important confirmation that compensation has correctly addressed fluorescence spillover into the detection channel of the probe.

5.4.7.3 Specific Challenges Intrinsic to Technology

Instrument-based variability over time is a well-known challenge for cytometry in general and for RO assay specially. This is because RO is often monitored at different time points, and sometimes over a long period of time, and samples cannot be stored for batch analysis. Daily instrument set-up and calibration are critical for the control of cytometers' performance. This typically involves instrument startup and laser performance monitoring based on the manufacturers' recommendations. Daily instrument bead-based calibration is also part of the process for characterizing and controlling variability (Shapiro, 2005).

5.4.7.3.1 Compensation and Nonspecific Binding

Compensation in flow cytometry refers to the need to compensate mathematically for fluorophore emission spectral overlap from one fluorophore into the detection channel of another. Conventional flow cytometers are equipped with filters to manage both emission wavelength and detection wavelength through compensation controls and FMO controls (Roederer, 2001). Nonspecific binding is managed through wash steps, buffer complexity, and experimental controls including unstained cells, and isotypes controls when appropriate.

FMO controls are also important to set gates (see below), help prevent overcompensation, and manage spill over. The latter is a significant issue for high-dimensional flow (Nguyen et al., 2013).

5.4.7.3.2 Gating

Flow cytometry defines a cell population expressing a particular marker based on measurement of the fluorescence intensity of an antibody against the marker. This is done graphically by drawing a gate or selecting a region which includes cells of interest. Thus, data analysis relies heavily on individual experience and expertise. This process can be straightforward as is the case for CD4 – a marker only found on a large subset of human T cells. CD4 cells can be easily selected on a scatter plot by drawing a gate around them due to the uniformity of CD4 expression. Therefore, if the same data set was manually gated by several individuals independently, the results can be very similar. However, for markers expressed at varying levels in multiple cell types, gating can be more subjective and lead to results that are difficult to reproduce. Inconsistency in results comes from the multiple ways for gating positive cells depending on the other markers used and how the gates were drawn for dim and bright populations. This challenge leads to efforts to develop autogating algorithms to improve consistency among different laboratories (Maecker et al., 2012). This is particularly important for high-dimensional cytometry which can measure 20–35 markers simultaneously, and for cytometry by time of flight (CyTOF) which measures even higher numbers of markers. In these cases, novel data visualization tools are required, since bivariate analysis is not feasible (Mair et al., 2016).

5.5 Section 5: Examples of Target Assay Development and Data Application

5.5.1 Target Assays and Target Engagement Assessment

In both preclinical and clinical studies of biotherapeutics, drug target level assessment is an important aspect of biomarker research. Levels of free and total ligand or drug target are monitored to assess target engagement – the most proximal measurement of PD effect. Measurement of free and total hormones (e.g., insulin, thyroglobulin, steroid hormone) dates back to the late 1950s (Yalow & Berson, 1959). These assays are used to monitor hormone levels during disease management and hormone replacement therapy. It is important to monitor free hormone levels because, typically, only the free form can interact with their receptors and is expected to be biologically active.

Similar thinking applies to free targets and drugs as the biologically active form. A key challenge is that the measurement of the free form of the ligand, target, or drug could be disrupted by the process of measurement (Midgley & Ekins, 1993).

A total target assay measures the level of a target both bound and unbound to a drug; while a free target assay measures only the level of a target not bound to a drug. These assays are important drug development tools. They are used frequently during early clinical development to measure target levels at baseline and post-drug treatment to provide proof of target engagement and mechanism of action, and to enable PK/PD modeling. Free and total targets can be assessed directly using reagents (e.g., antibodies, peptides, etc.) with the desired binding profile. In this section, we will focus on the specific challenges for free and total target assessment using LBA, LC-MS, and flow cytometry, and provide case studies on the development and use of total or free target assays.

For biotherapeutic targets, the level of *free target* decreases post-drug dosing, and reaches zero if all target proteins are engaged by the drug (100% target engagement). As such, free target measurement is the most proximal assessment for target engagement. When a target is involved in different disease indications, the free target measurement can also serve as a common target engagement biomarker among different indications. Free target concentrations can be measured directly, albeit with technical challenges (see later) or determined using data from total target levels and mathematical modeling (Wang et al., 2014).

Each biotherapeutic antibody can bind either one or two target molecules, shed or soluble targets, and form complexes. Typically, the bound form of the target is biologically inactive since it is unable to interact with its receptor or other binding partners. If the half-life of the target is shorter than that of the antibody, the level of *total target* would accumulate post-drug dosing (over 10,000 fold of the baseline level in some cases); it then reaches a plateau when the target is saturated by the biotherapeutics, and the target production and clearance has reached an equilibrium. For example, total IgE levels increase rapidly after dosing with omalizumab (Lachmann et al., 2009). Therefore, as such, a total target assay can be used to monitor the binding of a drug to a target, and target saturation. It is worth noting that knowledge of soluble target turnover rate can help understand and interpret free- and total-target data (Ciccimaro et al., 2017; Vahid et al., 2018).

Total target levels may change over time post-dosing through additional mechanisms. For example, Her2 is expressed at very high levels by certain tumor cells. Treatment with trastuzumab (Herceptin), an anti-Her2 antibody, reduces the constitutive shedding of the Her2 extracellular domain (Miller et al., 2017) and the levels of total Her2 in circulation. This can also occur to cell surface targets: treatment with rituximab kills cells expressing CD20 and reduces CD20+ cells to undetectable levels in the peripheral blood (Maloney et al., 1997).

Finally, there may be other molecules that bind to soluble receptors. These binding events are noncovalent and therefore reversible. The implications of these reversible interactions and changing target kinetics after dosing should be considered when measuring target and biotherapeutic drug levels. These issues are reviewed in detail by Lee et al. (2011) and Zheng et al. (2015).

5.5.2 Challenges and Mitigation Strategies for Free Target Assessment

Target engagement assessment through measuring free target or drug–target complex must evaluate and mitigate the consequences of equilibrium disruption. To measure free target, an immuno-based (ECL or LCMS) assay requires the capture reagent to bind the target at the same epitope with the study drug. One common approach is to use the drug itself as the capture reagent. The alternative is to use a capture reagent which binds the target at the same epitope as the drug. Quantification of free targets in the presence of drugs is technically challenging, because the drug–target complex may dissociate during sample processing and assay incubation, which would result in overestimation of the free target level. In addition, for low-level soluble targets, such as cytokines, a high-sensitivity assay is required to allow measurement of target level decrease from the baseline.

The challenge of overestimating free target levels due to drug–target dissociation has been well recognized in recent years (Lee et al., 2011; Zhang et al., 2017; Zheng et al., 2015). The drug–target dissociation rate is an important factor to consider during the feasibility stage of the free assay development. It is more challenging to develop a free target assay when the drug binding off-rate is high; and it is important to note that the binding off-rates obtained from BiaCore or KinExA studies (Drake et al., 2004) may differ from the true off-rates in a complex biological matrix. There are approaches to circumvent or minimize the drug–target dissociation issue. Colbert and coworkers (Colbert et al., 2014) showed that by carefully selecting assay reagents and implementing optimal assay conditions to minimize equilibrium shifts, it is possible to quantify free ligand with confidence. In fact, they demonstrated that under the carefully defined assay parameters, there was no evidence of complex dissociation over a range of sample dilution and assay incubation durations. Dysinger and Ma (2018) have also developed a free target assay for complement protein C5a using a Gyrolab platform. The authors and others have suggested that assay platforms which allow minimal sample dilution and abbreviated assay incubation time offer technical advantages for quantitative determination of free drug target.

Since free target assay development can be resource- and time-consuming, an alternative approach is to provide an estimation of free target levels based on measurement of total drug and total target levels using mechanism-based PK/PD modeling (Lee et al., 2011; Wang et al., 2014). However, this type of calculations relies on several assumptions, including that (1) the endogenous binding Kd between drug and target equals to what is measured in vitro, (2) the target production and elimination rates are constant, and (3) the binding reactions between the target and assay reagents have reached equilibrium. As discussed by Zheng et al. (2015), the predicted free target profile can deviate significantly from observed data if the model is not validated.

Example 1 LBA-Based Free- and Total-Target Assays to Inform Mechanism of Action and Enable PK/PD Modeling

Free- and total-target data to inform dosing decision for anti-PCSK9 antibody alirocumab

Alirocumab is a human mAb against proprotein convertase subtilisin/kexin type 9 (PCSK9). It binds PCSK9 and blocks its effect to accelerate the degradation of low-density lipoprotein (LDL) receptors, thereby preserving the function of LDL receptors and reducing circulating LDL cholesterol. In a randomized partial blinded study (Rey et al., 2016), both free- and total-PCSK9 were measured using LBAs to assess the relationships among the levels of LDL cholesterol, alirocumab, and free- and total-PCSK9. Subjects were randomized into three study groups: (1) alirocumab, (2) ezetimibe + alirocumab, and (3) fenofibrate + alirocumab. The study showed that treatment of ezetimibe or fenofibrate led to ~ 20%–25% of reduction in levels of LDL-cholesterol; treatment of alirocumab alone led to ~ 50% reduction in LDL-cholesterol. Coadministration of alirocumab with ezetimibe or fenofibrate produced significantly greater (up to 65%) LDL-cholesterol reduction. These results are consistent with the mechanism that alirocumab can further enhance standard LDL-cholesterol lowering therapies through reduction of PCSK9 and preservation of LDL receptor function. The study also showed that free-PCSK9 was reduced to very low concentrations (0–5 ng/mL) in all groups with alirocumab. Comparison of the concentration of PCSK9 and total alirocumab levels suggested that a sustained near maximum LDL-cholesterol lowering effect was achieved when free-PCSK9 concentrations remained less than ~ 20 ng/mL and alirocumab levels remained above ~ 10 mg/L. Together, these key measurements help to inform the dosing strategy and study design for alirocumab.

Example 2 Soluble and Membrane-Bound Target Engagement Workflows by LC-MS

When developing protein target engagement assays using a LC-MS/MS platform, a number of sample preparation approaches prior to MS detection have been successfully employed. The key strategy is to differentiate between the total-target and the drug-bound fraction and use these measurements to calculate the percentage *in vivo* target engagement. Alternatively, the level of free target can be measured in the presence of the therapeutic antibody drug candidate. Similar to LBA assays, the LC-MS-based free-target assay requires high sensitivity to measure the magnitude and duration of target engagement by the drug candidate, and careful design to minimize equilibrium disruption. For free- and total-target assays, the hybrid LBA–LC-MS/MS approach has proven to be very successful. This approach requires less stringent reagents for immunocapture, less development time, can make use of tissue biopsy samples, and can detect drug exposure and target antigen concentrations simultaneously (Ciccimaro et al., 2017; Sugimoto et al., 2018; Zhang et al., 2017). There follow two examples describing LBA–LC-MS/MS assay approaches for a soluble target and a membrane target.

Soluble Target Engagement

A recent example of the use of LBA–LC-MS/MS assays for the measurement of a soluble target is described by Zhang and coworkers (Zhang et al., 2017). In this example, free- and total-target assays were developed for IP-10, a low abundance and small chemokine protein target for the treatment of Crohn's disease.

Inhibition of IP-10 leads to suppression of various biological activities including recruitment of immune cells to the inflamed tissues in autoimmune diseases. The efficacy of anti-IP-10 mAb therapy is driven by the degree and duration of free IP-10 suppression. Measuring free IP-10 suppression directly informs the lead selection and development of anti-IP-10 mAbs. To build a high-quality free-target assay, the authors focused on having ultimate assay sensitivity and minimal drug–ligand dissociation.

Three steps were taken to achieve the desired sensitivity: (1) surrogate peptide selection; (2) immunocapture optimization to achieve the best recovery, minimize matrix effects, and to reduce background noise; and (3) the use of IP-10-depleted serum to generate calibration standards. Each step was methodically investigated: for example, the authors evaluated multiple proteolytic enzymes, and selected the peptides with the strongest signal as the surrogate peptide. They also attended to the physicochemical characteristics of the surrogate peptide, and used relatively high column temperatures to achieve appropriate chromatographic peak shape and resolution. Further, they incorporated strong reducing conditions prior to sample digestion to improve peptide yield from the capture antibody and IP-10 complex. Consequently, after immunocapture, elution of the captured analyte from the magnetic protein G beads resulted in a cleaner sample with significantly low background noise and excellent assay sensitivity.

The author's strategy to minimize drug–ligand complex dissociation was to use the therapeutic mAb as the capture reagent so that the drug and the capture antibody bind to the same epitope on the target with the same affinity and on/off rate. They also titrated the effect of therapeutic mAb concentration on the recovery of free IP-10. Other efforts included minimizing incubation time to reduce equilibrium shift; lowering incubation temperature to 7°C to reduce the binding exchange rate; minimal sample dilution to avoid dissociation of the complex; and minimum amount of capture antibody to avoid the avidity issue where the binding of one arm of the antibody facilitates the binding of the second arm.

Both free- and total IP-10 assays were validated in an analyte-depleted native matrix following the BMV PK-like guideline (FDA, 2018). The LOQ was 1 pM for the free assay and 10 pM for the total assay using 100 μL and 50 μL of serum, respectively. The quality of these assays were sufficient for lead anti-IP-10 mAb screening and candidate selection, for assessing the safety and efficacy via PK/PD correlations, and for dose projections for clinical development.

Membrane Target Engagement

Quantifying drug-bound targets in tissues is challenging given the need to solubilize the tissue, and to find appropriate extraction and binding conditions that ensure integrity of the drug-target complex. To investigate drug binding to membrane-bound CD40 receptors in cynomolgus monkey colon biopsies, Ciccimaro and coworkers (2017) used a strategy to directly measure both drug exposure and target engagement to enable assessment of the therapeutic antibody at the site of action.

CD40 is highly expressed in colon tissue with a low amount of soluble shed receptor in the circulation which is critical for immune regulation in the colon. Selective harvesting of the tissue biopsies by means of pinch biopsies ensured that the tissue sample is enriched in epithelial, lymphocytes, plasma, and mast cells from the lamina propria with minimal contamination from blood. For tissue lysis, the authors used bead mill processing of the biopsies at 4°C in the presence of tissue extraction buffer containing protease inhibitors. The supernatant portion was used to measure both total CD40 and bound CD40. Total CD40 was measured using mixed beads conjugated with a CD40 antibody with a different binding epitope from the anti-CD40 drug to capture free CD40 and CD40–drug complexes. A separate portion of tissue extract was then used to measure bound target captured with Protein G Magnetic Beads, where bound CD40 was measured by capturing CD40–drug complex and free drug. The affinity extracted protein in both samples was acid eluted and dried. The dried samples were resuspended, reduced, and alkylated and digested with trypsin. Stable isotope-labeled (SIL) tryptic peptides were added to each sample at a known concentration prior to LC-MS/MS analysis. The total CD40 assay achieved an LOQ of 0.78 ng/mL (26 pM) and was validated in a fit-for-purpose manner.

Percent target engagement was calculated from the ratio of bound target to total target. In these assays, the authors take advantage of MS which allows them to use matrix from a different species (surrogate matrix) to build calibration curves, and to use native matrix to build QCs by mixing ("ad-mixing") sample pools from healthy and diseased animals to generate low, mid, and high bound CD40 QCs. These QCs may be used to identify biases introduced during sample handling.

When compared to PK values measured in blood, the tissue target engagement profiles and drug exposure profiles in the colon biopsies were in agreement. For this study, RO was also measured by fluorescent activated cell sorting (FACS). RO by FACS and percent TE by hybrid LBA–LC-MS/MS in colon epithelial cells were in good agreement, although the magnitude of percent RO was greater than the percent TE measured by LBA–LC-MS/MS which requires additional experimentation to understand. Taken together, these data were used as direct confirmation that the drug reaches the site of action at therapeutically relevant concentrations, and that the drug is bound to the target. In addition, measurement of total-target and total-drug allowed determination of the stoichiometry as a function of dosing and was critical to understanding the PK/PD relationship.

Overall, relative to other approaches such as LBA or cytometry, the main advantages of using LBA–LC-MS/MS in target engagement assessment are: (1) percent TE is determined by direct measurement of bound target and total target via detection of tryptic peptides derived from the target protein and the therapeutic protein. This is a more direct approach than using a second detection antibody; and (2) soluble and membrane targets (regardless of cell type or surface expression) can potentially be measured in a tissue homogenate. Indeed, in recent years, more target engagement data using a LBA–LC-MS/MS approach have been used to guide human dose projections and patient stratification.

Example 3 Cytometry RO Assay

With an increase in programs targeting cell surface proteins there is a need to understand drug–RO preclinically to inform dose prediction using PK/PD modeling and to confirm sufficient target engagement. These assays can be quite challenging to develop especially with tight timelines. This is illustrated here in a preclinical study (unpublished) where the receptor occupancy of three biotherapeutic (BioTx) antibody molecules to a cellular target on blood cells were compared with each other and correlated with their PK parameters (see Figure 5.3). Since this was a single dose study with tight timelines, each BioTx antibody was used as a probe and was conjugated to Alexa Fluor 647. Specific and saturable binding was demonstrated for each probe, and a concentration above the EC80 was selected for detecting the cells. After demonstrating that each BioTx antibody probe was able to measure unoccupied target sites on cells preincubated with unlabeled drugs, a whole blood fix/lyze assay was developed which included an efficient phenotyping scheme, compensation, FMO controls, and controls for unstained cells and for nonspecific binding (NSB). Attempts to find an appropriate noncompeting antibody probe for generating a total-target assay were unsuccessful. Therefore, a decision was made to move forward without a total target assay. Pre-dose levels of target expression were used as a measure of total receptor number and post-dose median fluorescence intensity was used to calculate the percent free receptor (assuming total receptors = pre-dose levels). To minimize receptor internalization, all samples were immediately placed on wet ice after collection and sample processing and analysis were also conducted on wet ice or at 4–6°C. The cell population of interest in whole blood collected into sodium heparin tubes was considered relatively stable. However, to minimize the risk, samples were analyzed on the same day and as soon as possible after collection.

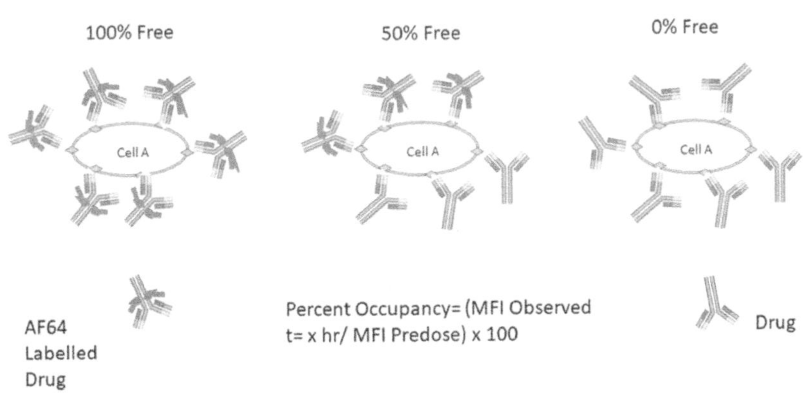

Figure 5.3 Receptor occupancy format.

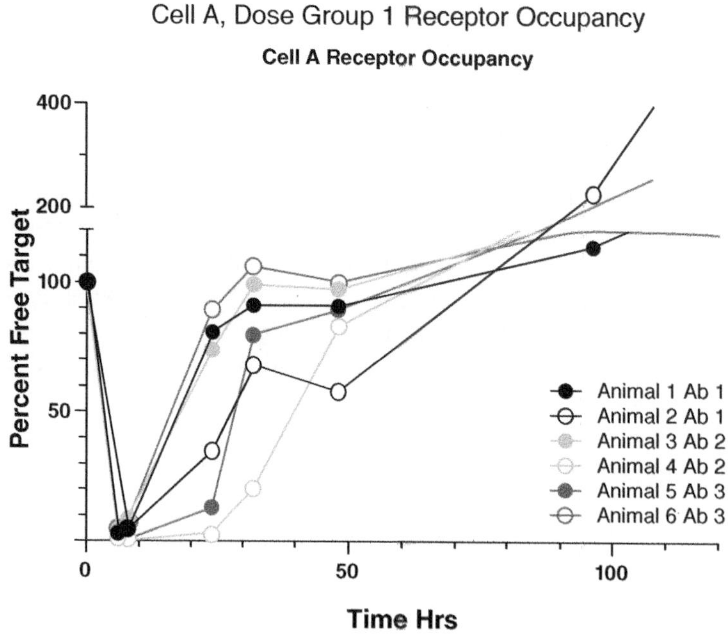

Figure 5.4 Receptor occupancy over time after dosing.

As shown in Figure 5.4, % of free target decreased after dosing in all groups, and returned to baseline between 24 and 48 hours. Higher than baseline (pre-dosing) binding was seen after 96 hours for two of the three testing groups. Binding above the linear range of detection was observed in these animals. These increases, observed in the presence of excess unlabeled antibody to reduce NSB, indicate that target expression was increased at later time points post-dosing. It is possible the amount of unlabeled drug added was insufficient to avoid NSB at these time points, although complete reduction of NSB was obtained with coincubation with 25 μg/mL of unlabeled drug in the pre-dose samples. Since assay imprecision was not well defined, small differences (less than 20%–30%) were not considered significant. However, in some animals, values of greater than 500%–1000% percent of pre-dosing levels were observed (data not shown). Since a total target assay was not available this increase could have been due to increased expression. However, the most likely reason for this increase was the presence of ADAs. Indeed, ADAs were found to be present in all the animals at 168–504 hours post-dosing using an ADA bridging assay.

Overall, the assay met the study needs by providing data for PK/PD modeling and enabling decision making by the team. This example also illustrates the value of NSB and FMO controls for data interpretation, the importance of including a total target sample, and the potential risk of ADA interference when using the drug as a probe.

5.6 Section 6: Summary

The pharmaceutical industry has significantly increased its use of biomarkers in recent years to support preclinical to clinical translation, and to monitor safety and efficacy. Consequently, biomarker bioanalysis has emerged as a special branch of the bioanalysis field. Since historically bioanalysis research has focused on PK measurement, much of the scientific practices and regulatory guidance for biomarker bioanalysis reflect those for PK analysis. In the last two decades, there has been significant effort among scientists in the academic, pharmaceutical communities, and regulatory agencies to raise awareness of the specific challenges intrinsic to biomarker assay validation, and to tackle these challenges through continued collaborations. Recent advances in the biomarker bioanalysis field are the results of such efforts and are epitomized by the emerging and evolving best practice for these unique challenge such as validation approaches for flow cytometric testing, reference calibrator characterization, parallelism assessment, biomarker specimen preservation and stability assessment, and validation of commercial kits and multiplexing assays (Cowan et al., 2017; Dakappagari et al., 2017; Jani et al., 2016; Khan et al., 2015; Lee et al., 2011).

Pharmaceutical scientists in the biomarker and clinical pharmacology fields have also directed more efforts toward generating data to enable PK/PD modeling, drug dose selection, and optimization. Changes in cell surface and soluble protein target levels pre- and post-drug treatment can be monitored using multiple analytical platforms, such as LBA, LC-MS, and flow cytometry. Recently, with a better understanding of off-rate kinetics (Cao et al., 2019) and application of rapid separations approaches and novel technology, there has been a renewed interest in the measurement of free target (Dysinger & Ma, 2018). However, developing free target assay can be resource-intensive; and the risk of overestimating free target remains a significant challenge. Therefore, the return on investment for this effort must be considered.

The use of LC-MS and, in particular, hybrid LBA–LC-MS methods to measure large molecule biomarkers has grown considerably in the past decade (Ciccimaro et al., 2017; Neubert et al., 2014, 2016, 2020; Palandra et al., 2013; Sugimoto et al., 2018; Zhang et al., 2017). Given that LC-MS is the most widely used bioanalytical tool for PK and TK analyses in support of drug discovery and development for the past 2 decades, this technology is also applied to biomarker bioanalysis. However, biomarker bioanalysis differs from PK bioanalysis in two major areas: the lack of analyte-free matrix to be used for assay development and validation, and the difficulty in procuring reference material which is identical to the endogenous analyte. These two differences have necessitated the use of innovative approaches such as combination of sample separation and enrichment approaches, hybrid LBA–LC-MS, and the use of surrogate matrices and surrogate analytes (Jones et al., 2012). Adoption of the use of LC-MS in biomarker bioanalytical laboratories can be limited by the complexity of the equipment, the high cost, the large sample requirement for low abundance molecules, and the lower throughput relative to LBA. However, there are also many advantages, including fast early assay development, good detection and selectivity in complex samples, direct measurement of the analyte, high level of multiplexing, and complementarity to LBA and flow cytometry, and these far outweigh the challenges. In recent years, more target engagement data for soluble and tissue targets have been generated using a LBA–LC-MS/MS approach for guiding human dose projections and patient stratification (Ciccimaro et al., 2017; Neubert et al., 2016, 2020).

The use of flow cytometry assays in industry has also increased due to interest in biotherapeutic approaches to modulate the immune system with targets on the surface of various cell types. Many unique challenges associated with developing flow cytometry assays have been addressed by a number of authors. It is time to integrate learning and experience from academia and clinical settings and apply them in drug development. This is an ongoing process with a particular focus on instrument control, specimen stability, and method validation (Brown et al., 2015; Davis et al., 2013; Du et al., 2015; O'Hara et al., 2011; Selliah et al., 2019; Wu et al., 2010). In addition, across the industry there is an increased effort targeting cell surface molecules and, thus, an increased need for RO measurement. New technology and approaches for RO measurement are emerging to enable the use of higher-dimensional phenotyping and better confidence in the results. With the routine use of a large number of markers in CyTOF panels, theoretical panels, and imaging cytometry, there is an increasing demand for high-dimensional data analysis. Numerous computational flow cytometry tools are being developed to aid this research (Saeys et al., 2016).

For biomarker analytical scientists, it is important to recognize that biomarker assay design and performance requirements are guided by the specific questions to be answered, which may change over the assay life cycle and among different users. Lee et al. (2006) devised the term "fit-for-purpose" to bring a unified approach to biomarker method validation. The recent effort on understanding the COU of each biomarker assay and its impact on assay development and validation (FDA-NIH Biomarker Working Group, 2018; Piccoli, 2019) is an extension of the FFP principle with more explicit focus on using the purpose of an assay as the single driver for all decisions during assay design and validation. There is no "one size fits all" approach for assay validation or for assay acceptance criteria. The performance of each assay must be viewed within the context of how the assay would be used, the decisions to be made with the data, and the stage of drug development. In addition, the COU framework also highlights the iterative nature of biomarker analytical validation and clinical utility.

Acknowledgment

The authors would like to thank Dr. Richard King for his careful review and insightful critique and comments.

References

Anderson, N. L., Anderson, N. G., Haines, L. R., Hardie, D. B., Olafson, R. W., & Pearson, T. W. (2004). Mass spectrometric quantitation of peptides and proteins using stable isotope standards and capture by anti-peptide antibodies (SISCAPA). *Journal of Proteome Research*, 3(2), 235–244. https:// DOI: 10.1074/mcp.M500331-MCP200

Arnold, M. E., Booth, B., King, L., & Ray, C. (2016). Workshop report: Crystal City VI-bioanalytical method validation for biomarkers. *The AAPS Journal*, 18(6), 1366–1372. https://doi.org/10.1208/s12248-016-9946-6

Azadeh, M., Sondag, P., Wang, Y., Raines, M., & Sailstad, J. (2019). Quality controls in ligand binding assays: recommendations and best practices for preparation, qualification, maintenance of lot to lot consistency, and prevention of assay drift. *The AAPS Journal*, 21, 89. https://doi.org/10.1208/s12248-019-0354-6

Barnidge, D. R., Goodmanson, M. K., Klee, G.G., & Muddiman, D.C. (2004a). Absolute quantification of the model biomarker prostate-specific antigen in serum by LC–MS/MS using protein cleavage and isotope dilution mass spectrometry. *Journal of Proteome Research*, 3(3), 644–652. https://doi.org/10.1021/pr049963d

Barnidge, D. R., Hall, G. D., Stocker, J. L., & Muddiman, D. C. (2004b). Evaluation of a cleavable stable isotope labeled synthetic peptide for absolute protein quantification using LC–MS/MS. *Journal of Proteome Research*, 3(3), 658–661. https://doi.org/10.1021/pr034124x

Barr, J. R., Maggio, V. L., Patterson, D. G., Jr., Cooper, G. R., Henderson, L. O., Turner, W. E., Smith, S. J., Hannon, W. H., Needham, L. L., & Sampson, E. J.. (1996). Isotope dilution–mass spectrometric quantification of specific proteins: model application with apolipoprotein A-I. *Clinical Chemistry*, 42(10), 1676–1682. https://doi.org/10.1093/clinchem/42.10.1676

Bernardi, D., Zaninotto, M., & Plebani, M. (2004). Requirements for improving quality in the measurement of bone markers. *Clinica Chimica Acta*, 346(1), 79–86. https://doi.org/10.1016/j.cccn.2004.02.036

Booth, B., Arnold, M. E., DeSilva, B., Amaravadi, L., Dudal, S., Fluhler, E., Gorovits, B., Haidar, S. H., Kadavil, J., Lowes, S., Nicholson, R., Rock, M., Skelly, M., Stevenson, L., Subramaniam, S., Weiner, R., & Woolf, E. (2015). Workshop report: Crystal City V--quantitative bioanalytical method validation and implementation: the 2013 revised FDA guidance. *The AAPS Journal*, 17(2), 277–288. https://doi.org/10.1208/s12248-014-9696-2

Brade, L., Engel, R., Christ, W. J., & Rietschel, E. T. (1997). A non-substituted primary hydroxyl group in position 6′ of free lipid A is required for binding of lipid A monoclonal antibodies. *Infection and Immunity*, 65, 3961–3965. https://doi.org/10.1128/iai.65.9.3961-3965.1997

Brown, L., Green, C. L., Jones, N., Stewart, J. J., Fraser, S., Howell, K., Xu, Y., Hill, C. G., Wiwi, C. A., White, W. I., O'Brien, P. J., & Litwin, V. (2015). Recommendations for the evaluation of specimen stability for flow cytometric testing during drug development. *Journal of Immunological Methods*, 418, 1–8. https://doi.org/10.1016/j.jim.2015.01.008

Cao, H., Zhang, G., Ciociola, K., Amaravadi, L., Qiu, Y., & Wu, J. (2019). An LC-MS/MS approach to assess total and free protein target in the serum of cynomolgus monkey. *Bioanalysis*, 11(5), 393–406. https://doi.org/10.4155/bio-2018-0294

Ciccimaro, E., Zhu, Y., Ostanin, D., Suchard, S., MacGuire, J., Xiao, Q., Dongre, A., Chimalakonda, A., Olah, T., & Shipkova, P. (2017). Antibody drug-target engagement measurement in tissue using quantitative affinity extraction liquid chromatography-mass spectrometry: method development and qualification. *Analytical Chemistry*, 89(9), 5115–5123. https://doi.org/10.1021/acs.analchem.7b00688

Colbert, A., Umble-Romero, A., Prokop, S., Xu, R., Gibbs, J., & Pederson, S. (2014). Characterization of a quantitative method to measure free proprotein convertase subtilisin/kexin type 9 in human serum. *mAbs*, 6(4), 1103–1113. https://doi.org/10.4161/mabs.28719

Cowan, K. J. Amaravadi, L., Cameron, M. J., Fink, D., Jani, D., Kamat, M., King, L., Neely, R. J., Ni, Y., Rhyne, P., Riffon, R., & Zhu, Y. (2017). Recommendations for selection and characterization of protein biomarker assay calibrator material. *The AAPS Journal*, 19(6), 1550–1563. https://doi.org/10.1208/s12248-017-0146-9

Dakappagari, N., Zhang, H., Stephen, L., Amaravadi, L., & Khan, M. U. (2017). Recommendations for clinical biomarker specimen preservation and stability assessments. *Bioanalysis*, 9(8), 643–653. https://doi.org/10.4155/bio-2017-0009

Davis, B. H. Wood, B., Oldaker, T., & Barnett, D. (2013). Validation of cell-based fluorescence assays: practice guidelines from the ICSH and ICCS part I rational and aims. *Cytometry Part B – Clinical Cytometry*, 84(5), 25–28. https://doi: 10.1002/cyto.b.21104

Davis, C., Wu, X., Li, W., Fan, H., Reddy, M. (2011). Stability of immunophenotypic markers in fixed peripheral blood for extended analysis using flow cytometry. *Journal of Immunological Methods*, 363(2), 158–165. https://doi.org/10.1016/j.jim.2010.09.029

Davis, K. A. Abrams, B., Iyer, S.B., Hoffman, R.A., & Bishop, J.E. (1998). Determination of CD4 antigen density on cells: role of antibody valency, avidity, clones, and conjugation. *Cytometry*, 33(2), 197–205. https://doi: 10.1002/(sici)1097-0320(19981001)33:2<97::aid-cyto14>3.0.co;2-p

der Strate, B. V., Longdin, R., Geerlings, M., Bachmayer, N., Cavallin, M., Litwin, V., Patel, M., Passe-Coutrin, W., Schoelch, C., Companjen, A., & Fjording, M. S. (2017). Best practices in performing flow cytometry in a regulated environment: feedback from experience within the European Bioanalysis Forum. *Bioanalysis*, 9(16), 1253–1264. https://doi.org/10.4155/bio-2017-0093

DeSilva, B., Garofolo, F., Rocc, I. M., Martinez, S., Dumont, I., Landry, F., Dicaire, C., Szekely-Klepser, G., Weiner, R., Arnold, M., Bansal, S., Bateman, K., Bauer, R., Booth, B., Davis, S., Dudal, S., Gouty, D., Grundy, J., Haidar, S., Hayes, R., Jemal, M., Kaur, S., Kelley, M., Knutsson, M., Le Blaye, O., Lee, J., Lowes, S., Ma, M., Mitsuoka, T., Neto, J. T., Nicholson, R., Ormsby, E., Sailstad, J., Stevenson, L., Tang, D., Welink, J., Viswanathan, C. T., Wang, L., Woolf, E., & Yang, E. (2012) White Paper on recent issues in bioanalysis and alignment of multiple guidelines. *Bioanalysis*, 4(18), 2213–26. https://doi: 10.4155/bio.12.205

Diks, A. M., Bonroy, C., Teodosio, C., Groenland, R. J., de Mooij, B., de Maertelaere, E., Neirynck, J., Philippé, J., Orfao, A., van Dongen, J. J. M., & Berkowska, M. A. (2019). Impact of blood storage and sample handling on quality of high dimensional flow cytometric data in multicenter clinical research. *Journal of Immunological Methods*, https://doi.org/10.1016/j.jim.2019.06.007

Drake, A. W., Myszka, D. G., & Klakamp, S. L. (2004). Characterizing high-affinity antigen/antibody complexes by kinetic- and equilibrium-based methods. *Analytical Biochemistry*, 328(1), 35–43. https://doi.org/10.1016/j.ab.2003.12.025

Du, L., Grover, A., Ramanan, S., & Litwin, V. (2015). The evolution of guidelines for the validation of flow cytometric methods. *International Journal of Laboratory Hematology*, 37(Suppl. 1), 3–10. https://doi: 10.1111/ijlh.12344

Dysinger, M., & Ma, M. (2018). A Gyrolab assay for the quantitation of free complement protein C5a in human plasma. *The AAPS Journal*, 20(6), 106. https://doi.org/10.1208/s12248-018-0266-x

Food and Drug Administration (FDA). (2001) Guidance for Industry: Bioanalytical Method Validation. http://www.labcompliance.de/documents/FDA/FDA-Others/Laboratory/f-507-bioanalytical-4252fnl.pdf

Food and Drug Administration (FDA). (2018, May). *Guidance for Industry: Bioanalytical Method Validation*. FDA, 1–22. http://www.fda.gov/Drugs/GuidanceComplianceRegulatoryInformation/Guidances/default.htm

Food and Drug Administration (FDA)-Biopharmaceutics-2001. (2013). *Guidance for Industry: Bioanalytical Method Validation, Food and Drug Administration Center for Drug Evaluation and Research (CDER) Center for Veterinary Medicine (CVM)*. http://www.labcompliance.de/documents/FDA/FDA-Others/Laboratory/f-507-bioanalytical-4252fnl.pdf

Food and Drug Administration (FDA)-National Institutes of Health (NIH) Biomarker Working Group. (2018). *BEST (Biomarkers, EndpointS, and other Tools) Resource*, 7–53. https://www.ncbi.nlm.nih.gov/books/NBK326791/pdf/Bookshelf_NBK326791.pdf

Fischer, S. K., Joyce, A., Spengler, M., Yang, T-Y., Zhuang, Y., Fjording, M. S., & Mikulskis, A. (2015) Emerging Technologies to Increase Ligand Binding Assay Sensitivity. *The AAPS Journal*, 17(1), 93. Http://doi: 10.1208/s12248-014-9682-8

Grant, R. P., & Hoofnagle, A. N. (2014). From lost in translation to paradise found: enabling protein biomarker method transfer by mass spectrometry. *Clinical Chemistry*, 60(7), 941–944. https://doi.org/10.1373/clinchem.2014.224840

Gratama, J. W., D'hautcourt, J-L., Mandy, F., Rothe, G., Barnett, D., Janossy, G., Papa, S., Schmitz, G., & Lenkei, R. (1998). Flow cytometric quantitation of immunofluorescence intensity: problems and perspectives. *Cytometry*, 33(2), 166–178. https://doi.org/10.1002/(SICI)1097-0320(19981001)33:2<166::AID-CYTO11>3.0.CO;2-S

Green, C. L., Stewart, J. J., Högerkorp, C. M., Lackey, A., Jones, N., Liang, M., Xu, Y., Ferbas, J., Moulard, M., Czechowska, K., & Mc Closkey, T. W. (2016). Recommendations for the development and validation of flow cytometry-based receptor occupancy assays. *Cytometry. Part B, Clinical Cytometry*, 90(2), 141–149. https://doi.org/10.1002/cyto.b.21339

Harth, S., Ten Haaf, A., Loew, C., Frisch, C., & Knappik, A. (2019). Generation by phage display and characterization of drug-target complex-specific antibodies for pharmacokinetic analysis of biotherapeutics. *mAbs*, 11(1), 178–190. https://doi.org/10.1080/19420862.2018.1538723

Honegger, A., Malebranche, A. D., Röthlisberger, D., & Plückthun, A. (2009). The influence of the framework core residues on the biophysical properties of immunoglobulin heavy chain variable domains. *Protein Engineering, Design & Selection*, 22(3), 121–134. https://doi.org/10.1093/protein/gzn077

Houghton R., Horro Pita C., Ward I., Macarthur R. (2009) Generic approach to validation of small-molecule LC-MS/MS biomarker assays. *Bioanalysis*, 1(8), 1365–1374. https://doi: 10.4155/bio.09.139

Houghton, R., Gouty, D., Allinson, J., Green, R., Losauro, M., Lowes, S., LeLacheur, R., Garofolo, F., Couerbe, P., Bronner, S., Struwe, P., Schiebl, C., Sangster, T., Pattison, C., Islam, R., Garofolo, W., Pawula, M., Buonarati, M., Hayes, R., Cameron, M., Nicholson, R., Harman, J., Wieling, J., De Boer, T., Reuschel, S., Cojocaru, L., Harter, T., Malone, M., & Nowatzke, W. (2012) Recommendations on biomarker bioanalytical method validation by GCC. *Bioanalysis*, 4(20), 2439–2446. https://doi: 10.4155/bio.12.197

Jager, E., van der Velden, V. H. J., te Marvelde, J. G., Walter, R. B., Agur, Z., & Vainstein, V. (2011). Targeted drug delivery by gemtuzumab ozogamicin: mechanism-based mathematical model for treatment strategy improvement and therapy individualization. *PLoS ONE*, 6(9): e24265. https://doi.org/10.1371/journal.pone.0024265

Jani, D., Allinson, J., Berisha, F., Cowan, K. J., Devanarayan, V., Gleason, C., Jeromin, A., Keller, S., Khan, M. U., Nowatzke, B., Rhyne, P., & Stephen, L. (2016). Recommendations for use and fit-for-purpose validation of biomarker multiplex ligand binding assays in drug development. *The AAPS Journal*, 18(1), 1–14. https://doi.org/10.1208/s12248-015-9820-y

Jenkins, R. G. (2016). Accuracy: a potential quandary in regulated bioanalysis of "endogenous" analytes. *Bioanalysis*, 8(23), 2393–2397. https://doi.org/10.4155/bio-2016-0247

Jian, W., Edom, R. W., Wang, D., Weng, N., & Zhang, S. W. (2013). Relative quantitation of glycoisoforms of intact apolipoprotein C3 in human plasma by liquid chromatography-high-resolution mass spectrometry. *Analytical Chemistry*, 85(5), 2867–2874. https://doi.org/10.1021/ac3034757

Jian, W., Kang, L., Burton, L., & Weng, N. (2016). A workflow for absolute quantitation of large therapeutic proteins in biological samples at intact level using LC-HRMS. *Bioanalysis*, 8(16), 1679–1691. https://doi.org/10.4155/bio-2016-0096

Jones, B. R., Schultz, G. A., Eckstein, J. A., & Ackermann, B. L. (2012). Surrogate matrix and surrogate analyte approaches for definitive quantitation of endogenous biomolecules. *Bioanalysis*, 4(19), 2343–2356. https://doi.org/10.4155/bio.12.200

Khan, M. U., Bowsher, R. R., Cameron, M., Devanarayan, V., Keller, S., King, L., Lee, J., Morimoto, A., Rhyne, P., Stephen, L., Wu, Y., Wyant, T., and Lachno, D.

R. (2015). Recommendations for adaptation and validation of commercial kits for biomarker quantification in drug development. *Bioanalysis*, 7(2), 229–242. https://doi.org/10.4155/bio.14.274

King, L. E., Farley, E., Imazato, M., Keefe, J., Khan, M., Ma, M., Pihl, K. S., & Sriraman, P. (2014) Ligand binding assay critical reagents and their stability: recommendations and best practices from the Global Bioanalysis Consortium Harmonization Team. *The AAPS Journal*, 16(3), 504–15. https//doi: 10.1208/s12248-014-9583-x

King, L. E. (2016). Parallelism experiments in biomarker ligand-binding assays to assess immunological similarity. *Bioanalysis*, 2387–2391. https://doi.org/10.4155/bio-2016-0245

Kuhn, E., Wu, J., Karl, J., Liao, H., Zolg, W., & Guild, B. (2004). Quantification of C-reactive protein in the serum of patients with rheumatoid arthritis using multiple reaction monitoring mass spectrometry and 13C-labeled peptide standards. *Proteomics*, 4(4), 1175–1186. https://doi.org/10.1002/pmic.200300670

Lacher, D. A., Hughes, J. P., & Carroll, M. D. (2005). Estimate of biological variation of laboratory analytes based on the Third National Health and Nutrition Examination Survey. *Clinical Chemistry*, 51(2), 450–452. https:// doi: 10.1373/clinchem.2004.039354

Lachmann, H. J., Lowe, P., Felix, S. D., Rordorf, C., Leslie, K., Madhoo, S., Wittkowski, H., Bek, S., Hartmann, N., Bosset, S., Hawkins, P. N., & Jung, T. (2009). In vivo regulation of interleukin 1β in patients with cryopyrin-associated periodic syndromes. *Journal of Experimental Medicine*, 206(5), 1029–1036. https://doi.org/10.1084/jem.20082481

Lanshoeft, C., Cianférani, S., & Heudi, O. (2017). Generic hybrid ligand binding assay liquid chromatography high-resolution mass spectrometry-based workflow for multiplexed human immunoglobulin g1 quantification at the intact protein level: Application to preclinical pharmacokinetic studies. *Analytical Chemistry*, 89(4), 2628–2635. https://doi.org/10.1021/acs.analchem.6b04997

Lassman, M. E., & Fernández-Metzler, C. (2014). Applications of low-flow LC–SRM for the analysis of large molecules in pharmaceutical R&D. *Bioanalysis*, 6(13), 1859–1867. https://doi.org/10.4155/bio.14.141

Lee, J. W., Devanarayan, V., Barrett, Y. C., Weiner, R., Allinson, J., Fountain, S., Keller, S., Weinryb, I., Green, M., Duan, L., Rogers, J.A., Millham, R., O'Brien, P. J., Sailstad, J., Khan, M., Ray, C., & Wagner, J. A. (2006). Fit-for-purpose method development and validation for successful biomarker measurement. *Pharmaceutical Research*, 23(2), 312–328. https://doi.org/10.1007/s11095-005-9045-3

Lee, J. W., Weiner, R. S., Sailstad, J. M., Bowsher, R. R., Knuth, D. W., O'Brien, P. J., Fourcroy, J. L., Dixit, R., Pandite, L., Pietrusko, R. G., Soares, H. D.,

Quarmby, V., Vesterqvist, O. L., Potter, D. M., Witliff, J. L., Fritche, H. A., O'Leary, T., Perlee, L., Kadam, S., Wagner, J. A. (2005). Method validation and measurement of biomarkers in nonclinical and clinical samples in drug development: a conference report. *Pharmaceutical Research*, 22(4), 499–511. https://doi.org/10.1007/s11095-005-2495-9

Lee, J. W., Kelley, M., King, L. E., Yang, J., Salimi-Moosavi, H., Tang, M. T., Lu, J. F. et al. (2011). Bioanalytical approaches to quantify "total" and "free" therapeutic antibodies and their targets: technical challenges and PK/PD applications over the course of drug development. *The AAPS Journal*, 13(1), 99–110. https://doi.org/10.1208/s12248-011-9251-3

Liang, M., Schwickart, M., Schneider, A. K., Vainshtein, I., Del Nagro, C., Standifer, N., & Roskos, L. K. (2016). Receptor occupancy assessment by flow cytometry as a pharmacodynamic biomarker in biopharmaceutical development. *Cytometry Part B*, 90(B), 117–127. https:// DOI: 10.1002/cyto.b.21259

Lizotte, P. H., Jones, R. E., Keogh, L., Ivanova, E., Liu, H., Awad, M. M., Hammerman, P. S., Gill, R. R., Richards, W. G., Barbie, D. A., Bass, A. J., Bueno, R., English, J. M., Bittinger, M., & Wong, K. K. (2016). Fine needle aspirate flow cytometric phenotyping characterizes immunosuppressive nature of the mesothelioma microenvironment. *Scientific Reports*, 6. 31745- https://doi.org/10.1038/srep31745

Lorentzon, M., Branco, J., Brandi, M. L., Bruyère, O., Chapurlat, R., Cooper, C., Cortet, B., Diez-Perez, A., Ferrari, S., Gasparik, A., Herrmann, M., Jorgensen, N. R., Kanis, J., Kaufman, J. M., Laslop, A., Locquet, M., Matijevic, R., McCloskey, E., Minisola, S., Pikner, R., Reginster, J. Y., Rizzoli, R., Szulc, P., Vlaskovska, M., & Cavalier, E. (2019). Algorithm for the use of biochemical markers of bone turnover in the diagnosis, assessment and follow-up of treatment for osteoporosis. *Advances in Therapy*, 36(10). https://doi.org/10.1007/s12325-019-01063-9

Maecker, H. T., McCoy, J. P., & Nussenblatt, R. (2012). Standardizing immunophenotyping for the Human Immunology Project. *Nature Reviews. Immunology*, 12(3), 191–200. https:// doi: 10.1038/nri3158

Mahnke, Y., Chattopadhyay, P., & Roederer, M. (2010). Publication of optimized multicolor immunofluorescence panels. *Cytometry Part A*, 77(9), 814–818. https://doi.org/10.1002/cyto.a.20916

Mair, F. et al. (2016). The end of gating? An introduction to automated analysis of high dimensional cytometry data. *European Journal of Immunology*, 46(1), 34–43. Wiley-VCH Verlag. https://doi.org/10.1002/eji.201545774

Maloney, D. G., Grillo-López, A. J., White, C. A., Bodkin, D., Schilder, R. J., Neidhart, J. A., Janakiraman, N., Foon, K. A., Liles, T. M., Dallaire, B. K., Wey, K., Royston, I., Davis, T., & Levy, R. (1997). IDEC-C2B8 (Rituximab) anti-CD20 monoclonal antibody therapy in patients with relapsed low-grade non-Hodgkin's lymphoma. *Blood*, 90(6), 2188–2195.

Marcelletti, J. F., Evans, C. L., Saxena, M., Lopez, A. E. (2015). Calculations for adjusting endogenous biomarker levels during analytical recovery assessments for ligand-binding assay bioanalytical method validation. *The AAPS Journal*, 17(4), 939–947. https:// DOI: 10.1208/s12248-015-9756-2

Mathews, J. A., Ni, Y. G., Wang, C., Peterson, J. E., Ray, C., Zhao, X., Duan, D., Hamon, S., Allinson, J., Hokom, M., Wegner, G. (2020). Considerations for soluble protein biomarker blood sample matrix selection. *The AAPS Journal*, 22(2), 38. https://doi.org/10.1208/s12248-020-0412-0

Midgley, J. E. M., & Ekins, R. (1993). The free thyroid hormone hypothesis and measurement of free hormones [1]. *Clinical Chemistry*, 39(6), 1342–1344. https://doi.org/10.1093/clinchem/39.6.1342

Miller, M. A., Sullivan, R. J., & Lauffenburger, D. A. (2017). Molecular pathways: receptor ectodomain shedding in treatment, resistance, and monitoring of cancer. *Clincal Cancer Research*, 23(3), 623–629. https://doi: 10.1158/1078-0432.CCR-16-0869

Neubert, H., Fan, Y.-Y., & Fernández Ocaña, M. (2016). Quantification of protein biomarkers in tissues: new capabilities with pellet digestion peptide immunoaffinity LC–MS/MS. *Bioanalysis*, 8(15), 1551–1555. https://doi.org/10.4155/bio-2016-0155

Neubert, H., Palandra, J., & Fernández Ocaña, M. (2014). Quantification of biotherapeutic targets: new opportunities with immunoaffinity LC–MS/MS. *Bioanalysis*, 6(13), 1731–1733. https://doi.org/10.4155/bio.14.138

Neubert, H., Shuford, C. M., Olah, T. V., Garofolo, F., Schultz, G. A., Jones, B. R., Amaravadi L., Laterza, O. F., Xu, K., & Ackermann, B. L. (2020). Protein biomarker quantification by immunoaffinity liquid chromatography–tandem mass spectrometry: current state and future vision. *Clinical Chemistry*, 66(2), 282–301. https://doi.org/10.1093/clinchem/hvz022

Nguyen, R., Perfetto, S., Mahnke, Y. D., Chattopadhyay, P., & Roederer, M. (2013). Quantifying spillover spreading for comparing instrument performance and aiding in multicolor panel design. *Cytometry Part A*, 83(3), 306–15. https:// DOI: 10.1002/cyto.a.22251

Neubert, H., Shuford, C. M., Olah, T. V., Garofolo, F., Schultz, G. A., Jones, B. R., Amaravadi, L., Laterza, O. F., Xu, K., & Ackermann, B. L. (2020). Protein biomarker quantification by immunoaffinity liquid chromatography–tandem mass spectrometry: Current state and future vision. *Clinical Chemistry*, 66(2), 282–301. https://doi.org/10.1093/clinchem/hvz022

Nowatzke, W., Cole, T. G., & Bowsher, R. R. (2010). Systematic analytical validation of commercial kits for the determination of novel biomarkers for clinical drug development. *Bioanalysis*, 2(2), 237–247. https://doi.org/10.4155/bio.09.191

Oberoi P., Suschak J., Stenglin M., Aghvanyan A., Nikolenko G., Roy D., Gardner L., Lockney D., Shukla A., Porter E., Miller T., Biesso A., Sardesai L., Sivakamasundar R., Mathew A., Glezer E. N., Wohlstadter J. N. (2016) True sensitivity of immunoassays: are concentrations of low abundance analytes real or artifacts? *The Journal of Immunology* 196(Suppl. 1), 138.9.

Oda, Y., Huang, K., Cross, F. R., Cowburn, D., & Chait, B. T. (1999). Accurate quantitation of protein expression and site-specific phosphorylation. *Proceedings of the National Academy of Sciences of the United States of America*, 96(12), 6591–6596. https://doi.org/10.1073/pnas.96.12.6591

O'Hara, D. M., Xu, Y., Liang, Z., Reddy, M. P., Wu, D. Y. & Litwin, V. (2011). Recommendations for the validation of flow cytometric testing during drug development: II assays. *Journal of Immunological Methods*, 363, 120–134. hhttps:// DOI: 10.1016/j.jim.2010.09.036

O'Hara, D. M., Theobald, V., Egan, A. C., Usansky, J., Krishna, M., TerWee, J., Maia, M., Spriggs, F. P., Kenney, J., Safavi, A., & Keefe, J. (2012). Ligand binding assays in the 21st century laboratory: Recommendations for characterization and supply of critical reagents. *The AAPS Journal*, 14(2), 316–328. https://doi.org/10.1208/s12248-012-9334-9

Palandra, J., Finelli, A., Zhu, M., Masferrer, J., & Neubert, H. (2013). Highly specific and sensitive measurements of human and monkey interleukin 21 using sequential protein and tryptic peptide immunoaffinity LC-MS/MS. *Analytical Chemistry*, 85(11), 5522–5529. https://doi.org/10.1021/ac4006765

Piccoli, S., & Sauer, J. M. (2019). Points to consider document: scientific and regulatory considerations for the analytical validation of assays used in the qualification of biomarkers in biological matrices. https://c-path.org/wp-content/uploads/2019/06/evidconsid-whitepaper-analyticalsectionv20190613.pdf

Reichard, A., & Asosingh, K. (2019). Best practices for preparing a single cell suspension from solid tissues for flow cytometry. *Cytometry. Part A*, 95(2), 219–226. https:// doi: 10.1002/cyto.a.23690

Rey, J., Poitiers, F., Paehler, T., Brunet, A., DiCioccio, T., Cannon, C. P., Surks, H. K., Pinquier, J-L., Hanotin, C., & Sasiela, W. J. (2016). Relationship between low-density lipoprotein cholesterol, free proprotein convertase Subtilisin/Kexin Type 9, and alirocumab levels after different lipid-lowering strategies. *Journal of the American Heart Association*, 5(6), e003323. https://doi.org/10.1161/JAHA.1,16.003323

Richards, S., Amaravadi, L., Pillutla, R., Birnboeck, H., Torri, A., Cowan, K. J., Papadimitriou, A., Garofolo, F., Satterwhite, C., Piccoli, S., Wu, B.,

Krinos-Fiorotti, C., Allinson, J., Berisha, F., Cocea, L., Croft, S., Frase, R. S., Galliccia, F., Gorovits, B., Gupta, S., Gupta, V., Haidar, S., Hottenstein, C., Ishii-Watabe, A., Jan, I. D., Kadavil, J., Kamerud, J., Kramer, D., Litwin, V., Lima Santos, G. M., Nelson, R., Ni, Y., Pedras-Vasconcelos, J., Qiu, Y., Rhyne, P., Safav, I. A., Saito, Y., Savoie, N., Scheibner, K., Schick E., Siguenza, P. Y., Smeraglia, J., Staack, R. F., Subramanyam, M., Sumner, G., Thway, T., Uhlinger, D., Ullmann, M., Vitaliti, A., Welink, J., Whiting, C. C., Xue, L., & Zeng, R.(2016). 2016 White Paper on recent issues in bioanalysis: focus on biomarker assay validation (BAV): (Part 3 – LBA, biomarkers and immunogenicity). *Bioanalysis*, 8(23), 2475–2496. https://doi.org/10.4155/bio-2016-4989

Rifai, N., Watson, I. D., & Miller, G. (2013). Commercial immunoassays in biomarkers studies: Researchers beware! *Clinical Chemistry and Laboratory Medicine*, 51(2), 249–251. https://doi.org/10.1515/cclm-2013-0015

Rissin, D. M., Piech, T., Patel, P. P., Chang, L., Rivnak, A. J., Ferrell, E. P., Randall, J. D., Provuncher, G. C., Wal, T. D. R., & Duffy, D. C. (2010). Single-molecule enzyme-linked immunosorbent assay detects serum proteins at sub-femtomolar concentrations. *Nature Biotechnology*, 28(6), 595–599. https://doi.org/10.1038/nbt.1641

Roederer, M. (2001). Spectral compensation for flow cytometry: visualization artifacts limitations, and caveats. *Cytometry*, 48(2), 113. https:// DOI: 10.1002/1097-0320(20011101)45:3<194::aid-cyto1163>3.0.co;2-c

Ruan, Q., Ji, Q.C., Arnold, M. E., Humphreys, W. G., & Zhu, M. (2011). Strategy and its implications of protein bioanalysis utilizing high-resolution mass spectrometric detection of intact protein. *Analytical Chemistry*, 83(23), 8937–8944. https://doi.org/10.1021/ac201540t

Saeys, Y., Van Gassen, S., & Lambrecht, B. N. (2016). Computational flow cytometry: Helping to make sense of high-dimensional immunology data. *Nature Reviews. Immunology*, 16(7): 449–462. https://doi.org/10.1038/nri.2016.56

Santockyte, R., Zeng, J., & Zheng, N. (2019). Critical reagent screening and characterization: benefits and approaches for protein biomarker assays by hybrid LC-MS. *Bioanalysis*, 11(8), 785–795. https://doi.org/10.4155/bio-2018-0277

Schultz, G. A., McCardle, K., & Neubert, H. (2016). Large-scale implementation of sequential protein and peptide immunoaffinity enrichment LC/nanoLC–MS/MS for human β-nerve growth factor. *Bioanalysis*, 8(8), 753–764. https://doi.org/10.4155/bio-2015-0022

Selliah, N., Eck, S., Green, C., Oldaker, T., Stewart, J., Vitaliti, A., & Litwin, V. (2019). Flow cytometry method validation protocols. *Current Protocols in Cytometry*, 87(1), e53. https://doi.org/10.1002/cpcy.53

Shapiro, H. M. (2005). Practical Flow Cytometry. John Wiley & Sons, Inc.

Shepherd, J. A., & Lu, Y. (2007). A generalized least significant change for individuals measured on different DXA systems. *Journal of Clinical Densitometry*, 10(3), 249–258. https:// DOI: 10.1016/j.jocd.2007.05.002

Staack, R. F., Stracke, J.O., Stubenrauch, K., Vogel, R., Schleypen, J., & Papadimitriou, A. (2011) Quality requirements for critical assay reagents used in bioanalysis of therapeutic proteins: what bioanalysts should know about their reagents. *Bioanalysis* 3(5):523–34. https:// doi: 10.4155/bio.11.16. DOI: 10.4155/bio.11.16

Stevenson, L. F., & Purushothama, S. (2014). Parallelism: considerations for the development, validation and implementation of PK and biomarker ligand-binding assays. *Bioanalysis*, 6(2), 185–198. https://doi.org/10.4155/bio.13.292

Sugimoto, H., Ghosh, D., Chen, S., Smith, M. D., Abu-Yousif, A. O., & Qian, M. G. (2018). Immunocapture-LC/MS-based target engagement measurement in tumor plasma membrane. *Analytical Chemistry*, 90(22), 13564–13571. https://doi.org/10.1021/acs.analchem.8b03726

Vahid, F., Chen, X., & Neubert, H. (2018). Protein turnover measurements in human serum by serial immunoaffinity LC-MS/MS. *Clinical Chemistry*, 62(2), 279. https:// DOI: 10.1373/clinchem.2017.272922

Varro, R., Chen, R., Sepulveda, H., & Apgar, J. (2007). Bead-based multianalyte flow immunoassays: the cytometric bead array system. *Methods in Molecular Biology*, 378, 125. https:// DOI: 10.1007/978-1-59745-323-3_9

Wang, L., Abbasi, F., Gaigalas, A. K., Vogt, R. F., & Marti, G. E. (2006). Comparison of fluorescein and phycoerythrin conjugates for quantifying CD20 expression on normal and leukemic B-cells. *Cytometry. Part B, Clinical Cytometry*, 70(B), 410–415. https:// DOI: 10.1002/cyto.b.20140

Wang, W., Wang, X., Doddareddy, R., Fink, D., McIntosh, T., Davis, H. M., & Honghui, Z. (2014). Mechanistic pharmacokinetic/target engagement/pharmacodynamic (PK/TE/PD) modeling in deciphering interplay between a monoclonal antibody and its soluble target in cynomolgus monkeys. *The AAPS Journal*, 16(1), 129–139. https://doi.org/10.1208/s12248-013-9545-8

White, A., Gray, C., & Corrie, J. E. T. (1985). Monoclonal antibodies to testosterone: The effect of immunogen structure on specificity. *Journal of Steroid Biochemistry*, 22(2), 169–175. https:// doi: 10.1016/0022-4731(85)90109-8

Wiener, M. C., Sachs, J. R., Deyanova, E. G., & Yates, N. A. (2004). Differential mass spectrometry: A label-free LC-MS method for finding significant differences in complex peptide and protein mixtures. *Analytical Chemistry*, 76(20), 6085–6096. https://doi.org/10.1021/ac0493875

Wisniewski, J. R. (2016). Quantitative evaluation of filter aided sample preparation (FASP) and multienzyme digestion FASP protocols. *Analytical Chemistry*, 88(10), 5438–5443. https:// DOI: 10.1021/acs.analchem.6b00859

Wisniewski, J. R., Zougman, A., Nagaraj, N., & Mann, M. (2009). *Nature Methods*, 6(5), 359–362. https:// DOI: 10.1038/nmeth.1322

Wu, D. Y., Patti-Diaz, L., & Hill, C. G. (2010). Development and validation of flow cytometry methods for pharmacodynamic clinical biomarkers. *Bioanalysis*, 2(9), 1617–1626. https:// DOI: 10.4155/bio.10.33

Yalow, R. S., & Berson, S. (1959). Assay of plasma insulin in human subjects by immunological methods. *Nature*, 184, 1648–1649. https:// DOI: 10.1038/1841648b0

Zhang, H., Gu, H., Shipkova, P., Ciccimaro, E., Sun, H., Zhao, Q., & Olah, T. V. (2017). Immunoaffinity LC–MS/MS for quantitative determination of a free and total protein target as a target engagement biomarker. *Bioanalysis*, 9(20), 1573–1588. https://doi.org/10.4155/bio-2017-0152

Zheng, S., McIntosh, T., & Wang, W. (2015). Utility of free and total target measurements as target engagement and efficacy biomarkers in biotherapeutic development – Opportunities and challenges. *Journal of Clinical Pharmacology*, 55(S3), S75–S84. https://doi.org/10.1002/jcph.357

6

Immunogenicity Risk Assessment for Biotherapeutics

Jochem Gokemeijer

Bristol Myers Squibb, Discovery Biotherapeutics

6.1 Introduction

Immunogenicity of biotherapeutics, characterized by the development of anti-drug antibodies (ADAs) can greatly affect the efficacy and safety of a biotherapeutic (Kierzek et al., 2019).

The sequela of immunogenicity is highly varied and can range from very mild (injection site reaction) to severe (anaphylaxis) and immunogenicity is therefore closely monitored during clinical testing of these therapeutics. Besides the potential safety impact of immunogenicity for patients, ADAs can impact the efficacy of a biotherapeutic if these antibodies are neutralizing antibodies which bind to the functional part of a biotherapeutic responsible for interacting with the therapeutic target.

Non-neutralizing ADAs can still have a significant impact on efficacy by altering the pharmacokinetics of the biotherapeutic; increased exposure can result in toxicities whereas reduced exposure can decrease target engagement and efficacy.

Since all biotherapeutics have the potential to elicit an immune response in humans, and most do to varying degrees, it is considered prudent to minimize the immunogenic characteristics of a novel biotherapeutic candidate. The non-human amino acid sequence "foreignness" of the therapeutic antibody, which is recognized as "non-self" by the immune system, is thought to be the major driver of a major histocompatibility complex (MHC) class II-mediated ADA

Bioanalytical Aspects in Biological Therapeutics, First Edition. Edited by Xiaohui Xu and Weifeng Xu.
© 2022 John Wiley & Sons, Inc. Published 2022 by John Wiley & Sons, Inc.

immune response. Thus, it is important for a safe and effective biotherapeutic to understand the level of "foreignness" and the potential of this "non-self" content to elicit an immune response.

Even fully recombinant endogenous human proteins that lack any "non-self" amino sequence have the potential to elicit an immune response (Abdolvahab et al., 2016; Cernea & Raz, 2020). This is thought to be the result of small changes in the characteristics of the proteins such as glycosylation and aggregation resulting in loss of immune tolerance and it is therefore prudent to assess the risk of recombinant endogenous biotherapeutics.

Recombinant endogenous proteins given to patients that have a genetic mutation resulting in either mutated, truncated, or complete abrogation of expression can be particularly susceptible to immunogenicity because these patients lack immune tolerance to the nonexpressed sections of the proteins. Enzyme replacement therapy for Pompe disease has shown that high levels of neutralizing ADAs to the recombinant endogenous enzyme can be especially high in patients that completely lack expression of the enzyme (Colella & Mingozzi, 2019).

Increased engineering of monoclonal antibody (mAb)-based biotherapeutics to minimize or maximize effector function, modulate pharmacokinetic characteristics, increase potency, or enable simultaneous binding to multiple targets (bispecific antibodies) increases the "non-self" content of these biotherapeutics and therefore the risk that one of these "non-self" peptides can be recognized by the human immune system and elicit an antibody-mediated immune response.

Determining the immunogenic potential and identification of the antigens in a biotherapeutic during the discovery phase of the drug development process allows one to select development candidates with reduced risk or apply protein engineering to reduce immunogenicity risk (Gokemeijer et al., 2017; Jawa et al., 2020; Sauna et al., 2018). Immunogenicity risk assessment tools and assays based on the mechanisms involved in a humoral MHC class-mediated immune response have made inroads in the drug discovery process and are routinely used to select drug candidates, assess immune liabilities, and guide protein engineering (Figure 6.1).

While it is possible to estimate the potential that a biotherapeutic will elicit an ADA response in patients, until now it has not been feasible to assess whether these ADAs have an impact on efficacy or exposure. Progress has been made in systems biology modeling that has the potential to model the impact of ADAs on efficacy and exposure (Kierzek et al., 2019).

Recent updates to the health authorities' recommendations for immunogenicity testing recommend an immunogenicity risk assessment report be part

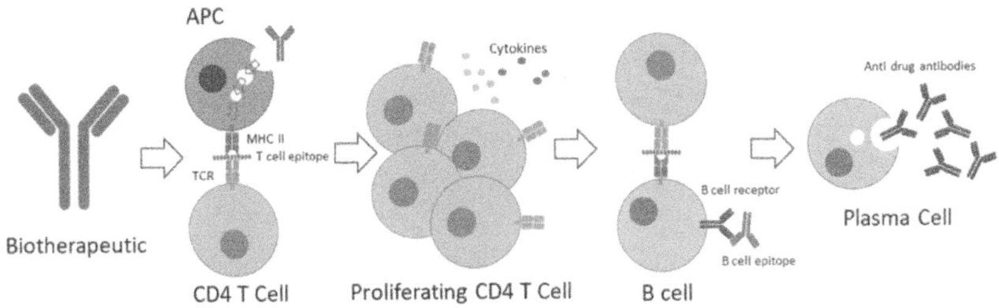

Figure 6.1 Overview of CD4 T cell-dependent major histocompatability complex (MHC) class II-mediated mechanisms involved in antidrug antibody development. Biotherapeutics are taken up by antigen-presenting cells (APC) through phagocytosis where they are processed and presented on MHC class II to naïve CD4 + T cells. Upon recognition of the MHC class II antigen complex through the T cell receptors (TCR), T cells are activated, characterized by the secreting of cytokines and proliferation. Migration of the activated T cells to the germinal center where interactions with B cells result in isotype switching and affinity maturation, resulting in the development of plasma B cells that secrete high-affinity IgG antidrug antibodies. Preclinical immunogenicity risk assessment tools and assays are built around the MHC class II antigen interaction (in silico human leukocyte antigen binding algorithms and MHC-associated peptide proteomics assay [MAPPS]) dendritic cell DC activation assay, and in vitro CD4 T cell activation and proliferation assay.

of the Investigational New Drug (IND) (Walsh et al., 2020) application or Biologics License Application (BLA) for a biotherapeutic (FDA, 2019).

Immunogenicity of biotherapeutics is a multifaceted process and many other parameters besides antigen content determine whether an ADA response develops.

Residual impurities such as endotoxin and host cell proteins can activate the innate immune system and provide a danger signal to the adaptive immune system driving it towards an activating versus tolerizing immune reaction. Protein aggregation and glycosylation levels and composition should also be considered as potential sources of immune activation (Moussa et al., 2016; Ross et al., 2000).

The target and mechanism of action of a biotherapeutic can also be a major factor in the potential immunogenicity of a biotherapeutic, particularly biotherapeutics that either inhibit the immune system such as anti-tumor necrosis factor alpha (TNFα) or anti-interleukin-6 (IL-6) mAbs or aim to activate the immune system such as the dendritic cell (DC) targets anti IL-21 R therapies (Xue et al., 2015) or anti-PD-1 or anti-PD-L1 mAbs checkpoint inhibitors (Davda et al., 2019; van Schouwenburg et al., 2013).

Patient-related factors such as the disease state impact on the immune system, that can result in either a suppressed or overactivated immune system, and prior treatment or co-dosed treatment can have a significant influence on the immunogenicity and development of ADAs in patients.

Lastly, it is important to weigh the potential immunogenicity risk of a biotherapeutic against the potential benefits the therapeutic can provide for patients (Chamberlain, 2019).

6.2 In Silico Immunogenicity Risk Assessment Tools

The binding and presentation of a linear peptide (T cell epitope) by antigen-presenting cells (APC) on an MHC class II receptor to a CD4 + helper cell is a critical step in the development of a high-affinity IgG ADA-mediated immune response. The T cell epitope consist of a core 9-mer that interreacts with the MHC class II receptor and variable flanking sequence that can increase the length of the peptide to 30 amino acids.

The binding pockets for the different MHC class II alleles present in the human population have been well characterized and binding studies have mapped the P1, P4, P7, and P9 positions of the core 9 sequence as the critical residues for interacting with the MHC class II (Grey et al., 1989).

Computer-based algorithms based on the MHC class II binding pocket and experimentally determined peptide binding data have been developed to calculate the potential binding properties of unknown peptides. These in silico immunogenicity risk assessment tools scan through a protein of interest by dividing the complete protein amino acid sequence into overlapping 9–15 residue peptides and calculating the predicted affinity or a ranking versus known peptides (Paul et al., 2015).

Human leukocyte antigen (HLA) alleles are highly variable in the human population and selecting a practical subset that is representative of the HLA distribution to make population-wide immunogenicity risk assessment can be challenging. Using a subset of 8–12 supertypes of HLA DRB1 alleles in these algorithms has been shown to encompass > 95% of the human variability. Inclusion of the DQ and DP alleles has been indicated in the immunogenicity of some biotherapeutics but is not included by all methods (Paul et al., 2015).

Cumulative MHC class II binding peptides across a biotherapeutic can be used as a proxy for population immunogenicity risk, in particular regions of a protein that show promiscuous binding across multiple HLA alleles are thought to pose immunogenic potential. Comparing total "non-self" T cell epitopes load within a biotherapeutic protein can then be compared to a series of biotherapeutics with documented clinical immunogenicity for comparison and ranking of immunogenicity risk.

The high throughput nature of these tools makes these in silico tools particular powerful methods to assess relative risk between sequence families during the lead selection phase of the biotherapeutics discovery process.

While for most biotherapeutics (mAbs and recombinant proteins) MHC class II binding potential is the relevant in silico immunogenicity risk assessment tool due to the extracellular localization of the protein, novel modalities such as gene therapy and cellular therapies have the potential to be presented on MHC class I due to their intracellular expression. In silico MHC class I algorithms are available and have been shown to be superior in predicting antigen binding compared to MHC class II algorithms. This is thought in part to be due to the much more constrained size of the binding pocket allowing almost exclusively 9-mer peptides in the MHC class I and less variability compared to MHC class II (Gorovits & Koren, 2019).

In general, not all MHC class II binding peptides identified utilizing in silico algorithms are antigens (Bui et al., 2007). There are several reasons for this. First, not every possible peptide contained within a biotherapeutic is generated by the intracellular proteolytic processing of proteins within the APC (Roche & Cresswell, 2016). Some algorithms are now incorporating the processing and presentation of antigens by utilizing the availability of in vitro eluted peptides from the MHC class II antigen complexes data into their risk assessment (Lill et al., 2018).

Second, most biotherapeutics consist mostly of "self" sequence; for example, most mAb-based biotherapeutics are fully human and the framework is fully "self" while the complementarity-determining regions (CDRs) due to the hypermutation during affinity maturation are much more likely to be recognized as "non-self" (Harding et al., 2010).

Endogenous proteins as part of the expression process are processed and presented on MHC class II and are tolerated by the immune system. The mechanisms for this self-tolerance or central tolerance is mediated through the T cell receptor (TCR). Once an MHC class II peptide complex is established and shuttled to the surface of the APC, it needs to be recognized by a TCR on a CD4 + T cell. During T cell thymic education T cells that display TCRs that recognize "self" antigens with high affinity are deleted.

This same mechanism results in tolerance for the "self" portions of biotherapeutics. This has been shown for fully human antibody-based therapeutics where in clinical testing the majority of the ADA epitopes are in the CDR in contrast to testing in non-human primates where the majority of the ADAs were found to bind the Ab framework (van Meer et al., 2013).

To assess whether a predicted T cell epitope will be tolerated as a "self" antigen many in silico algorithms and researchers filter out fully human T cell epitopes or assess whether and how much homology an identified T cell epitope has with an endogenous human amino acid sequence.

Other algorithms are leveraging the structure function knowledge of the TCR MHC class II interaction to predict the likelihood of tolerance. Since the MHC class II bound peptide docking residues are well established (P1, P4, P7, P9 positions) we also know which residues are interacting with the TCR (P2, P3, P6, P8 positions): peptides that are homologues to endogenous proteins at these positions are therefore more likely to be tolerated (Moise et al., 2013).

Recombinant endogenous protein-based biotherapeutics carry an immunogenicity impact risk for patients due to the potential of ADAs to cross-react with the endogenous counterpart of the biotherapeutic and neutralizing or depleting the endogenous counterpart. Immunogenicity to recombinant erythropoietin (EPO) has on rare occasions resulted in red cell aplasia in patients as a result of cross-reactivity of the ADAs to endogenous EPO (Casadevall et al., 2002).

Immunogenicity to recombinant endogenous protein is thought to occur due to a loss of tolerance as a result of low self-tolerance for the low plasma concentration of the endogenous counterpart (such as cytokines), slight differences in folding, and glycosylation of recombinant material versus endogenous counterpart and aggregation state of the protein. In silico analysis of the total T cell epitope content of recombinant endogenous biotherapeutics is therefore a viable immunogenicity risk assessment strategy.

Protein engineering of recombinant endogenous biotherapeutics to increase potency, stability, or pharmacodynamic properties can potentially introduce a "non-self" sequence through amino acid substitutions, linkers, or non-natural "junctions" between domains. Even very small changes to endogenous biotherapeutics can drive an immune response in patients. As an example, it was shown that a two amino acid change in recombinant factor VIIa, intended to increase potency of the proteins, was responsible for an increase in immunogenicity (Lamberth et al., 2017).

6.3 In Vitro Immunogenicity Risk Assessment Assays

6.3.1 MHC Binding Assays

Peptide MHC class I and II binding can be assessed in vitro with a plate-based assay utilizing recombinant MHC molecules. Fluorescently labeled model peptides that are prebound to the MHC molecule are challenged by a set of overlapping peptides spanning the biotherapeutic of interest.

Affinity (IC50) of a peptide is assessed by the displacement of the labeled peptide with a concentration series of the peptide of interest. Conversely, stability of the MHC–peptide complex can be assessed utilizing temperature or a chaotropic agent. Stability of the MHC–antigen complex has been reported to be a critical characteristic in determining the antigenicity of a peptide (Harndahl et al., 2012).

The affinity of the labeled model peptide should not be too high to allow for peptide exchange and prevent excessively long incubation times to reach equilibrium.

While the core binding motive for MHC class II is nine amino acids, most in vitro binding assay are performed with 15-mers or 12-mers with an overlap of three amino acids spanning the region of interest. This is mostly done to allow the flanking amino acids to interreact with the receptor which has been described as a stabilizing factor.

To make immunogenicity assessments for a patient population, a diverse set of HLA alleles that covers the diversity in the human population is required. Most MHC class II assays use a series of DRB1, DRB 3/4/5 while DQ and DP are considered optional (Paul et al., 2015).

Recombinant expression and purification of MHC molecules consisting of an alpha and beta chain can be challenging and might require a refolding step and a binding peptide to stabilize the complex. Recombinant MHC molecules can be sourced commercially and there are several contract research organizations that performin vitro MHC binding assays with the availability of large sets of HLA alleles.

6.3.2 T Cell Proliferation Assays

CD4 + T cell activation and proliferation is the next critical step in the development of a humoral immune response after antigen is presented on MHC class II and recognized by TCR. An in vitro T cell proliferation assay using a diverse set of human peripheral blood mononuclear cells (PBMC) donors can be used to assess the immunogenicity risk of biotherapeutics. These assays measure CD4 + T cell proliferation or activation after coincubation with a biotherapeutic in vitro.

While these assays are more labor intensive and time consuming than some of the other assays and methods described earlier, the main reason to use these assays is the incorporation of antigen processing and presentation, TCR recognition, T cell frequency, and cell-mediated tolerance mechanisms (Duke & Mitra-Kaushik, 2020). Additionally, one has the option to use cells from the patient's population that the biotherapy is targeted towards, and this can be particularly useful if there is the expectation that the patient's immune system was altered due to an underlying genetic disease.

6.3.3 PBMC Assay

The PBMC T cell proliferation assays use PBMC isolated from donor blood as the source for APCs and CD4 + T cells. The Ficoll gradient centrifugation method or bead-based isolation are used to isolate PBMC from whole blood. PBMC can be frozen in liquid nitrogen for an extended amount of time and thawed for an assay run. Blood "freshness" is a critical parameter for cell viability and PBMC should preferably be isolated either on the same day as the blood is drawn, or if shipping is necessary overnight shipping in phoshate-buffered saline at room temperature is an option. Additionally, optimization of freezing media and freezing and thawing protocols can yield rewards for assay performance.

The immune status of the blood donor can be assessed by testing a model antigen (keyhole limpet haemocyanin [KLH] or pan HLA DR-binding epitope [PADRE] peptides) as a quality control step and nonproliferating donors can be excluded from an assay run. KLH T cell-activating potency can have substantial batch-to-batch variability and several batches/manufacturers might have to be screened before choosing a reference batch.

It is recommended to use either human serum or serum-free media in these assays and avoid bovine serum to prevent possible reactivity to bovine antigens.

Some laboratories prefer to deplete CD8 + T cells from the PBMC to prevent cross-activation of CD4 + T cells by activated CD8 + T cells.

It has been demonstrated that the frequency of reactive CD4 + T cells to a specific antigen varies from approximate 1:1000 for highly immunogenic antigens to 1:1,000,000 for low immunogenic antigens (Spindeldreher et al., 2020). It is therefore critical to maximize the number of PBMC in these assays. Assuming 25%–60% of CD4 + T cells in PBMC, 1M–5M PBMC per donor using either multiple replicates of a 96- or 48-well plate are used in practice.

Biotherapeutics and controls are added directly to the PBMC and are incubated without cytokine support for 5–7 days before CD4 + T cell activation or proliferation is measured.

CD4 + T cell proliferation can be measured with radioactive tritiated thymidine or bromo-2′-deoxyuridine incorporation into the DNA of proliferating T cells. Alternatively labeling of the PBMC with carboxyfluorescein succinimidyl ester (CFSE), which is incorporated in the cell membrane and is diluted upon cell division, and CD4 staining allows proliferation to be measured by flow cytometry. Additionally, this method allows the other characteristics of the T cells to be measured as well as other T cells (CD8/Treg).

Activated CD4 + T cells produce IL-2 and interferon-gamma (INFγ) which can be measured using a bead-based or enzyme-linked immunosorbent assay (ELISA) format in the supernatant or intracellular staining using a flow cytometer. Expanding the cytokine profile to include innate immune activation markers in the supernatant can provide additional insights into the impact of the biotherapeutic on the immune system (Perry et al., 2008).

The percentage of donors or relative rate (stimulation index) of CD4 + T cell proliferation or activation as determined by a statistical method (ANOVA Student's t-test or others) versus media can be compared to the proliferation percentage of control biotherapeutics with established clinical ADA rates.

6.3.4 DC T Cell Assay

A modification of the PBMC assay format is the DC T cell assay in which monocyte-derived DC are generated and pulsed with the biotherapeutic of interest which are then added to the autologous PBMC. Increased sensitivity for immunogenic antigens has been reported for this assay format likely due to the number of APCs that can be generated and the activation of these APCs.

Another reason to utilize this assay format is to assess the immunogenicity risk of a biotherapeutic when it is desirable to circumvent the biotherapeutic from interacting with receptors and soluble factors expressed by PBMC. Utilizing B cells as APCs can be considered when the biotherapeutic target is expressed in DC as a strategy to target engagement influences.

In particular, biotherapeutics designed to either inhibit or enhance the immune system are likely to drive or block in vitro T cell proliferation and can obscure the amino acid sequence-driven immune response. Whether or not one would include or exclude the target engagement immune system effects of a biotherapeutic in an in vitro risk assessment assay remains an area of discussion.

Dendritic cells (DC) are generated from monocytes, which are isolated from PBMC of heathy donors using plate adherence protocols or bead-based anti-CD14 isolation kits. Isolated monocytes are cultured in hIL-4 and human granulocyte-macrophage colony-stimulating factor (hGM-CSF) for 3–5 days to induce differentiation to immature DC which can be characterized by CD14, CD86, and CD83 expression on flow cytometry. Assay culture media containing serum can be used although dedicated serum-free media formulated for DC can be used and has the potential to reduce background noise. DC are loaded with biotherapeutic proteins or peptides and added to the culture media so they can be taken up by the immature DC through phagocytosis, micropinocytosis, and receptor-mediated endocytosis.

Cells can be washed to remove the culture media containing the test articles and replaced with culture media containing one or more DC maturation agents for 1 or 2 days (TNFα, IFNγ, LPS, PGE, CD40L). Maturation of the DC results in the upregulation of MHC class II and activation markers such as CD86 and CD80 which have the potential to have an effect on the T cells in the next phase of the assay, and it is therefore important to carefully optimize the maturation protocol to minimize nonspecific T cell activation and assay background while maximizing antigen presentation.

Loaded mature DC are added to autologous PMBC or alternatively autologous CD4 T cells and incubated for 6–10 days to allow for antigen presentation and T cell activation to occur. Optimization of DC to PBMC or DC to CD4 + T cell ratios and numbers can increase the assay sensitivity and reduce background noise. Assay readout of activated CD4 + T cells can be measured as described for the PBMC assay format.

6.3.5 DC Internalization and Activation Assay

DC are the primary APCs for a humoral immune response and DC uptake and activation have been correlated with clinical immunogenicity. DC uptake assays have been developed to assess internalization of biotherapeutics. For this assay monocyte-derived DC can be generated in the same manner as described for in vitro DC T cell assays. Direct labeling of the biotherapeutic to monitor internalization is not recommended due to potential changes in biophysical characteristics as a result of the labeling procedure that could alter the uptake mechanism. Fluorescence resonance energy transfer (FRET)-based assays have been developed to assess DC uptake of biotherapeutics (Wen et al., 2020).Measuring the ability of a biotherapeutic to activate and mature DC using maturation markers CD40, CD80, CD86, and HLA-DR have also been used to assess immunogenicity risk (Xue et al., 2015).

6.3.6 Qualification and Validation of In Vitro Immunogenicity Risk Assessment

These in vitro assays are in general not validated and used in a nonregulated environment during the discovery and optimization stages of biotherapeutics development. Nevertheless, in order to generate reliable data it is important to understand the performance characteristics of these assays.

Documented qualification can be beneficial when the assay data are to be included in an integrated summary of immunogenicity section of the IND or

BLA (FDA, 2019). Assay validation and performance in a regulated environment have been reported as requested to monitor the potential immunogenicity of synthetic peptide impurities.

A fit-for-purpose validation approach focused on reproducibility and robustness can be applied to these assays to increase confidence in the assay results (Wickramarachchi et al., 2020).

Quality control proteins, preferably biotherapeutics with documented clinical immunogenicity rates, can be a good way to put assay results in context, track assay performance, and set acceptance criteria.

Repeatability can be a particularly challenging parameter due to the changing PBMC donor composition in the assay, and quality control proteins and trend analysis can be used to understand the repeatability and variability (Wullner et al., 2010).

6.3.7 MHC Peptide Elution Assays

The identification of T cell epitopes in a biotherapeutic is important to assess the immunogenicity risk of a biotherapeutic and to identify amino acid residues that can be substituted to reduce the antigenic load. In silico HLA binding algorithms have been demonstrated to be very good for determining whether a peptide can bind and approximate the affinity for MHC class II. Nevertheless, it is has been demonstrated that only a limited number of peptides contained within an immunogenic protein that can bind MHC class II are processed and presented on APCs (Paul et al., 2015).

APCs, presenting exogenous antigens to CD4 T cells through MHC class II, are sampling extracellular proteins through phagocytosis, micropinocytosis, and receptor-mediated endocytosis where they enter the endosomal/lysosomal system. In the endosome proteins are degraded by cathepsin proteases where individual peptides can bind MHC class II aided by HLA-DM, after which the MHC class migrates to the cell surface.

Moreover it is becoming clear that partial unfolding of an protein is followed by trimming of the exposed peptides by protases, making the stability of a protein and protein domains in the endosome environment an important factor in the likelihood of being presented by the APCs and subsequent recognition by a compatible TCR (Sadegh-Nasseri, 2016; Unanue et al., 2016).

Over the last several years ex vivo MHC-associated peptide proteomics (MAPPS) assays have been developed to identify the peptides presented on APCs in complex with MHC class II (Quarmby et al., 2018) (Figure 6.2).

Monocyte-derived DC isolated with anti-CD14 beads from PBMC or whole blood are mostly used for MAPPS assays in the same fashion as described for the DC T cell assay. Maturation agents such as lipopolysaccharide can be used

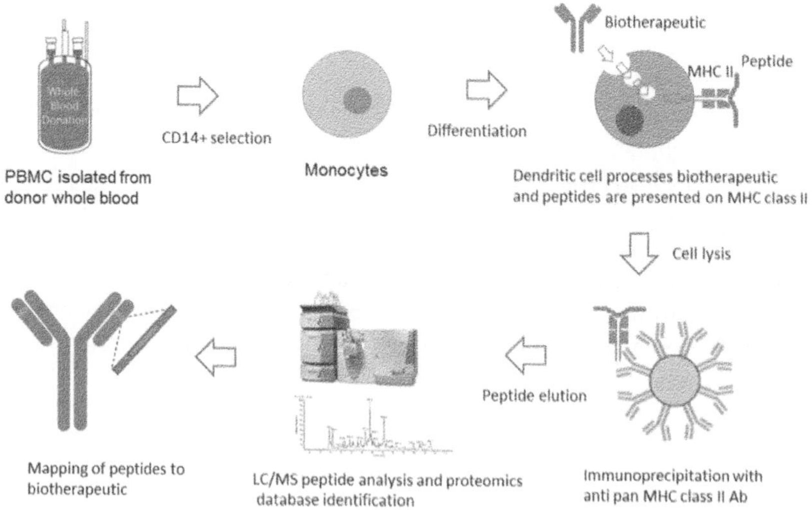

Figure 6.2 Overview of assay procedures in an MHC-associated peptide proteomics (MAPPS) assay. CD14 monocytes are isolated from peripheral blood mononuclear cells (PBMC) and differentiated to immature dendritic cells (DC) and incubated with a biotherapeutic. DC are lyzed after maturation and major histocompatability complex (MHC) class II-antigen complexes are isolated by immune precipitation using a pan MHC class II monoclonal antibody (mAb) and acid eluted antigens are identified by liquid chromatography/mass spectrometry (LC/MS).

for these assays since overactivation of the APCs is not considered a problem in contrast to a T cell assay proliferation assay where overactivated APCs can drive background noise.

Alternatively, adhesion protocols taking advantage of monocytes' characteristic cell culture plastic adherence can be used as a cost-effective and easily scalable solution for monocyte isolation. Frozen PBMC or monocytes stored in liquid nitrogen provide a convenient source of cells. Optimization of freezing conditions, recovery yields, and cell viability can yield big returns for assay sensitivity and the number of different peptides identified. Alternatively using freshly isolated monocytes circumvents cell loss due to freezing but does require keen logistics for leukopack sourcing.

During assay development it can be informative to check the phagocytosis function of the immature DC and the HLA expression and surface markers of the matured DC. Fluorescently labeled dextran gets internalized readily by immature DC and is measured using flow cytometry (Kato et al., 2000).

The sensitivity of the MAPPS assay is directly related to the number of mature DC and HLA–peptide complexes that can be isolated. Optimization of

DC yield (minimum 5–10 × 10^6), maturation conditions, pulsing conditions, and HLA isolation are recommended.

While DC are thought to be the main APC in the subcutaneous space one could consider utilizing B cells for antigen presentation that are mostly localized in the plasma compartment. This might be particularly relevant for biotherapeutics that are dosed intravenously. Additionally, differential expression of cathepsins in different cell types has been documented suggesting that antigen processing might be different in different cell types.

MHC class II peptide complexes are isolated by lyzing APC and immune precipitation (IP) with an anti-MHC class II antibody. Mild acid elution can be used to dissociate the peptide from the complex and analyze it with LC/MS. Identification of the eluted peptides can be done using the amino acid sequence of the biotherapeutic proteomic protein database algorithms such as PEAKS, Sequest, or MaxQuant.

Peptides identified by the MAPPS assay are in general qualitative since the differential ionization of peptides make quantitative comparisons challenging. Normalization against housekeeping proteins is a promising way to establish the relative abundance of identified antigens.

A large number of endogenous "self" peptides are presented on HLA by DC, and it is not unusual that the top 50–100 peptides identified all fall in this category before a peptide from a biotherapeutic is identified. It is therefore essential that the MAPPS assay is optimized for sensitivity and the number of peptides that can be identified. Most antigens identified with a MAPSS assay show a variety in lengths from 9 to 30 amino acids around the core binding 9-mer that will have to be consolidated to one antigen and mapped to the biotherapeutic. Moreover, not every peptide identified from a biotherapeutic is a T cell epitope capable of activating T cells. Validation of the antigenic potential of MAPPS-identified peptides can be assessed by T cell activation potential in vitro. MAPPS assay data and correlation with clinical immunogenicity are rapidly becoming available through individual studies and consortiums like the ABIRISK initiative (Hamze et al., 2017).

Novel modalities such as gene therapy and cellular therapy intracellular expression contain the potential to elicit an MHC class I immune response. To investigate the cellular immunogenicity risk of these modalities an anti-MHC class I antibody can be used to immune precipitate the MHC class I–peptide complex.

MAPPS assays for these modalities are much less developed than the MHC class II version and would need considerable development and optimization to ensure that DC or other cell types are appropriate for intracellular expression of the biotherapeutic and sufficient MHC class I expression.

Protein engineering to deimmunize a biotherapeutic can be initiated based on MAPPS-identified antigens. In silico algorithms can be used to identify the core MHC class II binding sequence and identify amino acid substitutions that abrogate MHC class II binding and reduce immunogenicity risk. Alternatively, direct HLA binding assays or in vitro T cell proliferation assays can be used to screen peptides containing amino acid substitutions. Deimmunization through protein engineering of biotherapeutics carries with it the risk of modifying or eliminating target engagement, in particular mAb engineering of the CDRs has been shown to be challenging. Structure–function relationship knowledge as well as in vitro display evolution can be used to probe deimmunizing amino acid substitutions while maintaining target engagement functionality.

MAPPS data to assess the immunogenicity risk of engineered endogenous proteins are particularly impactful. MAPPS data from the endogenous protein can identify regions that are likely to be processed and presented and can be avoided when considering protein engineering or the placement of linkers and junctions. MAPPS assay can then be used to ensure that none of the engineered sequence is in the final construct presented on MHC class II in silico and in vitro assays can be utilized to assess the immunogenicity risk of biotherapeutics in the preclinical stages of biotherapeutic drug development. Identification of the T cell epitopes that drive the recognition by the immune system allows use of protein engineering for deimmunization.

Utilization of orthogonal methods increases the confidence in the immunogenicity risk resulting in therapeutics that are less likely to elicit an immune response and are more likely to be safe and efficacious for patients.

References

Abdolvahab, M. H., Fazeli, A., Halim, A., Sediq, A. S., Fazeli, M. R., & Schellekens, H. (2016). Immunogenicity of recombinant human interferon beta-1b in immune-tolerant transgenic mice corresponds with the biophysical characteristics of aggregates. *Journal of Interferon & Cytokine Research*, 36(4), 247–257. https://doi.org/10.1089/jir.2015.0108

Bui, H. H., Peters, B., Assarsson, E., Mbawuike, I., & Sette, A. (2007). Ab and T cell epitopes of influenza A virus, knowledge and opportunities. *Proceedings of the National Academy of Sciences of the United States of America*, 104(1), 246–251. https://doi.org/10.1073/pnas.0609330104

Casadevall, N., Nataf, J., Viron, B., Kolta, A., Kiladjian, J. J., Martin-Dupont, P., ... Mayeux, P. (2002). Pure red-cell aplasia and antierythropoietin antibodies in patients treated with recombinant erythropoietin. *The New England Journal of Medicine*, 346(7), 469–475. https://doi.org/10.1056/NEJMoa011931

Cernea, S., & Raz, I. (2020). Insulin therapy: Future perspectives. *American Journal of Therapeutics*, 27(1), e121–e132. https://doi.org/10.1097/MJT.0000000000001076

Chamberlain, P. (2019). Effective presentation of immunogenicity risk assessments and related data in regulatory dossiers. *Bioanalysis*, 11(17), 1581–1592. https://doi.org/10.4155/bio-2018-0209

Colella, P., & Mingozzi, F. (2019). Gene therapy for Pompe disease: The time is now. *Human Gene Therapy*, 30(10), 1245–1262. https://doi.org/10.1089/hum.2019.109

Davda, J., Declerck, P., Hu-Lieskovan, S., Hickling, T. P., Jacobs, I. A., Chou, J., ... Kraynov, E. (2019). Immunogenicity of immunomodulatory, antibody-based, oncology therapeutics. *Journal for ImmunoTherapy of Cancer*, 7(1), 105. https://doi.org/10.1186/s40425-019-0586-0

Duke, B. R., & Mitra-Kaushik, S. (2020). Current in vitro assays for prediction of T cell mediated immunogenicity of biotherapeutics and manufacturing impurities. *Journal of Pharmaceutical Innovation*, 15(2), 202–218. https://doi.org/10.1007/s12247-019-09412-5

Food and Drug Administration (FDA). (2019). Immunogenicity testing of therapeutic protein products – Developing and validating assays for anti-drug antibody detection. Retrieved from https://www.fda.gov/media/119788/download

Gokemeijer, J., Jawa, V., & Mitra-Kaushik, S. (2017). How close are we to profiling immunogenicity risk using in silico algorithms and in vitro methods?: An industry perspective. *The AAPS Journal*, 19(6), 1587–1592. https://doi.org/10.1208/s12248-017-0143-z

Gorovits, B., & Koren, E. (2019). Immunogenicity of chimeric antigen receptor T-cell therapeutics. *BioDrugs*, 33(3), 275–284. https://doi.org/10.1007/s40259-019-00354-5

Grey, H. M., Buus, S., Colon, S., Miles, C., & Sette, A. (1989). Structural requirements and biological significance of interactions between peptides and the major histocompatibility complex. *Philosophical Transactions of the Royal Society of London B: Biological Sciences*, 323(1217), 545–552. https://doi.org/10.1098/rstb.1989.0034

Hamze, M., Meunier, S., Karle, A., Gdoura, A., Goudet, A., Szely, N., ... Maillere, B. (2017). Characterization of CD4 T cell epitopes of infliximab and rituximab identified from healthy donors. *Frontiers in Immunology*, 8, 500. https://doi.org/10.3389/fimmu.2017.00500

Harding, F. A., Stickler, M. M., Razo, J., & DuBridge, R. B. (2010). The immunogenicity of humanized and fully human antibodies: Residual immunogenicity resides in the CDR regions. *MAbs*, 2(3), 256–265. https://doi.org/10.4161/mabs.2.3.11641

Harndahl, M., Rasmussen, M., Roder, G., Dalgaard Pedersen, I., Sorensen, M., Nielsen, M., & Buus, S. (2012). Peptide-MHC class I stability is a better predictor than peptide affinity of CTL immunogenicity. *European Journal of Immunology*, 42(6), 1405–1416. https://doi.org/10.1002/eji.201141774

Jawa, V., Terry, F., Gokemeijer, J., Mitra-Kaushik, S., Roberts, B. J., Tourdot, S., & De Groot, A. S. (2020). T-cell dependent immunogenicity of protein therapeutics pre-clinical assessment and mitigation-Updated consensus and review. *Frontiers in Immunology*, 11, 1301. https://doi.org/10.3389/fimmu.2020.01301

Kato, M., Neil, T. K., Fearnley, D. B., McLellan, A. D., Vuckovic, S., & Hart, D. N. (2000). Expression of multilectin receptors and comparative FITC-dextran uptake by human dendritic cells. *International Immunology*, 12(11), 1511–1519. https://doi.org/10.1093/intimm/12.11.1511

Kierzek, A. M., Hickling, T. P., Figueroa, I., Kalvass, J. C., Nijsen, M., Mohan, K., ... van der Graaf, P. H. (2019). A quantitative systems pharmacology consortium approach to managing immunogenicity of therapeutic proteins. *CPT: Pharmacometrics & Systems Pharmacology*, 8(11), 773–776. https://doi.org/10.1002/psp4.12465

Lamberth, K., Reedtz-Runge, S. L., Simon, J., Klementyeva, K., Pandey, G. S., Padkjaer, S. B., ... Sauna, Z. E. (2017). Post hoc assessment of the immunogenicity of bioengineered factor VIIa demonstrates the use of preclinical tools. *Science Translational Medicine*, 9(372). https://doi.org/10.1126/scitranslmed.aag1286

Lill, J. R., van Veelen, P. A., Tenzer, S., Admon, A., Caron, E., Elias, J. E., ... Bassani-Sternberg, M. (2018). Minimal Information About an Immuno-Peptidomics Experiment (MIAIPE). *Proteomics*, 18(12), e1800110. https://doi.org/10.1002/pmic.201800110

Moise, L., Gutierrez, A. H., Bailey-Kellogg, C., Terry, F., Leng, Q., Abdel Hady, K. M., ... De Groot, A. S. (2013). The two-faced T cell epitope: Examining the host-microbe interface with JanusMatrix. *Human Vaccines & Immunotherapeutics*, 9(7), 1577–1586. https://doi.org/10.4161/hv.24615

Moussa, E. M., Panchal, J. P., Moorthy, B. S., Blum, J. S., Joubert, M. K., Narhi, L. O., & Topp, E. M. (2016). Immunogenicity of therapeutic protein aggregates. *Journal of Pharmaceutical Sciences*, 105(2), 417–430. https://doi.org/10.1016/j.xphs.2015.11.002

Paul, S., Lindestam Arlehamn, C. S., Scriba, T. J., Dillon, M. B., Oseroff, C., Hinz, D., ... Sette, A. (2015). Development and validation of a broad scheme for prediction of HLA class II restricted T cell epitopes. *Journal of Immunological Methods*, 422, 28–34. https://doi.org/10.1016/j.jim.2015.03.022

Perry, L. C., Jones, T. D., & Baker, M. P. (2008). New approaches to prediction of immune responses to therapeutic proteins during preclinical development. *Drugs in R&D*, 9(6), 385–396. https://doi.org/10.2165/0126839-200809060-00004

Quarmby, V., Phung, Q. T., & Lill, J. R. (2018). MAPPs for the identification of immunogenic hotspots of biotherapeutics; an overview of the technology and its application to the biopharmaceutical arena. *Expert Review of Proteomics*, 15(9), 733–748. https://doi.org/10.1080/14789450.2018.1521279

Roche, P. A., & Cresswell, P. (2016). Antigen processing and presentation mechanisms in myeloid cells. *Microbiology Spectrum*, 4(3). https://doi.org/10.1128/microbiolspec.MCHD-0008-2015

Ross, C., Clemmesen, K. M., Svenson, M., Sorensen, P. S., Koch-Henriksen, N., Skovgaard, G. L., & Bendtzen, K. (2000). Immunogenicity of interferon-beta in multiple sclerosis patients: Influence of preparation, dosage, dose frequency, and route of administration. Danish Multiple Sclerosis Study Group. *Annals of Neurology*, 48(5), 706–712. Retrieved from https://www.ncbi.nlm.nih.gov/pubmed/11079533

Sadegh-Nasseri, S. (2016). A step-by-step overview of the dynamic process of epitope selection by major histocompatibility complex class II for presentation to helper T cells. *F1000Research*, 5. https://doi.org/10.12688/f1000research.7664.1

Sauna, Z. E., Lagasse, D., Pedras-Vasconcelos, J., Golding, B., & Rosenberg, A. S. (2018). Evaluating and mitigating the immunogenicity of therapeutic proteins. *Trends in Biotechnology*, 36(10), 1068–1084. https://doi.org/10.1016/j.tibtech.2018.05.008

Spindeldreher, S., Karle, A., Correia, E., Tenon, M., Gottlieb, S., Huber, T., ... Kolbinger, F. (2020). T cell epitope mapping of secukinumab and ixekizumab in healthy donors. *MAbs*, 12(1), 1707418. https://doi.org/10.1080/19420862.2019.1707418

Unanue, E. R., Turk, V., & Neefjes, J. (2016). Variations in MHC class II antigen processing and presentation in health and disease. *Annual Review of Immunology*, 34, 265–297. https://doi.org/10.1146/annurev-immunol-041015-055420

van Meer, P. J., Kooijman, M., Brinks, V., Gispen-de Wied, C. C., Silva-Lima, B., Moors, E. H., & Schellekens, H. (2013). Immunogenicity of mAbs in non-human primates during nonclinical safety assessment. *MAbs*, 5(5), 810–816. https://doi.org/10.4161/mabs.25234

van Schouwenburg, P. A., Rispens, T., & Wolbink, G. J. (2013). Immunogenicity of anti-TNF biologic therapies for rheumatoid arthritis. *Nature Reviews Rheumatology*, 9(3), 164–172. https://doi.org/10.1038/nrrheum.2013.4

Walsh, R. E., Lannan, M., Wen, Y., Wang, X., Moreland, C. A., Willency, J., ... Malherbe, L. P. (2020). Post-hoc assessment of the immunogenicity of three antibodies reveals distinct immune stimulatory mechanisms. *MAbs*, 12(1), 1764829. https://doi.org/10.1080/19420862.2020.1764829

Wen, Y., Cahya, S., Zeng, W., Lin, J., Wang, X., Liu, L., ... Kaliyaperumal, A. (2020). Development of a FRET-based assay for analysis of mAbs internalization and processing by dendritic cells in preclinical immunogenicity risk assessment. *The AAPS Journal*, 22(3), 68. https://doi.org/10.1208/s12248-020-00444-1

Wickramarachchi, D., Steeno, G., You, Z., Shaik, S., Lepsy, C., & Xue, L. (2020). Fit-for-purpose validation and establishment of assay acceptance and reporting criteria of dendritic cell activation assay contributing to the assessment of immunogenicity risk. *The AAPS Journal*, 22(5), 114. https://doi.org/10.1208/s12248-020-00491-8

Wullner, D., Zhou, L., Bramhall, E., Kuck, A., Goletz, T. J., Swanson, S., ... Jawa, V. (2010). Considerations for optimization and validation of an in vitro PBMC derived T cell assay for immunogenicity prediction of biotherapeutics. *Clinical Immunology*, 137(1), 5–14. https://doi.org/10.1016/j.clim.2010.06.018

Xue, L., Hickling, T., Song, R., Nowak, J., & Rup, B. (2015). Contribution of enhanced engagement of antigen presentation machinery to the clinical immunogenicity of a human interleukin (IL)-21 receptor-blocking therapeutic antibody. *Clinical and Experimental Immunology*, 183(1), 102–113. https://doi.org/10.1111/cei.12711

7

Bioanalytical Strategy to Support Clinical Immunogenicity Assessment

Anti-drug Antibodies

Ying Wang[a,b] *and Michael Luong*[a]

[a]*BioMedicine Design, Pfizer Inc.*
[b]*Corresponding author*

7.1 Introduction

It is well established that protein biotherapeutics can elicit unwanted immune responses, even when the biotherapeutics have been engineered to have as close to 100% human sequence homology as possible. In addition, recent advancements in protein engineering have made it possible to produce molecules of increasing complexity, such as multidomain antibodies, to enhance their therapeutic potential. These new modalities carry heightened risks of immunogenicity. For example, CD3 bispecific antibody biotherapeutics that harness the power of the immune system to combat malignancies have demonstrated clinical efficacy (Bargou et al., 2008; Haense et al., 2016; Linke et al., 2010). However, due to its direct targeting of CD3 on T cells, this class of molecules presents a unique set of immunogenicity risks, including systemic T cell activation.

The consequences of immunogenicity range from mild to life-threatening with effects on both drug efficacy and patient safety. From the patient perspective, the generation of antidrug antibodies (ADA) can be extremely harmful in cases where the biotherapeutic has an endogenous counterpart, or in cases where the therapeutic is an essential life-saving enzyme replacement. Furthermore, a potent ADA response has the potential to cause immune complex formation and deposition, resulting in severe tissue damage and

Bioanalytical Aspects in Biological Therapeutics, First Edition. Edited by Xiaohui Xu and Weifeng Xu.
© 2022 John Wiley & Sons, Inc. Published 2022 by John Wiley & Sons, Inc.

subsequent organ failure (Leach et al., 2014). Finally, neutralizing antibodies (NAb) can completely inhibit the biotherapeutic, effectively negating its intervention in disease progression or life-saving potential. Therefore, a carefully crafted immunogenicity risk assessment and mitigation plan is essential, and a bioanalytical strategy to assess the immunogenicity of a given product during clinical development needs to reflect critical elements of the plan.

In this chapter, the authors focus on the bioanalytical strategy typically employed for the detection and characterization of ADA by means of various bioanalytical assays. Moreover, the authors discuss the challenges involved and viable solutions that can be implemented in the development of the ADA assays.

7.2 Immunogenicity Risk Assessment-Based Bioanalytical Strategy

To ensure patient safety, regulatory agencies mandate that sponsors have robust immunogenicity risk assessments and mitigation plans in place prior to clinical development of investigational biotherapeutics. In essence, regulatory guidance speaks to the importance of addressing immunogenicity in an integrated manner and illustrates approaches for defining risk levels and establishing risk management plans throughout the drug development cycle (Chamberlain, 2019; Koren et al., 2008; Rosenberg & Sauna, 2018). The generation of such a comprehensive plan requires expertise from multiple disciplines and involves scientists from many different fields. The integrated immunogenicity risk and mitigation plan is an important topic of discussion between sponsors and regulators during investigational new drug (IND) filings or consultations. As required by regulatory agencies, "a life-cycle management approach to immunogenicity through the creation of an integrated immunogenicity summary report that sponsors begin populating early in therapeutic protein product development and update at regular intervals as the individual product clinical program progresses through IND stages into the BLA [Biologics License Application] and even post approval stages" (FDA, 2019).

Regarding risk assessment, the process typically begins with in silico predictions of immunogenicity. The predictive tools, discussed in Chapter 6, are necessary to determine the potential of a biotherapeutic to induce ADA responses. These predictive tools provide critical information on whether a molecule carries the risk of inducing T cell- or B cell-driven antibody responses. The information provided by the predictive tools can guide efforts to de-risk a molecule, facilitate critical decision-making, and define a path forward for a given molecule.

Considerations of the integrated assessment during early stages of drug development are:

Basic information:

- Type of modality and structure
- Relationship between modality and structure to early predictions of immunogenicity
- Mode of action (MOA) and potential impact on innate and adaptive immune responses
- Review of immunogenicity data from preclinical studies

Although ADA incidence in animal studies does not predict the occurrence of ADA in humans, the impact of ADA on pharmacokinetic profiling could be interpreted for clinical study with the risk of unnecessary drug toxicity. Conversely, life-threatening ADA consequences caused by NAb in animal studies may be predictive in human studies, if drug targets are evolutionarily conserved between species (Rosenberg & Sauna, 2018).

Product indications:

- The therapeutic indications
- Consideration of a patient's genetic background or disease-related immune status
- Potential augmentation of immune responses if comedications are involved
- Dose prediction and routes of administration
- Drug disposition mechanisms and their impact on target depositions and consequences
- Pharmacodynamic markers that may predict the impact of immunogenicity on efficacy and safety as well as other pertinent factors that may provoke immunogenicity

Product manufacturing process:

- Protein structure and posttranslational modifications
- Formulation and packaging
- Potential formation of adducts, aggregation, or impurities

In summary, as immunogenicity assessment of biotherapeutics is an integral part of product development, the immunogenicity assessment paradigm should be based on the integrated risk assessments which drive risk mitigation plans. For example, while evaluation of adaptive immune responses that drive ADA development has become a required assessment, innate immune responses, which may also be elicited by certain drugs, should likewise be monitored by appropriate laboratory tests, especially when the MOA of the therapeutic involves modulation of the immune system.

7.2.1 Regulatory Considerations for Immunogenicity Assessment of Biotherapeutics

Regulatory guidance for the immunogenicity assessment of biotherapeutics in clinical product development has evolved over the past several decades.

In Europe, the first European Medicines Agency (EMA) draft guidance was published in 2006 ("Guideline on Immunogenicity Assessment of Therapeutic Proteins"). This draft was revised in 2015 and the finalized revision-1 was published in 2017. In the US, the Food and Drug Administration (FDA) published the first draft guidance in 2008 entitled "Immunogenicity Testing of Therapeutic Protein Products – Developing and Validating Assays for Anti-Drug Antibody Detection." This guidance was revised in 2018 and finalized in 2019. These revisions were made based on experiences from recent marketing authorization applications, scientific advice, emerging information, and consultation with the pharmaceutical industry. While the EMA and FDA took different approaches in issuing guidance on immunogenicity assessments, both agencies emphasized a risk-based approach and a bioanalytical strategy, in which the incidence of ADA/NAb is analyzed in the context of other clinical data (i.e., pharmacokinetics, pharmacodynamics, efficacy and safety) to fully understand the consequences of the induced ADA. Both the EMA and FDA guidance are aligned on the tiered approach of ADA testing and on the expectations for method development and validation. The EMA guidance also discusses in detail the factors that may influence the development of an immune response against a therapeutic protein, potential clinical consequences of immunogenicity, and nonclinical assessment of immunogenicity and its consequences. In contrast, the FDA guidance for industry finalized in 2019 is much more focused on the details of bioanalytical assay development and validation, including addressing assay interferences, fundamental assay performance characteristics to be validated, and anticipated acceptance criteria.

7.2.2 Typical ADA Testing Strategy

Bano et al. (2019) and Kurki et al. (2019) have summarized the comparability of regulatory guidance from EMA and FDA. In a summary for typical ADA testing strategy:

1) The tiered ADA testing approach, indicated in Figure 7.1, is typically required by both agencies.

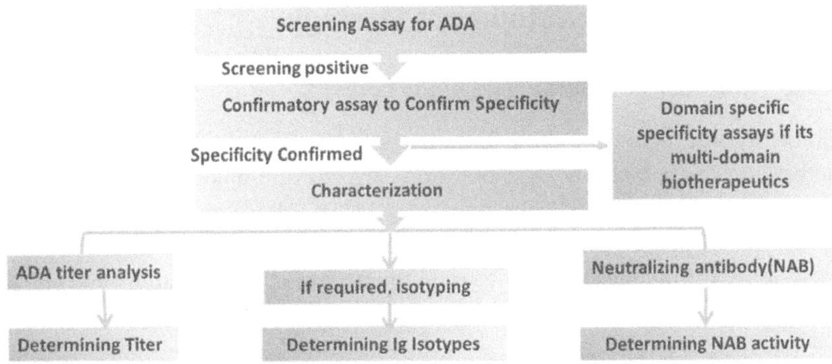

Figure 7.1 Tiered approach for antidrug antibody (ADA) testing and characterization.

2) Immunogenicity assays must be developed to be unaffected by interfering factors such as soluble targets, residual drug, and other matrix proteins (e.g., rheumatoid factor [RF]).
3) ADA screening assay false-positive rate is expected at 5%, while a 1% false-positive rate is expected for confirmatory assay.
4) Both agencies highlighted a need for the development and validation of NAb assays to detect NAb in study samples. Furthermore, the NAb assay is to be based on the MOA of the drug and the target and effector pathways of the biotherapeutic. When the drug target is on the cells or the MOA of the drug is associated with cellular responses or related to effector functions that trigger antibody-dependent cytotoxicity or complement-dependent cytotoxicity, a cell-based NAb assay is preferred.
5) The guidance from both agencies discussed a strategy to produce positive assay suitability controls, and both agree that animal-derived postive controls as suitability controls can be used in clinical ADA/NAb assays. However, the FDA emphasizes a preference for the use of affinity-purified PC from both monoclonal and polyclonal antibodies. For negative suitability controls, both agencies recommended the use of human matrix derived from drug-naïve study-relevant populations.
6) EMA and FDA require assay reagents to be characterized, while the FDA also emphasizes the need to evaluate reagent performance along with assay optimization to ensure specific assay performance.
7) For multidomain biotherapeutics, both agencies also require sponsors to determine the incidence of treatment-induced ADA to each of its functional domains that may be drivers of ADA production.

7.3 ADA Assay Development Strategy

7.3.1 ADA Assay Platforms and Assay Formats

Currently, the ligand binding format is still the dominant format for ADA detection. The commonly used platforms are enzyme-linked immunosorbent assay (ELISA) and Meso Scale Discovery (MSD). The assay formats on these platforms are diverse, including direct ADA binding assays or bridging assays. Each assay format has its benefits and limitations for ADA testing.

7.3.1.1 Direct ADA Binding Assay

The direct ADA binding assay can be performed using the ELISA or MSD platform. The assay works by capturing ADA using drug coated on the plate surface. In turn, the ADA is detected by an anti-human IgG/IgM conjugate. This type of assay can detect both IgG and IgM isotypes, and it is the simplest approach for ADA detection. In addition, because the drug coated on the plate is not chemically conjugated, there is less of a concern regarding the masking of epitopes on the drug.

However, this assay format has several inherent weaknesses:

- Potential interference from nonspecific matrix proteins that could interfere with the assay to cause high false-positive rates (e.g., anti-human IgG conjugate could capture nonrelevant IgG in samples).
- Susceptibility to specific matrix interference such as onboard drug or target interferences, even in instances where the interfering substance is at sub-microgram/milliliter levels.
- The stepwise procedures (plate coating, antibody binding, detection, and substrate addition) require multiple wash steps, which could cause the loss of low-affinity ADA, leading to inaccurate evaluation of ADA incidence.
- The postive control from animal sources as system suitability control requires the detection system to contain species-specific anti-IgG antibody conjugate (e.g., anti-human IgG and anti-rabbit IgG conjugate would be needed if the positive control host species is rabbit), which adds a level of complexity when it comes to evaluating assay performance characteristics such as sensitivity and drug tolerance levels.

7.3.1.2 Bridging Assay Format

The bridging assay format, which can be developed on the ELISA or MSD platform, offers enhanced specificity for ADA detection, due to its use of drug-derived capture and detection reagents.

The application of electrochemiluminescence (ECL) in bridging assays has become the most popular approach to ADA detection. The better tolerance of matrix interference, especially drug tolerance at the microgram/milliliter level, is driven by the homogenous reaction of samples with drug-derived reagents, in which an equilibrium favorable to the formation of ADA to drug can be established by ensuring a high stoichiometric ratio of labeled drug to residual drug. Further, the ECL bridging assay offers other benefits, such as the capability to detect low-affinity antibodies and a greater sensitivity for ADA detection, in part, due to the minimal wash steps involved in the method.

While the bridging assay has its advantages, the use of drug-derived reagents presents a concern regarding the potential of conjugation procedures to mask immunogenic epitopes, resulting in a subset of ADA going undetected. Therefore, in the development of drug conjugation procedures, it is critical to optimize the ratio of chemical moiety (e.g., biotin or ruthenium) to drug to yield specific and robust responses. Currently, common practice employs a challenge ratio of 8:1 (i.e., 8 or fewer chemical moieties to 1 drug molecule) in the generation of antibody drug-derived reagents. For antibody therapeutics, the specificity of the bridging assay could be impaired because it is susceptible to interferences, such as RF with human IgM isotypes and multimeric targets. In the presence of such interfering substances at levels beyond the tolerance of the assay, false-positive ADA results could be generated and even "confirmed" in confirmatory assays. The procedures used to eliminate assay interference are discussed in Section 7.4.

7.3.1.3 Other Platforms

Over recent years, newly developed technologies have been tested in the ADA assay setting to further improve assay sensitivity and enhance assay tolerance to interferences. The comparison of these new technological platforms to conventional platforms, such as ELISA and MSD, has provided insight into proper platform selection to suit intended purposes.

One of the emerging technologies that has been used for ADA assays is the Gyros platform, which is fully compatible with the bridging assay format. Relative to the MSD platform, the Gyros offers certain advantages, including the use of lower volumes (on the scale of nanoliters), automation, and specialized compact discs that allow for the incorporation of acid dissociation. Further, the Gyros limits processing time and minimizes sample consumption, which is critical to the support of nonclinical, rodent immunogenicity studies using microsampling (≤50 μL), or studies with pediatric populations. Collet-Brose et al. have compared the Gyros to MSD and ELISA platforms for bridging assays and confirmed that all three technologies were comparable regarding assay sensitivity, drug tolerance, and target tolerance, when sample pretreatment procedures were used (Collet-Brose et al., 2016).

While ligand binding assay (LBA) platforms are still predominantly used for ADA detection, during the past decade, liquid chromatography-mass spectrometry (LC-MS) technology has been evaluated and developed as an alternative, with the promise of superior tolerance to drug and target interferences due to the nature of the detection method. In addition to ADA detection, the LC-MS platform can also provide information on ADA isotypes, which enables the monitoring of isotype switching during clinical trials. As summarized by Chen et al., the specificity of the LC-MS platform for ADA detection across species has been strengthened by the addition of a specific form of sample pretreatment, resembling solid-phase extraction with acid dissociation (SPEAD), that allows for the dissociation of ADA from a solid support prior to LC-MS analysis (Chen et al., 2016). Chen et al. referred to this approach as immunocapture (IC)-LC-MS ADA analysis (Chen et al., 2016). The LC-MS platform can be a viable tool for ADA detection and characterization.

Other platforms, such as alpha-ELISA (Collet-Brose et al., 2016) and BIAcore (Barbosa et al., 2013), have also been explored by various groups, who have identified the pros and cons to these alternative platforms. The use of these platforms is limited to specific applications, which are not reviewed here.

7.4 Addressing Assay Interferences

Matrix interference is a common challenge to bioanalytical scientists during immunoassay development. Interference is mainly caused by matrix proteins present in study samples. In general, matrix interference can be categorized as nonspecific interferences (analyte-independent) or specific interferences (analyte-dependent) (Gorovits et al., 2014, 2016; Zhong et al., 2017). In ADA and NAb assay development, the major challenge involves interference from specific matrix proteins (e.g., residual drug at the time of sampling and soluble drug targets in circulation). These matrix proteins are either specific binding partners to the drug-derived reagents (such as target) or are competitors of drug-derived reagents for binding to ADA (such as residual drug in samples). This type of specific matrix interference impacts almost all known ADA assay formats, because the biotherapeutic itself is used as an assay component to capture and/or detect ADA. The consequence of this form of matrix interference can be either a false-positive ADA result, caused by a multimeric target bridging drug-derived reagent, or a false-negative ADA result, caused by residual drug forming an immune complex with ADA that competes with the

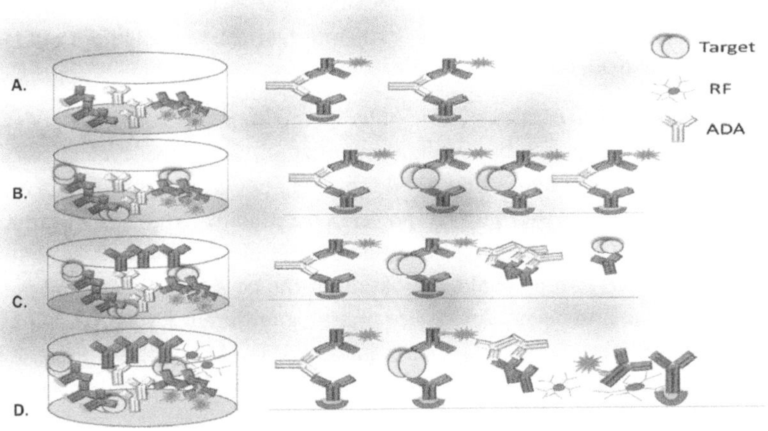

Figure 7.2 Examples of assay interferences and consequences. The figure illustrates several interference factors to an electrochemiluminescence (ECL) bridging antidrug antibody (ADA) assay using a monoclonal antibody biotherapeutic as an example; the same principle applies to other types of modality when using this assay format. (A) No interference, ADA binds to biotinylated and ruthenylated drug by its bivalent binding capability (or bridging both labeled drug) to form a complex, which can be detected by the ECL method. (B) Interference by the dimerized drug target, which also binds to labeled drugs inducing a false-positive signal that leads to false-positive results. (C) Besides target interference, onboard drug could form drug–ADA complexes in samples, which sequester ADA away from binding to labeled drugs and therefore diminish the signal that leads to false-negative results. (D) Besides target and onboard drug interference, rheumatoid factor (RF) could aggregate human IgG (labeled drug), therefore "bridging" the labeled drug to generate false-positive signals.

drug-derived reagent for ADA binding. Such interference directly impacts accurate immunogenicity assessment during clinical trials.

Nonspecific matrix interferences (such as RF) are also encountered in ADA assays, which likewise confound immunogenicity assessments. The potential interference and its consequences are illustrated in Figure 7.2.

To date, bioanalytical scientists have made considerable progress in addressing matrix interferences. The procedures which have proven successful in addressing these interferences are discussed in the next section.

7.4.1 Overcoming Onboard Drug Interference

7.4.1.1 General Considerations

The drug tolerance level of an ADA assay is defined as the level of residual drug at or below which the assay can reliability detect ADA in clinical study samples. It is an expectation of regulatory agencies that the drug tolerance level be carefully evaluated during assay development and is associated with assay sensitivity (FDA, 2019). While bioanalytical scientists have developed numerous procedures to overcome drug interference, several important factors should be considered:

1) Use of ADA surrogates for drug tolerance evaluations
Drug tolerance is generally evaluated using a positive control, typically a monoclonal antibody generated by hybridoma technology or phage display, as a surrogate for ADA. Alternatively, polyclonal antibodies from hyperimmunized animals can also be used. These antibodies are affinity purified, so that drug tolerance levels can be evaluated based on the concentration of the positive control in the absence and presence of drug at different levels. While positive controls from these sources are surrogates of ADA of varying affinities, studies have shown that the evaluation of drug tolerance is not solely driven by positive control affinity, rather by multiple factors such as cutpoint confidence level, stoichiometry of positive control–drug complexes, and presentation of epitopes (Egging et al., 2018).

2) Defining appropriate targeted drug tolerance levels
There are many factors that go into determining the drug tolerance levels for ADA and NAb assays: the drug modality; half-life; time in circulation; clinical study design (e.g., dosing regimen and dosing method); and residual drug concentrations. For example, in studies where drug safety and efficacy are commonly evaluated through dose escalations, the estimated trough levels (or levels at designated ADA sampling time points) should be carefully considered when defining drug tolerance levels. Therefore, it is critical to properly define clinically relevant drug tolerance levels and, subsequently, choose appropriate assay formats and sample pretreatment procedures to overcome onboard drug interference.

3) Drug modalities influence the strategy to overcome drug interference
Biotherapeutic modalities have become more diversified, which include monoclonal antibodies (from single to multidomain), antibody drug conjugates (ADCs), recombinant human proteins (e.g., growth hormones), fusion proteins, and peptides. While acid dissociation has become a common practice to dissociate immune complexes, more advanced drug depletion procedures, for example, SPEAD and affinity capture elution (ACE), are also frequently employed (Bourdage et al., 2007; Smith et al., 2007) and have been shown to be effective in removing residual drug from test samples. However, there are some concerns regarding the application of these procedures:

- The treatment of test samples with acid, as is common with SPEAD and ACE procedures, has the potential to impact ADA or NAb detection, for example, by destroying a subset of the total ADA repertoire.

- Certain drug modalities may not be compatible with acid treatment. In such cases, other procedures need to be developed to overcome drug interference (Xu et al., 2018).

Overall, the selection of a sample pretreatment procedure to deplete residual drug should be carefully considered after it has been deemed necessary. It is a regulatory expectation that bioanalytical scientists aim to achieve reliable detection of ADA through a rigorous method development process to ensure proper assessment of ADA incidence during clinical studies (FDA, 2019; Welink et al., 2018).

7.4.1.2 Summary of Sample Pretreatment Procedures to Overcome Drug Interference

7.4.1.2.1 Acid Dissociation Procedures with Antibody Detection in Bridging Assay

The bridging assay format, using biotin/streptavidin or avidin-labeled drug to capture ADA, is to some degree intrinsically tolerant to drug interference, because the concentration of labeled drug can be adjusted to favor the formation of ADA-labeled drug complexes over ADA-circulating drug complexes. The acid treatment of ADA samples has often been used to enhance the dissociation of immune complexes prior to incubation with drug-derived reagents, facilitating ADA detection in the bridging assay. The following factors are important for enhancing the drug tolerance level when acid dissociation is used as a sample pretreatment procedure:

1) The pH of the acid and the time samples are exposed to acid need to be optimized to achieve maximal reduction of residual drug interference, while ensuring minimal damage to ADA.
2) Following acid treatment, the neutralizing step may produce samples at neutral or slightly lower pH levels, such as 6.5 to 7, to maintain ADA and residual drug in a dissociated state.
3) Drug-derived reagents should be properly titrated to ensure that there is an excess amount of labeled drug to residual drug after acid dissociation, while preserving assay sensitivity.
4) Allow for adequate incubation time to drive equilibrium in the direction of ADA binding to the drug-derived reagent in a liquid phase reaction (e.g., overnight), prior to transferring samples to streptavidin/avidin coated plates for ADA detection.

7.4.1.2.2 Solid-Phase Extraction with Acid Dissociation (SPEAD)

SPEAD as a sample pretreatment procedure was originally reported by Smith and colleagues in 2007 (Smith et al., 2007). In brief, the procedure starts with or without acid dissociation of samples, followed by the addition of an excess amount of biotinylated drug to drive equilibrium in the direction of ADA binding to biotinylated drug. Subsequently, the ADA–biotinylated drug complexes are captured using streptavidin-coated ELISA plates. The bound ADA are then eluted by a short acid treatment. The eluted ADA are then detected by a direct antibody binding assay using a drug conjugate as detector. The use of SPEAD can result in a > 10–100-fold increase in residual drug tolerance. Over the years, SPEAD has evolved from a plate-based format to a beads-based format. One beads-based approach uses streptavidin-coupled magnetic beads to capture ADA complexed with biotinylated drug. Another approach is to use drug directly conjugated to activated magnetic beads (e.g., Dynabeads) as a solid support to capture ADA. Regardless of the solid support used for ADA capture, the goal of the procedure remains unchanged: to eliminate drug interference by enriching ADA. By applying SPEAD in ADA assays, drug tolerance levels in the sub-milligram/milliliter to milligram/milliliter range can be achieved at a positive control concentration of 100 ng/mL (sensitivity benchmark established by regulatory agencies). Drug tolerance can also be significantly improved in NAb assays. SPEAD procedures and ADA detection methods are illustrated in Figure 7.3.

It should be noted that a disadvantage of SPEAD is the concomitant enrichment of potential drug targets, resulting in enhanced target interference if not addressed in parallel.

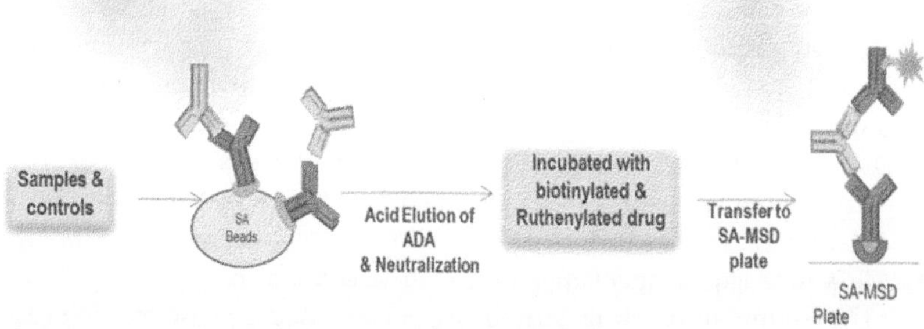

Figure 7.3 Illustration of solid-phase extraction with acid dissociation (SPEAD) procedures. This procedure includes four critical steps. (1) Acid dissociation of antidrug antibodies (ADA)–drug complex. (2) Immobilizing ADA in samples and controls on the solid support (drug-conjugated magnetic beads). (3) Elution of ADA from the beads. (4) Detection of ADA using an appropriate platform (here detection is achieved using an electrochemiluminescence bridging assay). MSD, Meso Scale Discovery; SA, streptavidin.

7.4.1.2.3 Polyethylene Glycol Precipitation and Acid Dissociation (PandA)

Polyethylene glycol (PEG) is a nontoxic, water-soluble, synthetic polymer. It is widely used in the chemical and biomedical industries. The use of PEG to precipitate protein molecules is influenced by both the size of the PEG polymer and the molecular weight of the protein itself. Regarding molecular weight, higher molecular weight proteins may be precipitated at lower concentrations of PEG (Atha & Ingham, 1981). Other conditions, such as salt concentration and pH level in PEG solutions, also play important roles in the ability of PEG to precipitate proteins. PEG precipitation of immune complexes has been previously described (Ohlson & Zetterstrand, 1985).

The PandA procedure was first reported by Zoghbi et al. (2015). This assay was performed sequentially, starting with saturation of ADA with an excess amount of mAb therapeutic to form drug–ADA complexes. Subsequently, the drug–ADA complexes were precipitated using PEG-8000 at a lower concentration. The precipitated drug–ADA complexes were dissociated by acid treatment and subsequently immobilized on high-binding MSD plates, which provide a large surface area for the capture of drug–ADA complexes, under acidic conditions (i.e., without neutralization). The ADA were detected using ruthenylated reagents. A critical element of the PandA procedure involves the use of an appropriate PEG concentration to ensure the precipitation of higher molecular weight complexes, such as drug–ADA complexes. Zhoghbi et al. showed that a drug tolerance level of at least 100 μg/mL could be consistently achieved with PandA at PC concentrations as low as 30 ng/mL.

Factors to consider when using PandA:

- Enhance assay robustness: PandA is a low-throughput process because it involves multiple experimental steps. Therefore, standardization of the procedures is critical to maintaining robust ADA detection and drug tolerance levels.
- Avoid under-recovery of ADA: carefully evaluate the percent PEG used to precipitate immune complexes and the pellet wash steps to ensure that ADA are recovered properly.

7.4.1.2.4 Alternative Sample Pretreatment Procedures to Address Drug Interference

Although the procedures described earlier have proven successful in overcoming drug interference, each procedure has its limitations. For example, all of the procedures involve acid treatment, which may be detrimental to assay reagents, such as positive control antibodies (Xu et al., 2018). Xu et al. reported such an incident when they applied acid dissociation in a procedure to overcome drug interference in a NAb assay for a drug that was a pegylated fab fragment. Xu et al. observed that the positive control lost its ability to bind to the drug-derived reagent due to the acid dissociation procedure. As an alternative, they developed a dissociation procedure involving the use of heat to disrupt ADA–drug complexes.

7.4.2 Addressing Target Interference

Circulating targets, including soluble cellular targets that are created from alternative mRNA splicing or from proteolytic cleavage of membrane-bound receptors (Heaney & Golde, 1998), retain the ability to bind to drug-derived reagents. When the concentration of soluble targets reaches the target tolerance threshold of ADA/NAb assays, the targets will likely interfere in ADA/NAb testing, regardless of assay formats. Target interference has the potential to produce both false-positive ADA responses (e.g., when a multimeric target bridges capture and detecting reagents) and false-negative ADA responses (e.g., when a target competes with ADA for binding to a drug-derived reagent). During the application of acid dissociation or a bead-based sample pretreatment procedure to overcome drug interference, the target can be released from its complex or enriched during elution of ADA. In both cases, target interference could be enhanced in the absence of mitigation procedures.

In the following sections, important considerations of target interference and procedures to address it are discussed.

7.4.2.1 General Considerations

The target tolerance level of an ADA assay refers to the concentration of circulating target above which the assay loses its ability to detect ADA. As with drug tolerance, target tolerance is evaluated using affinity-purified positive control in the absence and presence of clinically relevant concentrations of target. Three critical factors need to be considered in the evaluation of target tolerance:

1) Properly define target tolerance levels

As eloquently summarized by Zhong et al. (2017), the concentration of drug targets in the circulation is a highly complex matter. The concentration of a given target can be significantly influenced by disease biology, disease stage, drug mode of action, and even differences between subjects. Additionally, the biochemical–biophysical nature of the target (in terms of its ability to interact with drug and binding partners), and its degradation pathways also impact target concentrations.

Among the factors that may cause target concentrations to fluctuate, the status of targets after treatment is of lesser importance when determining a target tolerance level. For example, an increase in the concentration of a target after treatment may be due to the formation of drug–target complexes that extend the half-life of the target, or due to drug treatment (Chen et al., 2019). As mentioned earlier, acid dissociation procedures used to increase drug tolerance levels can also result in increases in soluble target interference (Dai et al., 2014). Therefore, the authors recommend two considerations in defining the target tolerance level: (a) evaluate target interferences using a range of concentrations spanning known target concentrations estimated in physiological/pathological states and (b) set the maximal target tolerance level, while considering potential increases in circulating target levels caused by drug administration, for example, ≥ 10 higher than estimated target concentration in pathological states depending on available pharmacokinetic and pharmacodynamic data.

2) Understand the nature of drug–target interactions and design fit-for-purpose procedures.

While the tools to overcome target interference in ADA assays have been well developed over the past several years, resolving target interference needs to be carefully considered, taking into account the biochemical nature of the target–drug interactions, the necessary elements for target–drug interactions, and the potential liabilities of target–drug binding under certain conditions. Biochemically, the binding of a drug to its target is noncovalent in nature, facilitated by ionic/electrostatic interactions, hydrogen bonds, van der Waals forces, and hydrophobic effects. As such, drug–target interactions are reversible and can be susceptible to certain detergents, or even slightly acidic or basic conditions (Chen et al., 2019).

During assay development, conditions that are detrimental to drug–target interactions, but do not impact ADA detection, can be incorporated in assay procedures. The following sections summarize some common procedures to overcome target interference in ADA assays.

7.4.2.2 Procedures to Overcome Target Interference

7.4.2.2.1 Depletion of a Soluble Target Through Its Scavenger(s)

1) Target depletion without solid-phase extraction

The ligand of the target, target-blocking protein, or anti-target antibodies are typical scavengers that can sequester the target from binding to drug-derived reagents in the ADA assay. In such cases, these target scavengers can be added to the sample diluent, allowing the target to form complexes with the scavengers without further depletion, and can be removed by washing steps. However, for NAb assay, target depletion (removal from samples) is necessary. It should be noted that for antibody biotherapeutics, the use of anti-target antibodies as a scavenger should be carefully assessed in the presence of blocking anti-idiotypic (anti-ID) positive controls that are against the complementarity-determining regions (CDRs), since some monoclonal antibody positive controls may be raised against the same or similar epitope as the drug, which could compete for the binding of ADA with drug-derived reagent and therefore cause false-negative ADA results.

2) Target depletion with solid-phase extraction
Target depletion using scavengers immobilized on a solid support has proven efficient for the removal of target from assay systems. Anti-target antibodies are often used in this depletion procedure. Plates or beads can be used as a solid support that binds to target–scavenger complexes. Bead-based procedures, in particular, are powerful, because the extent of target removal can be controlled by the quantity of beads and anti-target antibodies used. In comparison, plate-based procedures can be limited by the capacity of the plate surface area to accommodate high concentrations of anti-target antibodies. Two types of biochemical reactions are commonly used to immobilize anti-target antibodies on solid supports:

- Covalent coupling of anti-target antibodies on the preactivated magnetic beads
- Non-covalent binding by removal of target using biotinylated anti-target antibody binding to streptavidin beads

It should be noted that the depletion of target using anti-target antibodies could indirectly pull down the drug or drug–ADA complex. To avoid such potential negative impact, the procedures to dissociate the complex of drug–target should be applied. Furthermore, the use of polyclonal anti-target antibodies for target depletion is preferred (to increase the valency of the anti-target antibody to target). Ultimately, the assay controls should be set properly to monitor these potential negative impacts.

3) Combination of multiple procedures for target depletion
Sometimes, target depletion may need to be addressed using a combination of scavenger treatment and a subsequent procedure. For example, Chen et al. (2019) described a biotherapeutic protein target that exists as a homodimer and can form multiple inactive complexes with several inhibitory binding proteins. The target concentrations were increased significantly after acid dissociation. Chen et al. demonstrated that the drug–target binding can be interrupted under mild basic conditions (pH. 8.3), and complete target depletion at a clinically relevant level was achieved by applying a combination of target-blocking proteins to sequester the target and depletion by solid-phase extraction using anti-target antibodies immobilized on magnetic beads.

7.4.2.2.2 Other Procedures to Overcome Target Interference

1) PEG precipitation procedures
It is well known that immunoglobulins can be precipitated by PEG at 20%. At this PEG concentration, smaller proteins such as bovine serum albumin (66 kDa) are largely soluble (Creighton et al., 1973). PandA procedures introduced by Zoghbi et al. (2015) could be used as an alternative for target removal. It should be noted that if the target is tightly associated with drug, it could be precipitated with ADA as a complex, which would eventually result in poor efficiency for target depletion, and the precipitated target would be detected as a false-positive signal during the detection step.

2) Gel filtrations
Melon gel resin has long been used for IgG purification, because it binds and removes serum proteins, thereby allowing pure IgG to be collected in the flow-through fraction. The procedure is gentle, because no harsh elution conditions are involved. Therefore, it has been used to remove targets from serum samples. During the investigation of nerve growth factor (NGF) interference for an anti-NGF mAb therapeutic (fulranumab), Dai et al. (2014) compared several target removal procedures, including the use of melon gel resin, and showed that melon gel treatment can retain ADA positive signals over 50%. However, the use of melon gel resin was not effective at removing drug–target complexes, and target interference persisted. Therefore, proper dissociation of target–drug complexes, prior to melon gel treatment, is critical. Further, ADA recovery needs to be closely monitored between assays to confirm robust assay performance.

7.4.3 Addressing RF Interference

Rheumatoid factor (RF) can be found in the peripheral circulation of patients with autoimmune diseases; it is particularly prevalent in the rheumatoid arthritis patient population. In addition, studies have shown that the increase of RF has also been associated with certain cancer types (Ahn et al., 2017; Tian et al., 2014). The majority of RFs are of the IgM isotype and bind the Fc region of human IgG to form immune complexes. Therefore, RF can interfere with ADA detection, when the therapeutic contains a human Fc fragment (e.g., antibody or antibody-derived therapeutics and Fc fusion proteins) by bridging drug-derived reagents in the ADA assay. The result of RF interference is the potential to cause false-positive ADA results that cannot be resolved by confirmatory assays.

Several methodologies to overcome RF interference in immunogenicity assays have been considered, and examples are given below:

- Establishing ADA assay reagents with the Fab portion of the drug (Rispens et al., 2012) to avoid binding of RF with the Fc portion of antibody drugs.
- Size exclusion to identify the aggregation of drug-derived reagents and establish the assay conditions with low levels of aggregation between RF and drug-derived reagents (Tatarewicz et al., 2010).

These efforts could address RF interference to some degree. However, these methods can either be a challenge for routine clinical testing, or the procedures could underestimate ADA against the framework of antibody drug.

7.4.4 Method Optimization and Qualification

The procedures described earlier to resolve assay interferences need to be considered in the context of optimization of other ADA assay characteristics, as described here. Further, following development of ADA/NAb assays, a qualification process is needed to carefully examine assay characteristics and ensure successful method validation.

7.4.4.1 Method Optimization: Ensuring Assay Specificity While Maintaining Fit-for-Purpose Sensitivity

7.4.4.1.1 Assay Specificity is a Key Element to the Detection of Therapeutics-Elicited ADA

In program development, the ADA assay serves multiple purposes, but mainly it is intended for the evaluation of the potential immunogenicity of a drug and the impact of that immunogenicity on patient safety and drug efficacy. While drug competition assay can be used to demonstrate the specificity of the assay using ADA surrogate positive control as indicator, such procedures would be powerless to differentiate true ADA from those "specific" or nonspecific interference factors described previously, if not properly resolved. As discussed in previous sections, the ADA assay format and sample pretreatment procedures should be designed to overcome interference factors such as target and residual drug in study samples.

7.4.4.1.2 Maintaining Assay Sensitivity, While Eliminating Nonspecific Matrix Interferences

ADA assays need to achieve a certain level of sensitivity. As defined by the FDA, "Assay sensitivity is the lowest concentration at which the antibody preparation consistently produces either a positive result or a readout equal to the cutpoint determined for that assay" (FDA, 2019). Assay sensitivity is determined by using serial dilutions of a PC with sensitivity established at the method cutpoint. As required by regulatory agencies, the screening assay should retain approximately a 5% false-positive rate, while the confirmatory assay should yield a 1% false-positive rate. In practice, these requirements are impacted by multiple factors as described below.

1) Minimal required dilutions
Matrix interference is complex and can obscure ADA detection. Sample dilution is a common practice to reduce nonspecific matrix interference. The minimal required dilution (MRD) needs to be optimized based on matrix selectivity, using multiple samples from disease populations. However, in determining the MRD it is important that samples not be overly diluted to the degree of impacting assay sensitivity. Since Mire-Sluis et al. (2008) introduced this concept and provided procedures to define the MRD there have been different interpretations, resulting in inconsistencies in how the MRD is calculated between groups. Some groups calculate the MRD based entirely on sample dilutions, while other groups also take into consideration the addition of assay reagents. FDA guidance (FDA, 2019) has now clearly defined the calculation of MRD to indicate that the final dilution of the sample should be considered, with a range typically from 1:5 to 1:100 (1/5 to 1/100) to preserve sensitivity. For example, for a typical ECL bridging assay with acid dissociation, the MRD should be calculated as the sum of acid treatment (e.g., 1/10) and neutralization by sample diluent in the presence of biotinylated and ruthenylated drug (e.g., 1/3). In the preceding example, the MRD would be 1:30.

2) Adjusting assay stringency to eliminate nonspecific matrix interferences.

7.4.4.1.3 "Super Clean" ADA Assays Versus Assay Sensitivity

Ultimately, the assessment of immunogenicity in clinical studies is highly dependent on the suitability and quality of immunogenicity assays. Over the years, bioanalytical scientists have benefited greatly from emerging technologies, while accumulating significant experience in addressing matrix interference. Additionally, ADA positive controls, especially monoclonal antibodies, are highly specific with affinities for drug typically in the picomolar range. These factors have made it possible to develop "super clean" ADA assays with assay background levels close or equal to the instrument noise. A consequence of a "super clean" ADA assay would be a significant increase in false-positive rates, which may not align with other clinical data (e.g., pharmacokinetic and drug efficacy profiles). The high positive rates are largely due to unrealistically lower screening and confirmatory cutpoints, driven by low responses from cutpoint samples. To avoid such complications, the authors recommend that the optimization of ADA assays should not focus on the performance of the positive control in isolation, but rather in the context of study sample responses (Gorovits et al., 2019). The assay background should be around two to three times higher than instrument signals, which will give rise to meaningful cutpoints for ADA detection and confirmation, while distinguishing signal/noise ratio from a true ADA response. In addition, alternative statistical approaches, especially the procedures to remove outliers, should be explored for the purpose of establishment of meaningful cutpoint factors (Kubiak et al., 2018).

7.4.4.1.4 Assay Sensitivity in the Context of Specificity

The FDA has set a clear expectation for ADA assay sensitivity (100 ng/mL using a positive control) stating that "concentrations as low as 100 ng/mL may be associated with clinical events" (FDA, 2019). Regulatory agencies also dictate that the assay sensitivity be maintained in the presence of residual drug. Therefore, during ADA assay development, sensitivity should be evaluated and optimized in the context of drug and target tolerance levels. Further, in the case of IgE, regulators expect assay sensitivity to be even lower (nanogram/milliliter to picogram/milliliter range), due to the low circulating levels of this particular isotype (FDA, 2019).

7.4.4.1.5 Addressing High Prevalence of Positive Response in Drug-Naïve Populations

Clinical ADA assays are expected to be specific and to predominantly detect drug-induced antibodies. However, depending on various factors, ADA assays may also detect cross-reactive antibodies or preexisting antibodies that react with drug-derived reagents. For example, preexisting anti-PEG antibodies may cross-react with drug-derived reagents, when the drug in development is itself pegylated. van Schie et al. (2015) summarized the types of cross-reactive antibodies as anti-allotypes, anti-hinge antibodies, and anti-glycan antibodies. Gorovits et al. (2016) further discussed the types of preexisting antibodies associated with the therapeutic modality. Regardless of the source of cross-reactive and preexisting antibodies, they may affect the efficacy of the therapeutic and are often captured by ADA assays. A high prevalence of preexisting antibodies requires investigation during ADA assay development. Xue et al. (2017) summarized procedures that can be used in ADA assay development to investigate preexisting antibody responses. Regulatory agencies expect sponsors to identify preexisting antibodies and to understand their biology and their relationship to disease states, as preexisting antibodies can affect proper cutpoint analysis with consequences for accurate immunogenicity assessment (FDA, 2019).

7.4.4.1.6 Assay Suitability Controls

Assay suitability controls in ADA assays are important indicators of assay performance, which is governed by assay acceptance criteria.

1) Negative control

The negative control must be established using the same matrix type (serum or plasma) as drug-naïve populations and must be relevant to the targeted disease population. The negative control pool should be generated by screening samples collected from a minimum of 10 individual donors with equal representation of gender. The individual matrix samples must be ADA-negative, while the overall response of the negative control pool must be representative of the drug-naïve, disease-relevant population. The correlation between the negative control pool and the drug-naïve, disease-relevant population can be evaluated by statistical analysis, such as ANOVA in JMP and Pearson correlation analysis. The negative control pool should be prepared in large excess, with a volume that estimated to be enough for method validation and sample analysis.

2) Positive controls

Positive controls serve as surrogates for ADA in ADA detection and are used for assay sensitivity determinations. Positive controls are also used for the evaluation of other assay performance characteristics, such as drug and target tolerance levels. However, since the sources of positive controls are largely nonhuman, the positive controls do not truly reflect the ADA species generated against an administered drug.

The use of monoclonal ADA as positive controls is acceptable by regulatory agencies. In cases where multiple ADA clones are available for a given biotherapeutic, a thorough screening process would provide important information on the affinity of each clone and the epitope it recognizes. For antibody biotherapeutics, recommend choosing an anti-ID (blocking) ADA as a positive control.

The positive control should be spiked into negative control serum and stored properly. The performance of the positive control should be confirmed prior to method validation. During early stages of validation, the low positive control established in method development can be used; however, it should be further evaluated as method validation progresses. During method validation it is recommended to include the medium positive control (to monitor the linearity of the positive control; however, the medium positive control is not necessary in study support [FDA, 2019]).

For multidomain therapeutics, such as bispecific and trispecific antibodies, ADCs, and other antibody–drug conjugates, the positive controls should be specific for each domain. The specificity of each positive control for a given domain should be characterized during assay development, and the relevant positive controls should be used throughout assay validation.

References

Ahn, J. K., Hwang, J., Chang, Y., & Ryu, S. (2017). Rheumatoid factor positivity increases all-cause and cancer mortality: A cohort study. *Rheumatology International*, 37(7), 1135–1143. https://doi.org/10.1007/s00296-017-3738-x

Atha, D. H., & Ingham, K. C. (1981). Mechanism of precipitation of proteins by polyethylene glycols. Analysis in terms of excluded volume. *Journal of Biological Chemistry*, 256(23), 12108–12117. https://www.ncbi.nlm.nih.gov/pubmed/7298647

Bano, N., McKelvey, T., Spear, N., Yang, T. Y., Shankar, G., & Schantz, A. (2019). Analysis of regulatory guidance on antidrug antibody testing for therapeutic protein products. *Bioanalysis*, 11(24), 2283–2296. https://doi.org/10.4155/bio-2019-0241

Barbosa, M. D., Gokemeijer, J., Martin, A. D., & Bush, A. (2013). Altering drug tolerance of surface plasmon resonance assays for the detection of anti-drug antibodies. *Analytical Biochemistry*, 441(2), 174–179. https://doi.org/10.1016/j.ab.2013.07.019

Bargou, R., Leo, E., Zugmaier, G., Klinger, M., Goebeler, M., Knop, S., ... Kufer, P. (2008). Tumor regression in cancer patients by very low doses of a T cell-engaging antibody. *Science*, 321(5891), 974–977. https://doi.org/10.1126/science.1158545

Bourdage, J. S., Cook, C. A., Farrington, D. L., Chain, J. S., & Konrad, R. J. (2007). An Affinity Capture Elution (ACE) assay for detection of anti-drug antibody to monoclonal antibody therapeutics in the presence of high levels of drug. *Journal of Immunological Methods*, 327(1–2), 10–17. https://doi.org/10.1016/j.jim.2007.07.004

Chamberlain, P. (2019). Effective presentation of immunogenicity risk assessments and related data in regulatory dossiers. *Bioanalysis*, 11(17), 1581–1592. https://doi.org/10.4155/bio-2018-0209

Chen, J., Garlits, J., Dhulipala, G., Sirimanne, T., Xu, C., Lu, K., ... Sumner, G. (2019). Mitigating target interference in bridging immunogenicity assay with target-blocking reagents and mild basic pH. *Bioanalysis*, 11(17), 1569–1580. https://doi.org/10.4155/bio-2018-0187

Chen, L. Z., Roos, D., & Philip, E. (2016). Development of immunocapture-LC/MS assay for simultaneous ADA isotyping and semiquantitation. *Journal of Immunology Research*, 2016, 7682472. https://doi.org/10.1155/2016/7682472

Collet-Brose, J., Couble, P. J., Deehan, M. R., Nelson, R. J., Ferlin, W. G., & Lory, S. (2016). Evaluation of multiple immunoassay technology platforms to select the anti-drug antibody assay exhibiting the most appropriate drug and target tolerance. *Journal of Immunology Research*, 2016, 5069678. https://doi.org/10.1155/2016/5069678

Creighton, W. D., Lambert, P. H., & Miescher, P. A. (1973). Detection of antibodies and soluble antigen-antibody complexes by precipitation with polyethylene glycol. *Journal of Immunology*, 111(4), 1219–1227.https://www.ncbi.nlm.nih.gov/pubmed/4728683

Dai, S., Schantz, A., Clements-Egan, A., Cannon, M., & Shankar, G. (2014). Development of a method that eliminates false-positive results due to nerve growth factor interference in the assessment of fulranumab immunogenicity. *The AAPS Journal*, 16(3), 464–477. https://doi.org/10.1208/s12248-014-9581-z

Egging, D., Verhagen, J., Laat-Arts, K., Wit, B., Boekel, T. V., Buurman, M., ... Mulder, R. (2018). Sensitivity and drug tolerance of antidrug antibody assays in relation to positive control characteristics. *Bioanalysis*, 10(16), 1289–1306. https://doi.org/10.4155/bio-2018-0091

Food and Drug Administration (FDA). (2019). Immunogenicity Testing of Therapeutic Protein Products – Developing and Validating Assays for Anti-Drug Antibody Detection.

Gorovits, B., Clements-Egan, A., Birchler, M., Liang, M., Myler, H., Peng, K., ... Xue, L. (2016). Pre-existing antibody: Biotherapeutic modality-based review. *The AAPS Journal*, 18(2), 311–320. https://doi.org/10.1208/s12248-016-9878-1

Gorovits, B., McNally, J., Fiorotti, C., & Leung, S. (2014). Protein-based matrix interferences in ligand-binding assays. *Bioanalysis*, 6(8), 1131–1140. https://doi.org/10.4155/bio.14.56

Gorovits, B., Wang, Y., Zhu, L., Araya, M., Kamerud, J., & Lepsy, C. (2019). Anti-drug antibody assay conditions significantly impact assay screen and confirmatory cut-points. *The AAPS Journal*, 21(4), 71. https://doi.org/10.1208/s12248-019-0342-x

Haense, N., Atmaca, A., Pauligk, C., Steinmetz, K., Marme, F., Haag, G. M., ... Al-Batran, S. E. (2016). A phase I trial of the trifunctional anti Her2 x anti CD3 antibody ertumaxomab in patients with advanced solid tumors. *BMC Cancer*, 16, 420. https://doi.org/10.1186/s12885-016-2449-0

Heaney, M. L., & Golde, D. W. (1998). Soluble receptors in human disease. *Journal of Leukocyte Biology*, 64(2), 135–146. https://doi.org/10.1002/jlb.64.2.135

Koren, E., Smith, H. W., Shores, E., Shankar, G., Finco-Kent, D., Rup, B., ... Mire-Sluis, A. (2008). Recommendations on risk-based strategies for detection and characterization of antibodies against biotechnology products. *Journal of Immunological Methods*, 333(1–2), 1–9. https://doi.org/10.1016/j.jim.2008.01.001

Kubiak, R. J., Zhang, J., Ren, P., Yang, H., & Roskos, L. K. (2018). Excessive outlier removal may result in cut points that are not suitable for immunogenicity assessments. *Journal of Immunological Methods*, 463, 105–111. https://doi.org/10.1016/j.jim.2018.10.001

Kurki, P. (2019). Compatibility of immunogenicity guidance by the EMA and the US FDA. *Bioanalysis*, 11(17), 1619–1629. https://doi.org/10.4155/bio-2018-0243

Leach, M. W., Rottman, J. B., Hock, M. B., Finco, D., Rojko, J. L., & Beyer, J. C. (2014). Immunogenicity/hypersensitivity of biologics. *Toxicologic Pathology*, 42(1), 293–300. https://doi.org/10.1177/0192623313510987

Linke, R., Klein, A., & Seimetz, D. (2010). Catumaxomab: Clinical development and future directions. *MAbs*, 2(2), 129–136. https://doi.org/10.4161/mabs.2.2.11221

Ohlson, S., & Zetterstrand, K. (1985). Detection of circulating immune complexes by PEG precipitation combined with ELISA. *Journal of Immunological Methods*, 77(1), 87–93. https://doi.org/10.1016/0022-1759(85)90186-3

Rispens, T., de Vrieze, H., de Groot, E., Wouters, D., Stapel, S., Wolbink, G. J., & Aarden, L. A. (2012). Antibodies to constant domains of therapeutic monoclonal antibodies: Anti-hinge antibodies in immunogenicity testing. *Journal of Immunological Methods*, 375(1-2), 93–99. https://doi.org/10.1016/j.jim.2011.09.011

Rosenberg, A. S., & Sauna, Z. E. (2018). Immunogenicity assessment during the development of protein therapeutics. *Journal of Pharmacy and Pharmacology*, 70(5), 584–594. https://doi.org/10.1111/jphp.12810

Smith, H. W., Butterfield, A., & Sun, D. (2007). Detection of antibodies against therapeutic proteins in the presence of residual therapeutic protein using a solid-phase extraction with acid dissociation (SPEAD) sample treatment prior to ELISA. *Regulatory Toxicology and Pharmacology*, 49(3), 230–237. https://doi.org/10.1016/j.yrtph.2007.07.005

Tatarewicz, S., Miller, J. M., Swanson, S. J., & Moxness, M. S. (2010). Rheumatoid factor interference in immunogenicity assays for human monoclonal antibody therapeutics. *Journal of Immunological Methods*, 357(1–2), 10–16. https://doi.org/10.1016/j.jim.2010.03.012

Tian, G., Liang, J. N., Wang, Z. Y., & Zhou, D. (2014). Breast cancer risk in rheumatoid arthritis: An update meta-analysis. *BioMed Research International*, 2014, 453012. https://doi.org/10.1155/2014/453012

van Schie, K. A., Wolbink, G. J., & Rispens, T. (2015). Cross-reactive and pre-existing antibodies to therapeutic antibodies–Effects on treatment and immunogenicity. *MAbs*, 7(4), 662–671. https://doi.org/10.1080/19420862.2015.1048411

Welink, J., Xu, Y., Yang, E., Wilson, A., Henderson, N., Luo, L., ... Whale, E. (2018). 2018 White Paper on Recent Issues in Bioanalysis: "A global bioanalytical community perspective on last decade of incurred samples reanalysis (ISR)" (Part 1 – small molecule regulated bioanalysis, small molecule biomarkers, peptides & oligonucleotide bioanalysis). *Bioanalysis*, 10(22), 1781–1801. https://doi.org/10.4155/bio-2018-0268

Xu, W., Sank, M., Cummings, J., Carl, S., Juhel, M., Gleason, C., ... Pillutla, R. (2018). Bead-extraction and heat-dissociation (BEHD): A novel way to overcome drug and matrix interference in immunogenicity testing. *Journal of Immunological Methods*, 462, 34–41. https://doi.org/10.1016/j.jim.2018.08.003

Xue, L., Clements-Egan, A., Amaravadi, L., Birchler, M., Gorovits, B., Liang, M., ... Sung, C. (2017). Recommendations for the assessment and management of pre-existing drug-reactive antibodies during biotherapeutic development. *The AAPS Journal*, 19(6), 1576–1586. https://doi.org/10.1208/s12248-017-0153-x

Zhong, Z. D., Clements-Egan, A., Gorovits, B., Maia, M., Sumner, G., Theobald, V., ... Rajadhyaksha, M. (2017). Drug target interference in immunogenicity assays: Recommendations and mitigation strategies. *The AAPS Journal*, 19(6), 1564–1575. https://doi.org/10.1208/s12248-017-0148-7

Zoghbi, J., Xu, Y., Grabert, R., Theobald, V., & Richards, S. (2015). A breakthrough novel method to resolve the drug and target interference problem in immunogenicity assays. *Journal of Immunological Methods*, 426, 62–69. https://doi.org/10.1016/j.jim.2015.08.002

8

Bioanalytical Scheme for Antidrug Neutralizing Antibody Assays

Weifeng Xu,[a] Bonnie Wu[b] and Jim McNally[c]

[a]*PCD, Bioanalytics, Merck & Co., Inc. (corresponding author)*
[b]*Janssen R&D*
[c]*BioAgilytix*

8.1 Introduction

In standard tiered immunogenicity testing, antidrug antibodies (ADAs) positive clinical samples should be further characterized for their neutralizing capacity through in vitro neutralizing antibody (NAb) assays. NAb assessment monitors whether ADAs can inhibit the function of biotherapeutics by preventing the drug from binding to its target and blocking the downstream signaling events (Novick et al., 2000). Other than reduced drug efficacy, NAb can potentially lead to more severe, life-threatening side effects by neutralizing endogenous counterparts of the drug as exemplified by the pure red cell aplasia caused by anti-Epoetin antibodies that cross-react with and neutralize the endogenous erythropoietin in patients (Casadevall et al., 2002; Tovey & Lallemand, 2011). Consequently, NAb assessment is required prior to regulatory approval of biological license and in some cases even as a part of postmarking commitment. In this chapter we will first discuss the strategy for NAb assay format selection and when a NAb assay is needed, followed by important considerations and challenges associated with NAb assay development, validation, and method transfer.

Bioanalytical Aspects in Biological Therapeutics, First Edition. Edited by Xiaohui Xu and Weifeng Xu.
© 2022 John Wiley & Sons, Inc. Published 2022 by John Wiley & Sons, Inc.

8.2 Assay Format

Cell-based NAb assays use cells to detect NAb activity that inhibits the drug-mediated cellular response. This assay format was traditionally preferred by regulatory agencies (FDA, 2019). However, comparative analysis from industry sponsors has accumulated a large body of evidence demonstrating that non-cell-based assays are a viable assay platform for detection of NAbs to a variety of biotherapeutics with comparable or even superior assay performance (Finco et al., 2011; Hu et al., 2015; Wu et al., 2018). The most recent US Food and Drug Administration (FDA) Immunogenicity Guidance also explicitly states that "... ligand binding assays may be appropriate for antagonistic mAbs or receptor Fc fusion proteins that bind and inhibit the target" (FDA, 2019). Unlike cell-based assays, non-cell-based NAb assays usually employ immunoassays instead to measure drug–target interaction for NAb detection. Both cell-based and non-cell-based assay formats are expected to reflect the *in vivo* therapeutic mechanism of action (MOA) in order to generate clinically meaningful data.

The cell-based NAb assays are typically bioassays that measure a specific assay endpoint in a cellular signal transduction pathway. If the assay endpoint assesses the drug binding to the cell surface receptor, the first critical step that triggers the downstream cellular responses, a cell-based binding assay could be implemented for NAb detection through NAb-mediated inhibition of drug binding to target cells. Although this is just a binding assay, it is still considered a cell-based assay since the method involves the use of live cells.

The non-cell-based assays have been categorized into competitive ligand binding (CLB) assays and enzyme activity assays. The analytical procedure of the CLB NAb assays are based on the competition between NAbs and drug target for binding to the limited concentration of drug reagent used in the assay. NAb, when present in the sample matrix, is detected through its inhibition of the drug–target interaction. Conversely, the enzyme activity assays are biochemical assays uniquely designed for enzyme replacement therapies (Wang et al., 2008), for example, lysosomal enzymes for the treatment of lysosomal storage diseases. These assays measure the catalytic activity of the enzyme drug product based on its MOA.

Both cell-based and non-cell-based CLB NAb assays can be further classified into direct and indirect formats. The direct assay format typically monitors the assay response triggered by binding of drug, for example, STELARA or ustekinumab (Mojtahed Poor et al., 2019), to its corresponding target. Presence of NAbs inhibits drug–target interaction, reducing the assay signal. The indirect assay format monitors the ligand binding to its cognate receptor. Drug blocks the ligand–receptor interaction, inhibiting the assay response. An example is Yervoy which blocks the interaction between CTLA4 and CD80/CD86

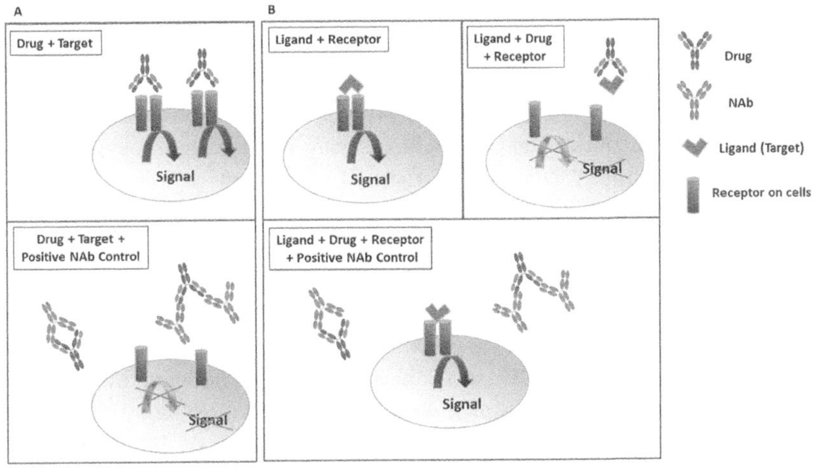

Figure 8.1 Direct (A) and indirect (B) cell-based neutralizing antibody assay formats (Wu et al., 2016).

(Xu et al., 2018a). When NAb is present, it reverses the drug-mediated inhibition of CTLA4–CD80/CD86 interaction, restoring the assay signal (Figure 8.1).

No matter which assay format is used, the method should be reliable to detect clinically relevant NAb responses with adequate precision, sensitivity, specificity, and selectivity. In particular, the NAb assay is expected to be tolerant to the drug and its target in matrix samples to avoid erroneous results and misclassification of NAb status for test samples.

8.3 Implementation of a Suitable NAb Assay Format to Support Immunogenicity Assessment

The advancement of technology platforms in drug discovery has resulted in drug products with a diverse complexity of MOAs. This presents an increasing challenge for NAb method development. The utility and suitability of both cell-based and non-cell-based assays for NAb assessment have further driven the development and establishment of consistent strategies that would guide industry sponsors on how to select a proper NAb assay format for their intended purpose with regulatory success.

Risk of immunogenicity is the primary guide for the NAb testing bioanalytical strategy. It informs the timing, frequency, and duration of sampling for bioanalysis and the scope of NAb assessment. For high-risk biotherapeutics, such as recombinant cytokines with nonredundant endogenous counterparts, the NAb assay needs to be ready for preclinical or phase I studies and may be used during interim sample analysis for real-time NAb assessment due to safety concerns. For low-risk biotherapeutics such as a monoclonal antibody (mAb) against a soluble ligand, the NAb assay is typically implemented at phase II and beyond. When the NAb method is not in place, it is advisable to bank the clinical samples for future analysis when the assay is available.

In the case of gene therapy programs, implementation of NAb assays may occur far earlier than other drug modalities. In particular, for adeno-associated viruses (AAV)-based gene therapies, it has been shown that there is a high incidence of preexisting antibodies in nonclinical species and in humans (Gorovits et al., 2020; Kumar et al., 2017; van Schie et al., 2015). Therefore, establishing NAb assays to prescreen animals and patients for either study inclusion/exclusion and/or stratification purposes may be necessary. For nonclinical studies, it could be important to identify animals that are either negative or low titer for NAb to minimize the impact of anti-AAV immunogenicity on the vector distribution and successful transduction. In these situations, NAb assays may be implemented as early as nonclinical biodistribution studies when needed.

Although the immunogenicity risk profile helps to determine the bioanalytical strategy (Shankar et al., 2007), the risk-based approach could only guide the NAb assay format selection for some high-risk molecules (e.g., cytokines or growth factors) instead of different varieties of biotherapeutics. This risk-based approach expects the use of cell-based assays for high-risk drug products and either cell-based or non-cell-based assays for low- or medium-risk biotherapeutics. However, this scheme cannot apply to enzyme replacement therapeutic products which are deemed high risk in immunogenicity due to their homology with endogenous counterpart(s) but require enzyme activity assays for NAb assessment. The same scenario can be applied to some mAb biotherapeutics that were initially predicted as low-risk molecules but attain a high-risk profile due to product aggregation or other critical biophysical attributes that make the drug products more immunogenic. Most importantly, for any given biotherapeutic, the risk level may change with experience and could vary with not only drug quality attributes but also patient immune status, genetic background, or route of drug administration. Despite the variation in the immunogenicity risk profiles during the clinical studies, the NAb assay format will likely remain unchanged and use the same method reflective of the drug's MOA.

Based on the best practice from industry sponsors, a consistent strategy has been formulated, proposing that the therapeutic MOA should serve as the primary guide for NAb assay format selection (Wu et al., 2016) (Table 8.1).

Table 8.1 Neutralizing antibody assay format selection based on mechanism of action (Wu et al., 2016).

MOA	Drug modality	Drug target	Drug–target interaction	Examples	Recommended assay format
Agonist	Recombinant protein or Ab	Cellular receptor	Drug binds and activates receptor	Cytokines, growth factors, EPO, agonists with no homology to endogenous protein	Cell-based assay as primary choice; non-cell-based assay as an alternative
Antagonist	mAb	Humoral target	Drug binds and inhibits the target	Golimumab, ustekinumab, adalimumab	Non-cell-based CLB assay
	mAb	Cellular receptor	Drug binds cellular receptor and competitively inhibits receptor–ligand interaction	Natalizumab, trastuzumab, tocilizumab	Cell-based assay or non-cell-based assay
	Soluble receptor	Ligand	Soluble receptor binds ligand and blocks receptor–ligand interaction	Etanercept, abatacept	Non-cell-based CLB assay recommended; cell-based assay possible with a suitable cell line
Targeted intracellular delivery of a potent cytotoxin mediated by Ab	ADC	Cellular receptor	ADC binds the cellular receptor and mediates the internalization of payload	Brentuximab vedotin, adotrastuzumab emtansine	Cell-based assay recommended; non-cell-based assay acceptable due to drug interference

(Continued)

Table 8.1 (Continued)

MOA	Drug modality	Drug target	Drug–target interaction	Examples	Recommended assay format
Target cell lysis through antibody effector function	mAb	Target cell receptor, FcγR, or complement	Ab binds to target cell receptor through variable region and FcγR or complement through Fc domain	Rituximab, cetuximab, alemtuzumab	Cell-based effector assay recommended; cell-based binding assay or non-cell-based CLB assay acceptable with justification
Enzyme replacement	Enzyme	Replace deficient protein in circulation or in target cells; may need cellular receptor for enzyme uptake	Enzyme functions in circulation or through cellular uptake	Human factor IX, imiglucerase, idursulfase, galsulfase	Enzyme bioactivity assay and/or cell-based assay; two assays may be needed

Abbreviations: Ab, antibody; ADC, antibody–drug conjugate; CLB, competitive ligand binding; FcγR, Fc-gamma receptors; mAb, monoclonal antibody; MOA, mechanism of action.

Together with MOA-based format selection, the assay performance characteristics (e.g., sensitivity, specificity, selectivity, precision, and robustness) should facilitate the selection of a suitable NAb assay format. The method is designed and optimized so that the NAb assay method with the best combination of performance characteristics would be used for reliable and sensitive NAb detection. In addition, the risk of immunogenicity helps to shape the expectation of the selected assay format. For example, for drug products with a high risk of severe clinical consequences, sensitive NAb assays with adequate drug tolerance are preferred so that NAbs can be assessed with adequate sensitivity at any sampling time points to facilitate clinical data analysis. In this circumstance, the NAb assay is expected to have a reasonable throughput and data turnaround time to support frequent sample bioanalysis.

It is important to note that assay format selection could be an evolving process that requires life cycle management especially for drugs investigated for multiple disease indications. In particular, the regulatory agencies should be consulted as soon as possible if there is a question relating to assay format selection or a major change of NAb assay format is necessary.

8.4 Current Regulatory Guidance and White Papers

The impact of immunogenicity, although often none or benign, could potentially impact drug efficacy and patient safety. When compared to ADAs, NAbs are more of a concern since they have an immediate and direct impact on loss of drug efficacy (e.g., Tysabri, IFN-beta) (Kivisakk et al., 2000; van Schie et al., 2017). When NAb neutralizes the endogenous counterpart of the drug it may even result in severe clinical consequences such as immunodeficiency syndrome (e.g., epoetin, PEG-rHuMGDF) (Casadevall et al., 2002; Wang et al., 2008). Therefore, detection of NAbs is an integral part of immunogenicity assessment and is required by regulatory agencies.

Health regulatory agencies and industry sponsors have published regulatory guidance (FDA, 2019) and white papers (Wu et al., 2016) to provide advice on how to select a suitable assay format for NAb assessment and to offer recommendations for the NAb assay method development and validation.

The United States Food and Drug Administration (FDA) released the first draft guideline "Assay Development for Immunogenicity Testing of Therapeutic Proteins" in 2009 (FDA, 2009) and a revised draft guidance "Assay Development and Validation for Immunogenicity Testing of Therapeutic Protein Products" in 2016 (FDA, 2016). The FDA 2009 draft guideline acknowledges that both cell-based bioassays and non-cell-based CLB assays have been used for the measurement of NAb activity. “While competitive ligand-binding assays may be the only alternative in some situations, generally FDA considers that bioassays are more reflective of the in vivo situation and are recommended.” “For NAb assays, the bioassay should be related to product mechanism of action, otherwise the assay will not be informative as to the effect of NAb on clinical results.” The recent FDA 2019 draft guidance provides more explicit recommendation on NAb assay format selection. “Selection of the appropriate assay format depends on various factors. These factors include, but are not limited to, the mechanism of action of the therapeutic protein product, its ability to reflect the in vivo situation most closely, and the selectivity, sensitivity, precision, and robustness of the assay. FDA recommends that neutralization assays use a cell-based bioassay format depending on the therapeutic protein product's mechanism of action because, frequently, cell-based bioassays more closely reflect the in vivo situation and therefore provide more relevant information than ligand-binding assays.”

The European Medicines Agency (EMA) published the first guideline on “Immunogenicity Assessment of Biotechnology-Derived Therapeutic Proteins” in 2006 (EMA, 2006), a revised draft in 2015 (EMA, 2015), and a finalized guideline on “Immunogenicity Assessment of Therapeutic Proteins” in 2017 (EMA, 2017). The EMA 2017 guideline states that “Understanding the mode of action, the target and effector pathways of the therapeutic is critical for identification of a suitable NAb assay format. Additionally, the risk of developing NAbs and the impact on clinical sequelae also needs to be considered. While cell-based assays are often employed for agonistic therapeutics, non-cell-based competitive ligand binding (CLB) assays are often considered for antagonistic molecules with humoral targets. For products that exert their activity only through direct binding to other molecules (e.g., some mAbs), CLB or other alternatives may be suitable. However, when these are used it must be demonstrated that they reflect neutralizing capacity/potential in an appropriate manner.” This statement aligns well with another separate EMA guideline issued in 2012 which recognizes that “...induced antibodies which block mAb binding to target are those which are mostly associated with reduced clinical efficacy. Therefore, competitive ligand binding assays may be the neutralizing assays of choice for mAbs rather than classic bioassays.”

In addition to the NAb format selection strategies, there have been series of regulatory guidelines (FDA, 2016) and white papers (Koren et al., 2008; Shankar et al., 2007; Wu et al., 2016) that recommend the best practice for developing a reliable NAb assay with adequate precision, sensitivity, specificity, and selectivity to detect clinically relevant NAbs. These publications address key issues during NAb method development and validation, for example, assay cutpoint assessment, drug activity curve and sensitivity, positive control selection, matrix interference including drug and its target interference, and NAb assay design for multispecific biotherapeutics (Civoli et al., 2012; Gupta et al., 2011; Jolicoeur & Tacey, 2012; Myler et al., 2019).

8.5 Key Considerations of NAb Assay Development

8.5.1 Selection of a Suitable Cell Line and Assay Endpoint for Cell-Based NAb Assays

For a cell-based NAb assay, a suitable cell line with a robust assay response reflecting the drug MOA is the most critical requirement. More often than not, a modified cell line overexpressing the drug target is needed to provide enough sensitivity. In addition, a stable clone with a high level of cell surface receptor would typically help to improve assay sensitivity and consistency. For example, CTLA-4, the target of Yervoy, is only transiently translocated from the intracellular compartment to the cell surface when T cells are activated under physiologic conditions. It is hard to find a cell line with stable and high surface CTLA-4 expression. Thus, a Jurkat cell line stably transfected with CTLA-4 containing a mutation in the transmembrane domain that prevents the internalization but maintains the physiologic function of CTLA-4 was constructed for a robust cell-based assay (Xu et al., 2018a).

A good starting point for cell-based NAb assay development is to evaluate the cell line used for drug potency assay, which measures the biological activity of drug product with adequate reproducibility and robustness. However, potency assays only assess drug function in cell culture media or assay buffer and do not typically need to be concerned about matrix interference and sensitivity. As opposed to potency assays, NAb assays measure NAbs in complex matrix samples such as serum and plasma. These biological matrices may contain interfering factors (e.g., free drug, drug target, lipid, nonspecific binding proteins) that would confound assay performance and result in erroneous results. The optimized NAb assays would overcome matrix interference to enable sensitive and reliable NAb detection.

Bispecific molecules usually bind two separate targets. One example is a mAb biotherapeutic with antibody-dependent cellular-mediated cytotoxicity (ADCC) effector function that binds to its target through the Fab portion and the Fc receptor through the Fc domain to engage in effector function.

Development of a cell-based ADCC assay requires two cell lines: a target cell line, and an effector cell line (e.g., natural killer cell line or an engineered Jurkat cell line expressing CD16 Fc receptor). When compared to a CLB assay (Wu et al., 2018), the cell-based ADCC assay allows the detection of NAbs to both functional domains of the drug molecule.

Lastly, it would be very helpful if "ready to use" frozen cells could be used in the cell-based NAb assay. This can not only improve the operational efficiency but also eliminate the variations introduced during continuous cell culture. The selection of a robust cell line is critical to allow the production of frozen cells that generate consistent and reproducible performance in the cell-based NAb assay.

Assay robustness and sensitivity of the cell-based assays are also related to the selected assay endpoint. For example, interleukin-2 (IL-2) secretion is the marker of T cell activation. To measure IL-2 production, multiple endpoints can be used: mRNA level, protein level, or IL-2 promoter-driven luciferase reporter expression. Monitoring IL-2 protein usually needs at least overnight accumulation before measurement with either enzyme-linked immunosorbent assay (ELISA) or wash-free AlphaLISA method. IL-2 mRNA level typically peaks around 4 hours after T cell activation and can be sensitively detected by either branched DNA or real-time polymerase chain reaction (PCR). Luciferase reporter expression can also be reliably detected approximately 4 hours poststimulation and enables sensitive NAb detection; however, this requires a reporter cell line with a stable and optimal level of luciferase gene expression.

Generally speaking, an endpoint supporting shorter assay time is desirable not only for increased operational efficiency but also for reduced exposure to cytotoxic matrix components or nonspecific cellular stimulation. The selection of the earliest and most specific signaling event as the assay endpoint relies on a thorough understanding of the signaling transduction pathways responsive to the drug–target interaction. For example, phosphorylation of the drug target or its immediate interactor (e.g., receptor tyrosine kinase), STAT protein phosphorylation, calcium flux, or cAMP level are all excellent candidates for an early assay endpoint.

Enzyme fragment complementation technology and other similar technologies have been applied to increase the cell-based assay specificity by constructing a cell line in which both the drug target and its proximal interacting protein are modified so that only the drug-specific pathway will induce the signal (Yan et al., 2002). With the advance of CRISPR technology, it is also possible to remove an interfering pathway from a parental cell line to reduce the matrix interference.

8.5.2 Assay Design for the CLB NAb Assays

Although non-cell-based CLB NAb assays are generally more sensitive and may be simpler to perform when compared to cell-based NAb assays, the method development is not always straightforward due to the unique consideration of reagents. The recombinant target used in the CLB NAb assays can be the extracellular domains of complex transmembrane receptor proteins which are usually fused with poly-histidine-tag or other protein framework (e.g., antibody Fc domain) to facilitate the purification. This may alter not only the configuration of the protein but also the binding affinity of the drug to the recombinant target protein. Therefore, the comparability between recombinant and cellular target protein needs to be evaluated by assessing the binding affinity of drug to both forms of target proteins using surface plasmon resonance (SPR) or a cell-based binding assay, respectively. In cases where recombinant drug target protein is unavailable, a cell-based binding assay can be implemented for NAb assay development. In addition, both the labeled drug and target used in the CLB assays need to be conjugated with suitable challenge ratio to preserve their biological activity and generate optimal assay signal in the assay.

8.5.3 Sample Pretreatment to Overcome Drug and Target Interference

When compared to ADA assays, NAb assays usually generate an inadequate drug tolerance. To obtain sufficient sensitivity, a relatively low level of drug reagent is used for NAb detection. This renders the NAb assay more susceptible to interference from an excess amount of circulating drug in the samples, In addition, drug from ADA samples will be competed by the labeled drug, while in NAb assays drug from the sample cannot be distinguished from drug reagent, resulting in erroneous results. Most of the mAb biotherapeutics have a relatively long half-life in the sample matrix, leading to high trough concentrations of drug that would render NAb assays inadequate for drug tolerance. This is particularly the case for biotherapeutics that serve as immune checkpoint inhibitors in oncological indications. These therapeutic drugs are typically dosed at a very high level. As a result, the immunogenicity test samples may contain more than 100 μg/mL of the circulating drug and would need sample pretreatment to dissociate drug–ADA immune complexes or NAb purification to improve the sensitivity of NAb detection in the presence of circulating drug. Various methods to overcome drug interference are summarized together with their pros and cons in Table 8.2.

Table 8.2 Case studies to overcome drug interference.

Method and outline	Pros and Cons	Reference
Affinity capture elution (ACE): acid dissociation of sample and incubation with plate-coated drug	**Pros:** no need to biotinylate drug; no need of magnetic beads; easy to perform and low cost **Cons:** low binding capacity and low NAb recovery; purified NAbs would be eluted with second acid; may only be suitable for CLB assays	Bourdage et al. (2007)
Solid-phase extraction with acid dissociation (SPEAD): acid dissociation of sample and incubation with biotin-drug coated on SA plate	**Pros:** no need of magnetic beads; easy to perform and low cost **Cons:** low binding capacity and low NAb recovery; may only be suitable for CLB assays	Smith et al. (2007)
Bead extraction with acid dissociation (BEAD): acid dissociation of sample and incubation with SA-coated magnetic beads	**Pros:** High binding capacity on the SA-coated magnetic beads and high NAb recovery, suitable for CLB as well as cell-based assays **Cons:** high cost, tedious to handle beads	Lofgren et al. (2006)
PEG precipitation and acid dissociation (PandA): use PEG to selectively precipitate drug/ADA complex and remove free drug	**Pros:** no need for magnetic beads **Cons:** MSD plate-based assay, only suitable for CLB assays	Zoghbi et al. (2015)
Acid dissociation and use drug target to remove free drug	**Pros:** same assay format for both ADA and NAb **Cons:** MSD plate-based assay, only suitable for CLB assays	Sloan et al. (2016)
Bead extraction with heat dissociation (BEHD): immune complex dissociation through heat treatment and incubation with SA-coated magnetic beads	**Pros:** avoid using harsh acid in the first step so that NAb activity can be better preserved **Cons:** need to find the best temperature for each drug; may only be suitable for biotherapeutics (e.g., domain Ab) with lower thermal stability	Xu et al. (2018b)
Double acid dissociation to sequentially remove free drug and drug target	**Pros:** can remove drug and its target interferences simultaneously; suitable for both ADA and NAb assays **Cons:** one more acid dissociation step; may compromise NAb recovery	Xiang et al. (2019)

Abbreviations: Ab, antibody; ADA, antidrug antibody; CLB; competitive ligand binding; MSD, Meso Scale Discovery; NAb, neutralizing antibody; PEG, polyethylene glycol; SA, streptavidin.

Table 8.3 Case studies to mitigate matrix interference.

Drug	Interfering factors and mechanism	Pretreatment and MOA	Reference
Anti-PDGFR-α mAb	Cell-based NAb assay measures the phosphorylation of drug target: PDGFR. Natural ligands PDGF-AA and PDGF-AB bind to the receptor and induce phosphorylation of PDGFR-α	Add inhibitory Ab against PDGF-A to block the binding of both PDGF-AA and PDGF-AB to PDGEF-α	Schwickart et al. (2014)
IgG1 mAb to deplete target cells via CDC	Addition of individual naïve serum completely inhibited CDC activity • Delipidation, desalting, and ammonium sulfate precipitation followed by reconstitution of the IgG fraction did not help • Particular set of cytokines, RF level, or preexisting serum complement immune complexes was not related to serum interference, suggesting other interfering factors	Protein A/G resin to purify IgGs, including ADAs and NAbs, thus excluding any other serum factors	McCutcheon et al. (2010)
BCMAXCD3 bispecific Ab	Soluble BCMA	Biotinylated anti-BCMA polyclonal Ab conjugated to SA-coated magnetic beads for sBCMA removal	Wang et al. (2019)

Abbreviations: Ab, antibody; ADA, antidrug antibody; BCMA; B cell maturation antigen; CDC, complement-dependent cytotoxicity; mAb, monoclonal antibody; MOA, mechanism of action; NAb, neutralizing antibody; PDGFR, platelet-derived growth factor receptor; RF, rheumatoid factor; SA, streptavidin; sBCMA, serum B cell maturation antigen.

In addition to circulating drug, other matrix factors in samples (e.g., cytokines, rheumatoid factors, circulating drug, or soluble drug target) could either increase or inhibit the endpoint signal, leading to false-negative or false-positive NAb results. The sample pretreatment may include heat inactivation, enrichment of ADAs by a capture reagent (e.g., protein A/G or drug), and inhibition of drug target by an anti-target antibody (see case studies in Table 8.3) to mitigate the matrix interference.

Using biotinylated drug to enrich and extract ADA is a commonly used sample pretreatment procedure which removes not only circulating drug but also other interfering matrix factors in test samples. The plate-based sample pretreatment such as affinity capture elution (ACE) and solid-phase extraction with acid dissociation (SPEAD) as shown in Table 8.2 have been applied successfully to both CLB and some sensitive cell-based NAb assays. Conversely, bead-based methods (e.g., bead extraction with acid dissociation [BEAD]) would extract more NAb due to the higher surface area and binding capacity of streptavidin-coated magnetic beads compared to that of a 96-well plate, thereby generating a higher assay drug tolerance.

Several steps in the BEAD procedure need to be properly optimized to facilitate successful method development (Figure 8.2). The pH for the acid dissociation steps should be acidic enough to dissociate NAb–drug complex but not too harsh to denature NAb or prevent the detection of low-affinity NAb. In addition, the acid and base buffer used during the second acid dissociation and

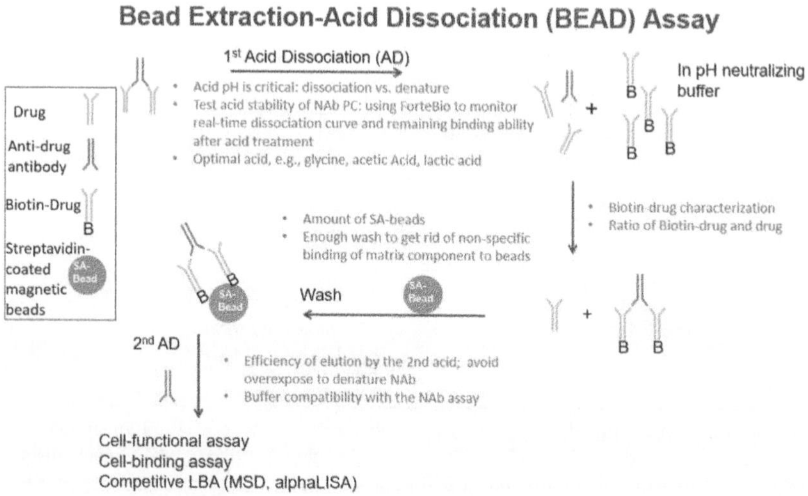

Figure 8.2 Challenges for bead extraction with acid dissociation (BEAD) optimization. AD, acid dissociation; LBA, ligand binding asssay; MSD, Meso Scale Discovery; NAb, neutralizing antibody; PC, positive control; SA, streptavidin.

pH neutralization need to be compatible with the downstream assay steps. It has been observed that some cells, and even AlphaLISA methods, are sensitive to buffer conditions (Wu et al., 2019). It is also important to characterize the biotinylated drug to ensure that there is no free biotin, unlabeled drug, or aggregates to interfere with the assay; and an appropriate amount of SA-conjugated magnetic beads would allow the maximum capture of NAb bound biotin–drug in the BEAD method.

One concern for acid dissociation is that harsh acid used in the sample treatment may destroy ADA/NAb in test samples. Indeed, it was found that as high as 40% of NAb monoclonal positive controls were denatured after acid treatment (Xu et al., 2018a). There have been discussions on how to overcome drug or drug target interference for NAb assay while minimizing the loss of NAb during sample pretreatment (Wang et al., 2019; Wu et al., 2019; Xiang et al., 2019). One report utilized heating instead of acid to break apart drug/ADA complexes, taking advantage of the small size and low thermal stability of the drug so that only drug is denatured at an elevated temperature but not ADA (Xu et al., 2018b). A combination of a less harsh acid and mild heating could facilitate the dissociation of drug/ADA complexes while minimizing the denaturation of ADA/NAb in test samples (Figure 8.2).

8.5.4 Optimization of Key Assay Performance Characteristics

Once the assay platform and endpoint have been determined, the assay needs to be further optimized for suitable sensitivity and robustness. The key assay performance characteristics such as sensitivity, selectivity, and robustness should be optimized according to recommendations from regulatory guidelines and white papers (FDA, 2019; Gupta et al., 2011).

It is advisable that the sensitivity and robustness of the cell-based assays could be improved by optimizing critical factors, for example, cell number, reagent drug concentration and/or ligand concentration, incubation time, and minimal required dilution (MRD). Furthermore, the miniaturized 384-well assay format may require fewer cell numbers, and a lower concentration of drug and/or ligand and therefore would provide a better sensitivity.

The FDA guidance recommends to "choose a therapeutic protein product concentration whose activity readout is sensitive to inhibition" (FDA, 2019). It is preferable that the selected therapeutic protein concentration is in the linear range of the drug dose–response curve (Figure 8.2). However, for some assays with a broad range of assay windows and high signal-to-noise (S/N) ratio (e.g., MSD assays or cell-based assays highly responsive to drug treatment) the lower portion of the drug dose–response curve still generates an assay window adequate for robust assay performance. Also when this seemingly nonlinear region is plotted in the log scale, it could be perfectly linear (Rusnak, 2020). In this circumstance, the assay could be performed at a lower drug concentration outside the linear portion of the dose–response curve to achieve a better sensitivity without compromising the assay performance.

The NAb method development could be further facilitated by design of experimental methods. For example, in an indirect cell-based NAb assay, a reporter cell line is stimulated by its ligand which in turn is inhibited by a therapeutic drug. The drug-mediated ligand inhibition would be restored by antidrug NAb in sample matrix. The method development involves the optimization of cell number, ligand concentration that generates a suitable assay window, and drug concentration that would block the ligand activity by approximately 70%. The complicated interplay of different factors should be evaluated to optimize assay sensitivity. The traditional one-factor-a-time approach may not be effective enough to assess the trends and identify the best combination of conditions for assay improvement. To overcome this limitation, statistical design of experiments can be employed to simplify the assay optimization and evaluate multiple assay parameters simultaneously (Chen et al., 2012).

8.6 Assay Validation

Once an assay has been developed, it usually needs to be validated prior to sample testing. The parameters for a NAb assay validation generally include cutpoint, accuracy and precision, sensitivity, selectivity, and specificity. The consideration and expectation for NAb assay validation have been delineated in FDA guidance on immunogenicity testing of therapeutic protein products and several industry white papers (FDA, 2019; Gouty et al., 2017; Gupta et al., 2011), Recently, the American Association of Pharmaceutical Scientists (AAPS) Immunogenicity Working Group has been working on the harmonization of assay validation and data reporting for both ADA and NAb assays. There are some specific considerations for cell-based NAb assay method transfer that warrant further discussion.

Unlike CLB NAb assays, an analyst experienced in cell biology and cell-based assays is critical not only for cell-based assay development but also for method transfer. This is mainly due to the fact that cells are living organisms and special care and handling are needed. In addition, cell-based NAb assays usually comprise multiple steps, often beginning with a sample pretreatment such as BEAD to improve drug tolerance, followed by a cell-based assay and a final readout assay. Due to high number of projects and tight timeline, it may be challenging for a contract research organization (CRO) to have both the experienced analysts with a troubleshooting mindset and the time needed for assay optimization. Therefore, most of the major biopharmaceutics tend to develop and prevalidate a cell-based NAb assay in house and then transfer to CROs with experienced cell biologists to perform the assay methods. To ensure the success of method transfer, it is helpful to provide on-site training on assay method for the CRO analysts. The periodic site visit would also provide opportunities to observe the method execution by the analysts and assess whether suitable reagents and instruments are being used to enable seamless method transfer and partial validation.

8.7 Future Challenges and Direction: NAb Assays for New Modalities

The advancement of technology platforms for drug discovery has spawned many new treatment modalities (e.g., multispecific biotherapeutics and probodies) to offer novel solutions to combat a wide range of indications. Therefore, new assay platforms and strategies beyond traditional approaches may be needed for the assessment of NAb to these biotherapeutics with novel drug MOA.

For a multispecific biotherapeutic, NAb assays may need to detect NAbs for each functional domain. This can be achieved in one single assay or separate assays. One assay could be acceptable to assess NAb responses to the entire drug molecule. However, separate NAb assays, each specific for one functional domain, may be required to measure immune responses to different domains of a multispecific molecule as recommended by the FDA guideline (FDA, 2019). The increased workload warrants the exploration of multiplex NAb assay formats that offer an effective method that enables the detection of domain-specific NAbs in one single assay for further characterization of ADAs to multispecific biotherapeutics.

A probody is a mAb with antigen binding sites preblocked with masking peptides so that it cannot bind to its target in circulation or normal tissue. Once in the tumor microenvironment, however, high protease activity will digest the linker peptide and release the masking peptides to activate the therapeutic mAb (Autio et al., 2020). The design of probody is particularly useful when the drug target is also expressed on healthy cells. A regular mAb drug would lead to severe side effects and may also require a much higher dosing level. A probody NAb assay should detect NAbs to both the variable region of mAb and the linker peptide (Figure 8.3). The ADA assay specific to the linker peptide is important since antibodies binding to the linker peptide could prevent the protease cleavage, thus blocking the drug function and compromising drug

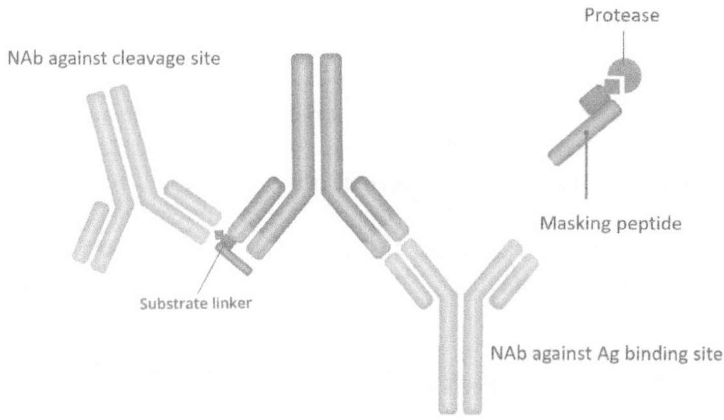

Figure 8.3 Possible neutralizing antibodies (NAbs) against probody.

efficacy (Lopez et al., 2019). While NAb against the variable region can be tested using the traditional CLB assay, NAb binding to the linker peptide could be detected with a direct binding assay (Heyduk et al., 2018). Ideally the anti-linker peptide NAb assay would detect NAbs that not only block the protease binding but also its subsequent cleavage of the linker peptide.

The aforementioned example indicates that the new therapeutic modalities with complicated drug MOA present increasing challenges for NAb method development. It calls on innovation in detection technologies to enable monitoring clinically relevant NAb responses with adequate robustness, sensitivity, and throughput.

References

Autio, K. A., Boni, V., Humphrey, R. W., & Naing, A. (2020). Probody therapeutics: An emerging class of therapies designed to enhance on-target effects with reduced off-tumor toxicity for use in immuno-oncology. *Clinical Cancer Research*, 26(5), 984–989. https://doi.org/10.1158/1078-0432.CCR-19-1457

Bourdage, J. S., Cook, C. A., Farrington, D. L., Chain, J. S., & Konrad, R. J. (2007). An Affinity Capture Elution (ACE) assay for detection of anti-drug antibody to monoclonal antibody therapeutics in the presence of high levels of drug. *Journal of Immunological Methods*, 327(1–2), 10–17. https://doi.org/10.1016/j.jim.2007.07.004

Casadevall, N., Nataf, J., Viron, B., Kolta, A., Kiladjian, -J.-J., Martin-Dupont, P., Michaud, P., Papo, T., Ugo, V., Teyssandier, I., Varet, B., & Mayeux, P. (2002). Pure red-cell aplasia and antierythropoietin antibodies in patients treated with recombinant erythropoietin. *New England Journal of Medicine*, 346(7), 469–475. https://doi.org/10.1056/NEJMoa011931

Chen, X. C., Zhou, L., Gupta, S., & Civoli, F. (2012). Implementation of design of experiments (DOE) in the development and validation of a cell-based bioassay for the detection of anti-drug neutralizing antibodies in human serum. *Journal of Immunological Methods*, 376(1–2), 32–45. https://doi.org/10.1016/j.jim.2011.11.004

Civoli, F., Kroenke, M. A., Reynhardt, K., Zhuang, Y., Kaliyaperumal, A., & Gupta, S. (2012). Development and optimization of neutralizing antibody assays to monitor clinical immunogenicity. *Bioanalysis*, 4(22), 2725–2735. https://doi.org/10.4155/bio.12.239

European Medicines Agency (EMA). (2006). Concept Paper on guideline on immunogenicity assessment of therapeutic proteins.

European Medicines Agency (EMA). (2015). Guideline on immunogenicity assessment of biotechnology-derived therapeutic protein.

European Medicines Agency (EMA). (2017). Guideline on immunogenicity assessment of therapeutic proteins.

Finco, D., Baltrukonis, D., Clements-Egan, A., Delaria, K., Gunn, G. R., 3rd, Lowe, J., Maia, M., & Wong, T. (2011). Comparison of competitive ligand-binding assay and bioassay formats for the measurement of neutralizing antibodies to protein therapeutics. *Journal of Pharmaceutical and Biomedical Analysis*, 54(2), 351–358. https://doi.org/10.1016/j.jpba.2010.08.029

Food and Drug Administration (FDA). (2009). Guidance for industry assay development for immunogenicity testing of therapeutic proteins.

Food and Drug Administration (FDA). (2016). Assay development and validation for immunogenicity testing of therapeutic protein products; revised draft guidance for industry.

Food and Drug Administration (FDA). (2019). Immunogenicity testing of therapeutic protein products – Developing and validating assays for anti-drug antibody detection.

Gorovits, B., Fiscella, M., Havert, M., Koren, E., Long, B., Milton, M., & Purushothama, S. (2020). Recommendations for the development of cell-based anti-viral vector neutralizing antibody assays. *The AAPS Journal*, 22(2), 24. https://doi.org/10.1208/s12248-019-0403-1

Gouty, D., Cai, C. C., Cai, X. Y., Kasinath, A., Kumar, V., Alvandkouhi, S., Yang, J., Pederson, S., Babbitt, B., Peritt, D., Rudy, A., Koppenburg, V., Dasilva, A., Ullmann, M., Liu, S., & Satterwhite, C. (2017). Recommendations for the development and validation of neutralizing antibody assays in support of biosimilar assessment. *The AAPS Journal*, 20(1), 25. https://doi.org/10.1208/s12248-017-0181-6

Gupta, S., Devanarayan, V., Finco, D., Gunn, G. R., 3RD, Kirshner, S., Richards, S., Rup, B., Song, A., & Subramanyam, M. (2011). Recommendations for the validation of cell-based assays used for the detection of neutralizing antibody immune responses elicited against biological therapeutics. *Journal of Pharmaceutical and Biomedical Analysis*, 55(5), 878–888. https://doi.org/10.1016/j.jpba.2011.03.038

Heyduk, E., Hickey, R., Pozzi, N., & Heyduk, T. (2018). Peptide ligand-based ELISA reagents for antibody detection. *Analytical Biochemistry*, 559, 55–61. https://doi.org/10.1016/j.ab.2018.08.012

Hu, J., Wala, I., Han, H., Nagatani, J., Barger, T., Civoli, F., Kaliyaperumal, A., Zhuang, Y., & Gupta, S. (2015). Comparison of cell-based and non-cell-based assay platforms for the detection of clinically relevant anti-drug neutralizing antibodies for immunogenicity assessment of therapeutic proteins. *Journal of Immunological Methods*, 419, 1–8. https://doi.org/10.1016/j.jim.2015.02.006

Jolicoeur, P., & Tacey, R. L. (2012). Development and validation of cell-based assays for the detection of neutralizing antibodies to drug products: A practical approach. *Bioanalysis*, 4(24), 2959–2970. https://doi.org/10.4155/bio.12.285

Kivisakk, P., Alm, G. V., Fredrikson, S., & Link, H. (2000). Neutralizing and binding anti-interferon-beta (IFN-beta) antibodies. A comparison between IFN-beta-1a and IFN-beta-1b treatment in multiple sclerosis. *European Journal of Neurology*, 7(1), 27–34. https://doi.org/10.1046/j.1468-1331.2000.00002.x

Koren, E., Smith, H. W., Shores, E., Shankar, G., Finco-kent, D., Rup, B., Barrett, Y. C., Devanarayan, V., Gorovits, B., Gupta, S., Parish, T., Quarmby, V., Moxness, M., Swanson, S. J., Taniguchi, G., Zuckerman, L. A., Stebbins, C. C., & Mire-Sluis, A. (2008). Recommendations on risk-based strategies for detection and characterization of antibodies against biotechnology products. *Journal of Immunological Methods*, 333(1–2), 1–9. https://doi.org/10.1016/j.jim.2008.01.001

Kumar, S. C., Delcarpini, J. A., Qu, Q., Kane, M., & Gorovits, B. (2017). Mitigation of pre-existing antibodies to a biotherapeutic in non-clinical species when establishing anti-drug antibody assay cutpoint. *The AAPS Journal*, 19(1), 313–319. https://doi.org/10.1208/s12248-016-0011-2

Lofgren, J. A., Wala, I., Koren, E., Swanson, S. J., & Jing, S. (2006). Detection of neutralizing anti-therapeutic protein antibodies in serum or plasma samples containing high levels of the therapeutic protein. *Journal of Immunological Methods*, 308(1–2), 101–108. https://doi.org/10.1016/j.jim.2005.10.007

Lopez, T., Mustafa, Z., Chen, C., Lee, K. B., Ramirez, A., Benitez, C., Luo, X., Ji, R. R., & Ge, X. (2019). Functional selection of protease inhibitory antibodies. *Proceedings of the National Academy of Sciences of the United States of America*, 116(33), 16314–16319. https://doi.org/10.1073/pnas.1903330116

McCutcheon, K. M., Quarmby, V., & song, A. (2010). Development and optimization of a cell-based neutralizing antibody assay using a sample pre-treatment step to eliminate serum interference. *Journal of Immunological Methods*, 358(1–2), 35–45. https://doi.org/10.1016/j.jim.2010.03.016

Mojtahed Poor, S., Ulshofer, T., Gabriel, L. A., Henke, M., Kohm, M., Behrens, F., Geisslinger, G., Parnham, M. J., Burkhardt, H., & Schiffmann, S. (2019). Immunogenicity assay development and validation for biological therapy as exemplified by ustekinumab. *Clinical and Experimental Immunology*, 196(2), 259–275. https://doi.org/10.1111/cei.13261

Myler, H., Gorovits, B., Phillips, K., Devanarayan, V., Clements-Egan, A., Gunn, G. R., Kirshner, S., Desilva, B., & Shah, V. P. (2019). Report on the AAPS immunogenicity guidance forum. *The AAPS Journal*, 21(4), 55. https://doi.org/10.1208/s12248-019-0328-8

Novick, D., Nabioullin, R. R., Ragsdale, W., McKenna, S., Weiser, W., Garone, L., Burkins, C., Kim, S. H., Rubinstein, M., Tepper, M. A., & Arulanandam, A. R. (2000). The neutralization of type I IFN biologic actions by anti-IFNAR-2 monoclonal antibodies is not entirely due to inhibition of Jak-Stat tyrosine phosphorylation. *Journal of Interferon & Cytokine Research*, 20(11), 971–982. https://doi.org/10.1089/10799900050198417

Rusnak, D. (2020). Managing sensitivity in neutralizing antibody assays. Land O'Lakes Annual Bioanalytical Conference. https://www.bioagilytix.com/resources/webinars/managing-sensitivity-in-nab-assays-when-wide-dynamic-range-makes-yes-mean-no-and-no-mean-yes

Schwickart, M., Mehrzai, F., Pearson, J., Shaghasi, N., Chavez, C., Schneider, A., Wu, S., Roskos, L., & Liang, M. (2014). Identification and elimination of target-related matrix interference in a neutralizing anti-drug antibody assay. *Journal of Immunological Methods*, 403(1–2), 52–61.https://doi.org/10.1016/j.jim.2013.11.018

Shankar, G., Pendley, C., & Stein, K. E. (2007). A risk-based bioanalytical strategy for the assessment of antibody immune responses against biological drugs. *Nature Biotechnology*, 25(5), 555–561. https://doi.org/10.1038/nbt1303

Sloan, J. H., Conway, R. G., Pottanat, T. G., Troutt, J. S., Higgs, R. E., Konrad, R. J., & Qian, Y. W. (2016). An innovative and highly drug-tolerant approach for detecting neutralizing antibodies directed to therapeutic antibodies. *Bioanalysis*, 8(20), 2157–2168. https://doi.org/10.4155/bio-2016-0161

Smith, H. W., Butterfield, A., & Sun, D. (2007). Detection of antibodies against therapeutic proteins in the presence of residual therapeutic protein using a solid-phase extraction with acid dissociation (SPEAD) sample treatment prior to ELISA. *Regulatory Toxicology and Pharmacology*, 49(3), 230–237. https://doi.org/10.1016/j.yrtph.2007.07.005

Tovey, M. G., & Lallemand, C. (2011). Immunogenicity and other problems associated with the use of biopharmaceuticals. *Therapeutic Advances in Drug Safety*, 2(3), 113–128. https://doi.org/10.1177/2042098611406318

van Schie, K. A., Kruithof, S., van Schouwenburg, P. A., Vennegoor, A., Killestein, J., Wolbink, G., & Rispens, T. (2017). Neutralizing capacity of monoclonal and polyclonal anti-natalizumab antibodies: The immune response to antibody therapeutics preferentially targets the antigen-binding site. *The Journal of Allergy and Clinical Immunology*, 139(3), 1035–1037 e6. https://doi.org/10.1016/j.jaci.2016.09.014

van Schie, K. A., Wolbink, G. J., & Rispens, T. (2015). Cross-reactive and pre-existing antibodies to therapeutic antibodies–Effects on treatment and immunogenicity. *MAbs*, 7(4), 662–671. https://doi.org/10.1080/19420862.2015.1048411

Wang, J., Lozier, J., Johnson, G., Kirshner, S., Verthelyi, D., Pariser, A., Shores, E., & Rosenberg, A. (2008). Neutralizing antibodies to therapeutic enzymes: considerations for testing, prevention and treatment. *Nature Biotechnology*, 26(8), 901–908. https://doi.org/10.1038/nbt.1484

Wang, Y., Luong, M., Guadiz, C., Zhang, M., & Gorovits, B. (2019). Addressing soluble target interference in the development of a functional assay for the detection of neutralizing antibodies against a BCMA-CD3 bispecific antibody. *Journal of Immunological Methods*, 474, 112642. https://doi.org/10.1016/j.jim.2019.112642

Wu, B., Chung, S., Jiang, X. R., McNally, J., Pedras-Vasconcelos, J., Pillutla, R., White, J. T., Xu, Y., & Gupta, S. (2016). Strategies to determine assay format for the assessment of neutralizing antibody responses to biotherapeutics. *The AAPS Journal*, 18(6), 1335–1350. https://doi.org/10.1208/s12248-016-9954-6

Wu, B., Schnarr, M., Devlin, J. L., Brown, S., & Yang, T. Y. (2019). Approaches to improve drug tolerance and target tolerance in the assessment of neutralizing anti-drug antibodies. *Bioanalysis*, 11(22), 2061–2074. https://doi.org/10.4155/bio-2019-0184

Wu, Y., Akhgar, A., Li, J. J., Yu, B., Chen, C., Lee, N., White, W. I., & Roskos, L. K. (2018). Selection of a ligand-binding neutralizing antibody assay for benralizumab: Comparison with an antibody-dependent cell-mediated cytotoxicity (ADCC) cell-based assay. *The AAPS Journal*, 20(3), 49. https://doi.org/10.1208/s12248-018-0207-8

Xiang, Y., Parng, C., Olson, K., Seletskaia, E., Gorovits, B., Jani, D., Caiazzo, T., Joyce, A., & Donley, J. (2019). Neutralizing antibody assay development with high drug and target tolerance to support clinical development of an anti-TFPI therapeutic monoclonal antibody. *The AAPS Journal*, 21(3), 46. https://doi.org/10.1208/s12248-019-0320-3

Xu, W., Cummings, J., Sank, M., Juhel, M., Li, X., Gleason, C., Desilva, B. S., Dodge, R. W., & Pillutla, R. (2018a). Development and validation of a functional cell-based neutralizing antibody assay for ipilimumab. *Bioanalysis*, 10(16), 1273–1287. https://doi.org/10.4155/bio-2018-0109

Xu, W., Sank, M., Cummings, J., Carl, S., Juhel, M., Gleason, C., Dodge, R., Desilva, B. S., Kolaitis, G., & Pillutla, R. (2018b). Bead-extraction and heat-dissociation (BEHD): A novel way to overcome drug and matrix interference in immunogenicity testing. *Journal of Immunological Methods*, 462, 34–41. https://doi.org/10.1016/j.jim.2018.08.003

Yan, Y. X., Boldt-houle, D. M., Tillotson, B. P., Gee, M. A., D'eon, B. J., Chang, X. J., Olesen, C. E., & Palmer, M. A. (2002). Cell-based high-throughput screening assay system for monitoring G protein-coupled receptor activation using beta-galactosidase enzyme complementation technology. *Journal of Biomolecular Screening*, 7(5), 451–459. https://doi.org/10.1177/108705702237677

Zoghbi, J., Xu, Y., Grabert, R., Theobald, V., & Richards, S. (2015). A breakthrough novel method to resolve the drug and target interference problem in immunogenicity assays. *Journal of Immunological Methods*, 426, 62–69. https://doi.org/10.1016/j.jim.2015.08.002

9

Critical Reagents in Bioanalysis

Yang Xu[a], Agostinho Gomes Rocha[b], Shannon Chilewski[c], Krisna C. Duong-Ly[d], Kun Yang[e] and Jonathan Haulenbeek[f]

[a]*Regulated Bioanalysis, Pharmacokinetics, Pharmacodynamics, and Drug Metabolism (PPDM), Merck & Co., Inc.*
[b]*Bioanalytical Services, Smithers*
[c]*Translational Sciences, Bristol Myers Squibb*
[d]*Regulated Bioanalysis, PPDM, Merck & Co., Inc.*
[e]*Regulated Bioanalysis, PPDM, Merck & Co., Inc. present affiliation: Bioanalytical Science, Regeneron Pharmaceuticals Inc.*
[f]*Translational Sciences, Bristol Myers Squibb, Princeton, NJ, USA; present affiliation: Bioanalysis, Early Phase, Syneos Health*

Defining Critical Reagents

Bioanalysis in support of discovery and development of biotherapeutics would not be possible without the use of what have been termed "critical reagents." While the term can be used to encompass many parts of a bioanalytical assay, this chapter is aimed at providing some guiding principles for the development, characterization, and life cycle management of biological critical reagents. For the purposes of the following discussions we will stick to well-defined and characterizable biological reagents that are used to capture and detect relevant biological species from patient samples. Excluded from this discussion will be reagents such as matrix, buffers, analytical plates, and equipment.

In bioanalytical assays, biological species (e.g., drugs, antidrug antibodies, biomarkers, etc.) are measured either directly or indirectly. These assays include pharmacokinetics, immunogenicity, and numerous pharmacodynamic assays. To be successful, they require the use of specific, selective, and well-characterized biological reagents. Often the results or readouts from the assay are deemed to be the ultimate merit of the assay. However, it is the critical reagents that define the readout and require attention as the foundation of the method development process. Obtaining these critical biological reagents is generally time consuming with lead times of up to a year or more. Once generated, critical reagents require characterization to quantify the critical quality attributes (CQAs) that can have profound effects on assay results.

Bioanalytical Aspects in Biological Therapeutics, First Edition. Edited by Xiaohui Xu and Weifeng Xu.
© 2022 John Wiley & Sons, Inc. Published 2022 by John Wiley & Sons, Inc.

A well-coordinated reagent generation and characterization plan is paramount for successful assay development and life cycle management. While the current regulatory guidance focuses on characterizing the CQAs of therapeutic proteins through the determination of physicochemical and immunochemical properties, biological activity, purity, and impurities, and so on, such guidance for critical reagent characterizations is lacking.

In this chapter we will seek to (1) lay out a comprehensive reagent strategy; (2) define the fundamentals of reagent identity; (3) suggest tools and techniques for determining reagent identity and quality; and (4) discuss key elements of a reagent development plan, including early planning for reagent generation, reagent considerations in functional assay development strategy, and reagent life cycle management recommendations. At the end, case studies will be provided for reagent characterization in assay development and troubleshooting using a tiered approach.

It is the hope of the authors that, on reading this chapter, the reader will have a better understanding of how critical reagent characterization plays a role in method development and life cycle management. Applying the learning as suggestions herein to daily routines and programs can not only decrease assay downtime but save valuable resources in drug development.

9.1 Comprehensive Reagent Strategy

A plan that incorporates reagent development and characterization as well as functional assay testing is required for successful and timely bioanalytical assay development. This plan involves the production process, the confirmation of function, the biophysical characteristics of the material (purity, concentration, form, chemical modification, etc.), and a life cycle management strategy. While the proposition of such a comprehensive plan at first appears daunting, it can be broken down into stages as shown in Table 9.1.

Table 9.1 Stages of reagent development and characterization plan.

Stages	Considerations	Examples
Identification of functional characteristics	Specificity	Binding to antigen X but not antigens Y and Z (e.g., free vs. total drug measurements); lack of cross-reactivity among closely related species, etc.
	Type of reagent	Monoclonal antibody, polyclonal antibody, aptamer, etc.

Stages	Considerations	Examples
	Intended assay platform	Enzyme-linked immunosorbent assay (ELISA), electrochemiluminescence (ECL), flow cytometry, immunohistochemisty (IHC), etc.
	Chemical modification	Labeling with biotin, ruthenium, fluorophore, chromophore, etc.
Predicted long-term needs	Projected length of use for assay	Will the reagent be used for short-term exploratory projects, lengthy clinical development programs, and/or multiple therapeutic programs?
	Commercial vs. custom	Freedom to operate, single-source vendor, production/development time
	Recombinant production	Species, expression system, molecular tags, etc.
Documentation of initial reagent source	Define identity	Source information and documentation of production
	CPF	Concentration, purity, function
Life cycle management	Recombinant production	Larger lots, lot-to-lot variability
	Formulation	Long-term stability, ease of use
	Storage	Container type, storage temperature, single use vs. bulk storage
	Testing plan	What are indicators of quality? What type of monitoring and retesting plan is needed to ensure consistent reagent quality over time?

9.2 Fundamentals of Reagent Identity

CQAs include characteristics that may affect the assay robustness and can be used as performance predictors. The need for detailed characterization of CQAs has been recognized by the industry (King et al., 2014; O'Hara et al., 2012; Staack et al., 2011).

In a recent white paper, the concept of minimal characterization was discussed by many industry leaders in varied bioanalytical disciplines. A consensus on what constitutes proper characterization was described as concentration, purity, and functional testing (CPF) (Fandozzi et al., 2019; Piccoli et al., 2019). The CPF characteristics serve as a fingerprint for a reagent and determine a reagent's identity. Concentration and purity are assessed through several available tools to be described in the next section. Functional testing involves analysis not only on activity strength but also on specificity/cross-reactivity to ensure the critical reagents can be used without interference from endogenous species in biologics matrix. This can include assay performance testing in the intended assay format but may also describe additional orthogonal biochemical testing such as binding activity and potency studies, where popular approaches include surface plasmon resonance (SPR) and functional enzymatic methods (Geist et al., 2013).

In addition, other important attributes include (1) isoelectric points of proteins, which may impact the selection of an adequate buffer formulation to mitigate stability issues (Lehermayr et al., 2011); (2) protein structure (e.g., molecular weight, sequence, and posttranslation modifications) that may impact the antibody recognition; and (3) conjugation with different molecules (e.g., biotin, ruthenium, etc.) for capture or detection, where the number of labels/protein molecule (incorporation ratio) is an important metric of a conjugated critical reagent and is particularly useful when conjugating additional batches of the same reagent to ensure assay consistency. Several approaches, including high-resolution mass spectrometry (HRMS), fluorescence, or absorbance allow an estimate of the degree of labeling (Haulenbeek & Piccoli, 2014; Vira et al., 2010). Additional insights on impact of incorporation ratio of labeled reagents to assay performance will be discussed in Section 9.4.2.2.

9.3 Tools for Determining Reagent Identity and Quality

Appropriately assessing reagent identity and quality attributes requires using a series of complementary analytical tools, depending on the purpose of investigation. In most cases, conclusions may not be possible without consideration of multiple attributes from analytical studies. This section will focus on different analytical tools for assessing reagent identity and comparability from molecular and binding characteristics, respectively.

9.3.1 Tools for Determining Molecular Characteristics

9.3.1.1 Tools for Assessing Protein Concentration and Purity

The most routine quality attributes provided in a certificate of analysis are concentration and purity (percent aggregation). One method of determining total protein involves measuring ultraviolet (UV) absorbance at a wavelength of 280 nm and applying Beer's law, which relates absorbance to the product of extinction coefficient of the given protein, concentration, and path length through which absorbance is measured. If the samples are labeled, contain UV-absorbing additives, or lack tryptophan residues, the absorbance at 280 nm will not yield an accurate measurement of protein concentration (Simonian & Smith, 2006).

In these cases, the colorimetric method initially described by Bradford is preferable (Bradford, 1976). In this method, protein samples and calibration standards are incubated with a Coomassie blue dye. After measurement of absorbance values, the total protein concentrations of the samples can be back-calculated using their absorbance aligned within the calibration curve. This method relies on careful selection of the appropriate protein standard; if the Coomassie blue dye does not interact with the test protein in the same manner as the standard, inaccurate test protein concentrations will be calculated.

A popular alternative to the Bradford assay is the bicinchoninic acid (BCA) assay. This colorimetric assay works via the Biuret reaction. Briefly, when the protein is an alkaline environment a reduction of copper ions occurs. A complex is then formed between the copper ions and the BCA which results in a purple color (Brady & MacNaughtan, 2015). As with the Bradford assay, the standard used must be carefully selected, as the different amino acid composition of the test protein compared to the standard can lead to inaccurate results. However, the BCA assay offers some advantages over the Bradford as it tends to have less variability and is compatible with detergents (Goldring, 2019).

Protein purity/impurity assessment allows for the elucidation of whether a molecule is undergoing aggregation, degradation, or has traces of impurities. Protein purity/impurity is generally measured using size-exclusion chromatography (SEC) with UV detection. SEC distinguishes different species based on their molecular weight. Large aggregates or high molecular weight (HMW) species do not enter the pores of the SEC column matrix and instead elute at the void volume of the column. In contrast, smaller protein oligomers and low molecular weight (LMW) species (e.g., degradation products, etc.) enter the pores of the SEC column matrix and elute based on size. Generally, the retention times of any species of interest can be analyzed against those of standard proteins of known size to estimate the molecular weight of the species. Although these attributes alone do not provide sufficient information for determination of the reagent identity, measuring aggregation is critical because highly aggregated reagents can affect reagent stability and assay performance. Alternative chromatographic methods such as ion-exchange chromatography (IEX), which separates different species by charge, can also be used to assess the level of impurities.

9.3.1.2 Tools for Assessing Charge Variant and Isoelectric Point

The isoelectric point (pI) of a protein indicates the pH at which the reagent carries no net electrical charge. The pI of a protein affects its solubility at a given pH, and therefore is an important piece of information to obtain to ensure adequate reagent formulation (Lehermayr et al., 2011). The pI can often be estimated based on its amino acid sequence (Bjellqvist et al., 1993, 1994) and several online servers can facilitate these calculations. These calculations, however, cannot account for the heterogeneity of a protein sample due to charge variants. Charge variants commonly occur as a result of both chemical and posttranslational modifications (PTMs) including deamidation, oxidation, glycosylation, and glycation, during reagent generation, purification, and storage (Mack & Darling, 2014). These modifications can result in species with different pIs, and potentially affect reagent solubility, binding, and/or stability.

The charge variant profile can be experimentally determined by traditional ion-exchange high-performance liquid chromatography (IEC or IEX) and/or capillary electrophoresis (CE)-based techniques, such as capillary isoelectric focusing (cIEF) and/or imaging cIEF (icIEF), which allow the characterization of the species' net charge (Andersen et al., 2004; Felten et al., 2011; Sosic et al., 2008). In icIEF, pI-based separation can be detected with whole-column detection in the capillary, which eliminates the need for a lengthy mobilization step as in cIEF. Common interference from sample matrix can be resolved by adding solubilizers, dilution, or buffer exchange, and separation resolution can be improved by addition of narrow-range ampholytes (Mack & Darling, 2014).

The major limitation associated with IEC is that it often requires intensive method development for different types of proteins, and thus it is challenging to have an IEC-based platform method to characterize all critical reagents. In contrast, the icIEF technology could be used as a universal and quick tool to provide information on protein charge heterogeneity and pI, and consequently allow assessing of protein identity, charge variants, and stability with increased throughput (Mack & Darling, 2014).

9.3.1.3 Tools for Assessing Primary Structure/Sequence

Another important attribute of a protein/antibody reagent is its primary structure. Confirmation of sequences and evaluation of PTMs, especially those in the complementarity-determining region (CDR) that could impact reagent performance, provide further evidence of reagent identity and comparability. HRMS has been recognized as an important tool for quality control of protein products. Different front-end separation instruments, such as liquid chromatography or CE coupled with MS, separate proteins from impurities and thus facilitate characterization of quality attributes.

The assessment using MS can be performed at three different levels: top-down (intact protein), middle-down (subunit analysis), or bottom-up (peptide mapping) (Giorgetti et al., 2020; Huttenhain & Hess, 2010). Selecting whether to use single-level or a combination of three levels of characterization will depend on the purpose of the investigation.

- The intact mass allows high level evaluation of total molecular weight with different glycosylation structures. Often, when reagent sequence information is the focus of investigation (e.g., assessing incorporation ratio [IR] of labeled reagents [biotin or ruthenium tags]), deglycosylation may be necessary to minimize the heterogeneity of the study reagents (see case study at the end of this chapter).
- For reagents with similar total MW, middle-down analyses could be used to investigate structural differences in different fragments/subunits of monoclonal antibodies (mAbs). Digestion with IdeS protease (an enzyme from *Streptococcus pyogenes*) can specifically cleave IgGs in the hinge region between two glycine residues at the specific amino acid sequence CPPCPAPELLG/GPSVF in the Fc chains to generate a single-chain Fc fragment (scFc, ~ 25 kDa) and a $F(ab)'_2$ fragment (~100 kDa) (An et al., 2014). Optional reduction of intrachain and interchain disulfide bonds cleaves the $F(ab)'_2$ fragment and releases the light chain (~25 kDa) and Fd fragment (~25 kDa) (An et al., 2014). Using a reversed-phase liquid chromatography (LC) coupled with HRMS, the subunit polypeptides (scFc, light chain, and Fd) with different accurate masses can be detected and analyzed after deconvolution of the spectra. Using high-resolution settings for the MS analysis, it is possible to observe isotopically resolved signals that distinguish between protein variants with very close masses.
- Peptide mapping can provide insight into specific sequence information for a reagent, including sequence variants due to PTMs and/or site-specific glycosylation. Samples are digested via a chemical or enzymatic reaction and then subjected to separation and identification of the resultant peptides. Good coverage of the peptide map, especially around the CDR region, facilitates assessment of reagent comparability and identification of important changes to reagent primary structure, including chemical modifications that may occur during production and/or storage.

9.3.2 Tools for Determining Binding Characteristics

Biosensor technologies based on SPR and bio-layer interferometry (BLI) are common tools for determining binding characteristics of molecular interactions. As complementary techniques to methods discussed in the previous section, SPR and BLI are valuable tools for assessing reagent quality. Common applications include but are not limited to ranking of candidate reagents for binding ability (affinity and binding kinetics) to a biomolecule, determination of specificity (cross-reactivity) toward one protein versus other closely related proteins, determination of competitive or noncompetitive binding between different molecules for a binding partner, and comparison of different reagent production lots (Duo et al., 2018). SPR and BLI have also been used as tools to assess conditions for acid dissociation of drug/antidrug antibody (ADA) complexes, often a necessary step in ADA and neutralizing antidrug antibody (NAb) assays (Sickert et al., 2008; Wu et al., 2019).

9.3.2.1 Surface Plasmon Resonance (SPR)

In an SPR experiment, one molecule, the ligand, is immobilized to a sensor chip and the binding of a second molecule, the analyte, is flowed over the sensor chip, which is coated with a thin layer of metal, often gold. Binding of the analyte increases the mass on the surface of the chip and alters the refractive index. This leads to a change in the resonance angle, reported as a change in the resonance units (ΔRU). The ΔRU is proportional to the mass on the surface and thus the number of analyte molecules bound to the surface. The resultant data can be fitted to either steady-state affinity or kinetics models to derive parameters such as affinity (equilibrium binding constants K_A and K_D), association rate (k_a or k_{on}), and dissociation rate (k_d or k_{off}).

While SPR provides a wealth of useful information, SPR assay development can also be time consuming and labor intensive. A typical SPR experiment can be divided into four parts: ligand immobilization, analyte interaction, ligand regeneration, and data analysis. One of the earliest decisions in studying an interaction is the choice of binding partner to immobilize as the ligand. Once the ligand has been chosen, the immobilization can be achieved through several possible chemistries including amine coupling of unlabeled reagents and streptavidin capture of biotinylated reagents, among others. Once ligand is immobilized, a dose response of analyte is flowed over the biosensor and the ΔRU as a function of time is recorded by the SPR instrument as a sensorgram. At this step, it is very important that the analyte buffer and the running buffer are well matched since differences in the buffers can result in bulk shifts in the SPR signal. Also, since analytes in biotherapeutic development frequently exhibit very slow dissociation rates, it is often advantageous to regenerate the surface between analyte injections rather than wait for complete dissociation in running buffer. Once the dose responses have been completed, commercially available SPR evaluation software packages can be used to fit the data to a binding model and derive kinetic parameters.

SPR instruments are commercialized under a variety of trade names by different companies. Examples include Biacore from Cytiva (formerly GE Healthcare Life Sciences), ProteOn XPR36 from Bio-Rad Laboratories, Molecular Affinity Screening System (MASS)-1 from Sierra Sensors GmbH and Sierra SPR-32 from Bruker, MP-SPR Navi family from BioNavis, Pioneer from Pall ForteBio, OpenSPR from Nicoya, Carterra LSA from Carterra, Inc., and so on. Features of SPR instruments that are important to consider are sensitivity, kinetic range, throughput, flexibility of experimental design, and automation.

9.3.2.2 Bio-Layer Interferometry (BLI)

Like SPR, BLI is an optical technique for measuring kinetics and affinity of biomolecular interactions (Petersen, 2017). In BLI, the ligand is immobilized on a sensor tip. Shifts in the wavelength of white light occur when tip thickness changes due to binding of analyte in solution (ForteBio, 2020; Sultana & Lee, 2015). BLI method development is analogous to that of SPR and involves immobilization of one binding partner, interaction with the second binding partner, surface regeneration, and data analysis. The primary difference between BLI and SPR is that in BLI there is no fluidic system. Instead, one binding partner is coated onto a "dip and read" biosensor and brought to a solution containing the second binding partner. The plate containing the second binding partner is then agitated to facilitate flow and the wavelength shift measured. The ForteBio Octet HTX is a popular label-free BLI-based system that measures molecular interactions in real time to provide antibody quantitation and kinetic data (ForteBio, 2020; Takkar et al., 2020). The high throughput ability of the HTX makes it ideal for use in screening and epitope binning assays (Takkar et al., 2020).

While SPR and BLI can be used to measure similar parameters, there are advantages and disadvantages of each method. Generally, SPR produces more consistent results and, in one study, was demonstrated to resolve high-affinity interactions with slow dissociation rates better than BLI (Yang et al., 2017). One possible reason for this discrepancy is that evaporation is a major factor in BLI since samples are exposed to the environment during data collection; in contrast, SPR platforms permit sealing of samples (Yang et al., 2017). Limiting experimental time can mitigate evaporation effects; however, proper analysis of slow dissociation interactions usually requires lengthy dissociation times.

BLI, however, has a few advantages over SPR. In BLI there is no need for any kind of buffer matching and thus BLI instruments are more tolerant of crude samples such as serum-containing samples as well as samples that contain glycerol, dimethyl sulfoxide (DMSO), and other common additives (Petersen, 2017). Another advantage is that many BLI instruments are configured for high-throughput sample testing in parallel, which vastly reduces total experimental time (Petersen, 2017).

Given the available analytical tools discussed in this section, a tiered approach has been proposed for reagent characterization, evaluation, and troubleshooting (Yang et al., 2020). A case study will be detailed later.

9.4 Reagent Development Plan

9.4.1 Early Planning for Reagent Generation

Critical reagent generation for biologics development is a time-consuming process. While it may be possible to incorporate some commercial "off the shelf" reagents or generic reagents such as anti-human Fc antibodies (Bautista et al., 2012), several custom reagents are often required in pharmacokinetics (PK) and ADA/immunogenicity (IMG) assays. Historically, custom reagents often result from either hybridoma technology (Parray et al., 2020; Salimi-Moosavi et al., 2015; Zaroff & Tan, 2019) or phage display (Adamson et al., 2019; Chin et al., 2015; Harth et al., 2019; Tornetta et al., 2007). Regardless of the mode, reagent generation encompasses the identification of reagent needs, immunogen design, screening to isolate reagent candidates, testing and characterization of those candidates, assessment of selectivity and other desired features, large-scale production of selected reagents, and confirmation that the final reagents produced are suitable for the bioanalytical assay(s). Depending on the platform(s) used in the bioanalytical assays and the purpose of the reagents, critical reagent generation may also involve incorporation of labels such as biotin, ruthenium, fluorescent, or radioactive tags and subsequent incorporation ratio assessment. Given these complexities, early planning is of the utmost importance to ensure timely support to good laboratory practice and clinical studies during biologics drug development.

9.4.1.1 Identification of Reagent Needs and Bioanalytical Assay Platform Considerations

Reagent needs for bioanalytical studies depend on the assay platform(s) and the purpose of the studies. In general, ligand binding assays (LBA) are considered the gold standard to support PK, IMG, and biomarker quantification for therapeutic antibodies. Typical LBA platforms include radioimmunoassay (RIA), traditional enzyme-linked immunosorbent assay (ELISA), Meso Scale Discovery (MSD), and Gyros Protein Technologies, among others (Li et al., 2011; Mikulskis et al., 2011; Mora et al., 2010; Nencini et al., 2014; Soderstrom et al., 2011; Spriggs et al., 2012). All these platforms require several critical reagents including positive controls, capture, and detection antibodies against the analytes of interest (biotherapeutic drugs). In addition, liquid chromatography coupled with mass spectrometry (LC-MS) is gaining increased attention and is being used as a complementary tool to LBA. In LC-MS, the critical reagents include capture antibodies that facilitate affinity purification and enrichment of the analytes of interest.

9.4.1.2 Anti-Idiotypic Antibodies (Anti-IDs) in Bioanalytical Studies

Regardless of the assay platform(s), critical reagents that can specifically differentiate the analyte of interest from biologic matrix is desired. For example, a generic anti-human Fc antibody can recognize the epitope in human Fc and differentiate humanized antibody drugs from endogenous IgGs in preclinical species; therefore, such generic reagents can be used as capture antibodies to support LBA in preclinical studies (Bautista et al., 2012). However, such anti-human Fc reagents cannot be used to capture humanized drugs in human serum, because the drug and endogenous human IgGs share nearly identical antibody backbones. Therefore, reagent specificity must be considered along with assay matrix and format design, often necessitating custom reagent generation.

Anti-idiotypic antibodies (anti-IDs) are a class of custom reagents extensively used in biotherapeutic development. These molecules recognize the unique paratope or CDR of an antibody, binding specifically to that antibody and distinguishing it from others with similar antibody backbones. In biologics development, anti-IDs can be selected to bind to the target/drug complex, the free drug, or the total drug (Kelley et al., 2013; Talbot et al., 2015; Thway et al., 2013). Anti-IDs against the total drug and target/drug complex are useful as capture and/or detection reagents in PK assays after labeling with appropriate tags (e.g., biotin or ruthenium, etc.) to quantitate drug concentrations in biologic matrices (e.g., serum, blood, urine, or other biologic fluid or tissue) following drug administration. In IMG assessments, anti-IDs that bind selectively to the free drug only are neutralizing and can serve as positive controls when analyzing the presence/absence of ADA due to the induced immune response toward the investigational drugs in patient samples (Talbot et al., 2015).

9.4.1.3 Immunogen Design for Anti-IDs Generation

While anti-IDs could theoretically be generated through a number of different techniques, those used in biotherapeutic development are most often generated using hybridoma technology (Parray et al., 2020; Salimi-Moosavi et al., 2015; Zaroff & Tan, 2019) or phage display (Adamson et al., 2019; Chin et al., 2015; Harth et al., 2019; Tornetta et al., 2007). For both techniques, an immunogen is required to generate anti-IDs candidates. Depending on the purpose and therapeutic modality, immunogens can include whole antibodies, antibody fragments, hormones, cytokines, or peptides. Proper immunogen design is critical to ensuring that adequate numbers of anti-IDs candidates are identified in the screening process and that these candidates are specific for the desired region of the immunogen.

Smaller biomolecules such as cytokines, peptides, and hormones may not be immunogenic on their own and thus require conjugation to a carrier protein. Commonly used carrier proteins include keyhole limpet hemocyanin (KLH) and bovine serum albumin (BSA). KLH is derived from the giant keyhole limpet *Megatura crenulata* and is ideal for mammalian hosts. BSA is derived from plasma protein of cattle. For both carrier proteins, conjugation to the biomolecule occurs in a two-step reaction where the carrier is first activated through free amine groups and subsequently conjugated to the biomolecule via cysteine residues on the biomolecule surface. As the immune response for such a conjugate is against the entire biomolecule/linker/carrier protein complex, some of the initial anti-IDs candidates may bind specifically to the linker or carrier protein instead of the biomolecule. Conjugation to a different carrier protein (e.g., ovalbumin) can be performed similarly to KLH or BSA. The resultant immunogen can be used to confirm whether anti-IDs candidates are specific for the biomolecule of interest.

Larger biotherapeutic molecules such as antibodies tend to be immunogenic without the need for carrier proteins; however, steps must be taken during immunogen design to ensure specificity against the epitope of interest. Most commonly, the desired anti-IDs bind specifically to the paratope or CDR region. In this case, an $F(ab)_2$ fragment that lacks the Fc region may be used as an immunogen. This approach enhances the screening efficiency by eliminating any candidates that would bind to the Fc region. Another approach would be to incorporate a deselection strategy (see next section) using an antibody that has a similar backbone to the biotherapeutic but a different CDR region.

The dependence on animals can be greatly reduced through the use of phage display technologies (Adamson et al., 2019; Chin et al., 2015; Harth et al., 2019; Tornetta et al., 2007). In phage display, the immunogen (bait) is immobilized to a solid support; this often occurs using a biotinylated version of the immunogen and employing capture with a streptavidin-coated support. Once immobilized, the immunogen is screened against a library containing phage particles that display a wide diversity of proteins/peptides. The high-affinity binders can be identified after multiple rounds of selection and amplification.

9.4.1.4 Screening Paradigms During Reagent Generation

Immunogen design must be carried out with consideration to the screening strategy for identification of the initial pool of anti-IDs candidates. An ideal screening strategy will heavily favor identification of anti-IDs candidates with the desired characteristics such as high specificity and affinity for the analyte(s) of interest. As mentioned previously, anti-IDs could be screened for binding to the target/drug complex, the free drug, or the total drug; depending on the goal of the study, screening strategies need to take into consideration the circulating levels of the soluble target as this target will compete with binding of potential anti-IDs candidates in the LBA (Verch et al., 2010). Other circumstances may warrant certain specificity requirements. For example, if the analyte of interest is a mutein, the anti-ID candidates must exhibit selectivity for the mutein over its wild-type cognate. Another common situation is to select anti-IDs that bind to the intact drug rather than a metabolite or degradation product.

One common method of improving specificity is to incorporate a deselection target in the anti-ID identification process. For example, in the case of anti-ID generation against a mutein target, deselection with the wild-type could be performed on any initially identified candidates. Upon deselection, only those candidates that bind selectively to the mutein should remain. In general, several rounds of selection and deselection are performed to generate the initial pool of anti-IDs candidates. Usually, ~ 20–30 candidates will advance to method development screening and biophysical characterization.

9.4.2 Reagent Considerations in Functional Assay Development Strategy

9.4.2.1 Reagent Selection as Part of Assay Development

Antibody recognition and kinetics form the fundamental basis of all LBAs. As such, the importance of antibody selection in the development of LBAs is crucial to the success and reliability of the assay (Duo et al., 2017; Rup & O'Hara, 2007). While the final antibody selection depends upon the individual assay goals, the binding affinity, epitope coverage, and selectivity characteristics of potential candidates are key considerations in this process (Mayer & Hottenstein, 2016). Therefore, molecular understanding of these critical reagents during assay development is vital to the performance, quality, and reliability of LBAs (Rup & O'Hara, 2007).

Several methods can be used for screening of reagent candidates. Conventional, immunoassay-based antibody screening methods can be labor intensive, time consuming, and costly, particularly if large antibody reagent panels are to be processed. If only a small number of candidates are remaining at this screening stage, however, functional assay testing in the intended assay format is ideal. BLI (described earlier in this chapter) is commonly used to characterize molecular binding and is a quicker alternative to screening large antibody panels for LBA assay development (Sultana & Lee, 2015). BLI or SPR experiments can also provide orthogonal measurements for assessing the suitability of the reagents (Geist et al., 2013; Rup & O'Hara, 2007; Santockyte et al., 2019; Staack et al., 2011).These orthogonal measurements can be performed using synthetic proteins or markers (Duo et al., 2018). These biophysical data can be particularly useful when troubleshooting potential assay issues that may arise. For example, SPR and BLI can be used to assess discrepancies in activity between different lots of reagents with identical physical concentrations.

9.4.2.2 Critical Attributes to Labeled/Conjugated Critical Reagents

Conjugated critical reagents are vital for the development of LBAs. Conjugation tags can include biotin, sulfo-tag, digoxigenin, fluorophores, enzymes such as alkaline phosphatase or horseradish peroxidase, and so on. For any labeled reagents, purity (percent unlabeled and/or percent aggregation) and degree of labeling (DOL) or incorporation ratio (IR) are important attributes to assess.

For biotinylated reagents, purity of critical reagents is particularly important (Rocha et al., 2019; Rup & O'Hara, 2007). The presence of unconjugated biotin in the reagent can caused the signal-to-noise ratio in the assay to drop significantly. This is likely due to the unconjugated, excess biotin material competing for space on the streptavidin surface, reducing the binding of the biotinylated reagent and thereby reducing the overall signal. Early assessment using SEC could help to identify the reagent purity (percent unlabeled and percent aggregation) after labeling and avoid the issue in LBAs.

When designing LBAs, it is important to consider the assay platform as well as the desired DOL or IR for any conjugated reagents. Many groups have reported that DOLs can affect performance in functional assays (Jani et al., 2018; Kubiak et al., 2016; Rocha et al., 2019, 2020; Yang et al., 2020). There is no well-defined answer to the question "What is the best DOL for my assay?". While "under-conjugating" a molecule might not provide enough labels/tags for the desired purpose, conversely "over-conjugating" can promote epitope masking, which can also negatively impact the assay (Rocha et al., 2019). Rather than empirically conjugating at a random DOL, systematic DOL screenings should be planned. One approach is to perform multiple small-batch conjugations at different DOLs and identify the optimal conjugated reagent DOL (Rocha et al., 2020). Subsequently, a larger reagent batch can be conjugated at this optimal DOL. Such an upfront and rational approach to identify the optimal conjugated reagent will undoubtedly accelerate assay development and improve assay robustness.

Additionally, these observations facilitate decisions regarding potential DOL modifications to improve assay performance. For example, if a high drug concentration is needed to support a toxicology study, sensitivity may not be needed and so a low DOL can be used in a chemiluminescent ELISA. However, as the drug moves forward into the clinic, greater sensitivity may be needed, thus this may be achievable by increasing the DOL of the detector.

The case studies detailed later in this chapter will seek to further promote the need for appropriate characterization of reagents to ensure high quality from the start of assay development. The impact of purity as it pertains to aggregated protein and degree of labeling in ELISA platforms has also been discussed (Haulenbeek & Beaver, 2021).

9.4.3 Reagent Life Cycle Management Recommendations

Given the importance of critical reagents to the assay performance and the lengthy drug development process, developing a life cycle management plan is essential (O'Hara et al., 2012; Pihl et al., 2018, 2019).

A critical reagent life cycle management plan is a tool that outlines the key aspects of a reagent context. Several situations related to the critical reagent might impact the course of a study, such as: lack of stability or poor performance, limited reagent availability, issues with reagent handling, large batch-to-batch variability, and so on. An unenviable scenario is that, in the middle of a pivotal study, critical reagent is found out to be nonfunctional and thus new reagents need to be made.

There are two distinct phases of a reagent life cycle: the reagent production and development and the maintenance of that reagent supply for a certain study (O'Hara & Theobald, 2013). Systematic characterization of critical reagents becomes crucial not only for better understanding of their physicochemical characteristics, but also monitoring trends in stability and anticipating shifts in reagent performance. Defining a life cycle management plan will mitigate assay troubleshooting by identifying physicochemical performance trends and avoid undesirable reagent problems, such as stability issues or insufficient stock. Such a plan includes important quality attributes and other parameters such as storage, stocks, resupply and de novo reagent preparation, qualification of different lots, and timelines.

9.4.3.1 Storage Considerations

When discussing life cycle management of critical reagents, storage-related best practices are essential to maintain long-term supply of critical reagents. Storage conditions can impact the assay performance (Kubiak et al., 2016; Rocha et al., 2019). Recommendations on storage conditions have been critically discussed by regulators and industry (EMA, 2011; FDA, 2018; Japanese Ministry of Health, Labour and Welfare, 2014; King et al., 2014).

During early phase reagent development, the adequate buffer formulation should be defined. Different proteins present different higher-order structures because of their differences in amino acid compositions, charge and hydrophobicity profiles, so therefore a "one size fits all" concept for buffer formulation is not quite applicable. TheCQAs and historical data should be considered when formulating the storage buffer. Proteins can be stabilized by adjusting salt concentration, pH, and presence of amino acids, sugars, glycerol, carrier proteins, and antimicrobial agents (Zbacnik et al., 2017).

The container type should be adequate to the context of the reagent. For example, the reagents might need to be shielded from light, stored in specific material types (plastic, glass, etc.) due to stability concerns, or need to be processed under sterility conditions. The containers/vials should be adequately labeled with unique identifiers and clear information traceable by a LIMS system. In addition, the volume of each subaliquot should be reflected upfront. For example, sample evaporation can occur in vials with very low volumes, but if the molecule is not stable to multiple freeze–thaw cycles, smaller volume subaliquots might be necessary.

To prolong the shelf time of a critical reagent, and increase long-term stability, a suitable storage temperature should be used (Pihl et al., 2019). Recommended storage temperatures and expiration dates for critical reagents have been proposed (O'Hara & Theobald, 2013). The quality attributes previously discussed can be affected during storage periods and might be predictors of stability issues (Kueltzo et al., 2008; Mahler et al., 2009; Zheng & Janis, 2006). However, functional testing or assay performance should ultimately determine the reagent stability and dictate the expiration dates.

Additional attention should be given to the storage unit itself, to understand if adequate temperature monitoring systems and backup units exist. In addition, when required, shipping and transportation details should be considered (Murphy et al., 2013).

9.4.3.2 Lot and Batch Qualification

When conjugating a new lot of capture and detection reagent, it is crucial to understand the DOL and aggregation levels. One must compare the new lot of material to the old lot, as changes can shift the sensitivity and assay range resulting in further assay development to successfully bridge the two lots of reagent. If the DOL and aggregation is understood from the beginning, new lots can be generated with the same characteristics, which will lead to easier bridging of new lots into validated assays, resulting in less work and better assay performance.

9.4.3.3 Stability Considerations

The relationship between observed physicochemical changes in a critical reagent and the stability of critical reagents can be confounded by assay variables that are not part of testing plans. The current practice is to assess the suitability of a reagent by performance in assay facilitating extension of retest dates based on continued acceptable assay performance. In practice, this continues to be an appropriate manner to handle this retest/stability question. The degree to which a critical reagent undergoes physicochemical changes before it will impact the performance of a method is dependent on the type of reagent/labeling being utilized and the assay conditions. Systematic explorations of these questions should be undertaken, with the goal of setting specific criteria. However, to avoid the potential impact of unknown changes, we recommend the appropriate handling of critical reagents such as the use of single-use aliquots and the avoidance of repeated freeze–thaw cycles and when possible the creation of large lots to avoid lot-to-lot changes over time.

Looking forward, critical reagent characterization outside the framework of the bioanalytical assay is necessary to determine the appropriate quality attributes to be measured for stability. As demonstrated by the case studies in the next section, key quality attributes of the reagents (DOL and percent HMW species) have profound consequences on the final assay performance. Poor characterization of the initial quality of the reagents makes it hard to identify and correct performance issues during the life cycle of an assay. Implementing a testing plan and identifying key quality attributes of the reagents can greatly reduce assay downtime. Decisions regarding storage buffers, containers, or temperatures should only be made once key attributes have been determined or empirically determined in the assay. Knowing the quality attributes leads to a testing plan that is meaningful to the assay performance.

9.5 Tiered Characterization Approach for Troubleshooting with Case Studies of Critical Reagents in Clinical PK and ADA Method Development

In a recent publication, a tiered approach was proposed to implement appropriate characterization tools for the evaluation of protein-based critical reagents to ensure the consistency of bioanalytical assays in support of clinical studies (Yang et al., 2020). The approach contains three tiers based on outcomes of reagent characterization. Case studies employing this tiered approach will be summarized here and have been discussed more extensively elsewhere (Yang et al., 2020).

Tier 1: SEC, IEX, and icIEF

Case Study #1

In this case, a pair of hybridoma anti-IDs were selected as capture and detection reagents in a PK assay using the MSD platform. After production as recombinant proteins in large-scale, both small-scale and large-scale anti-IDs were subjected to routine analytical characterization including SDS-PAGE, analytical SEC,and binding ELISA. The small-scale and large-scale lots showed comparable purity (>95%) and EC_{50}; however, the assay using the large-scale anti-IDs showed unacceptable soluble target interference (Table 9.2).

To investigate the assay performance differences, further analytical characterization was performed to compare the small-scale and large-scale anti-IDs. Analytical SEC demonstrated that the purity of each anti-IDs was above 95% (Figure 9.1A, 9.1B), suggesting no significant differences in percent aggregation. Notably, a 10-second difference in the retention times of the small-scale and large-scale capture anti-IDs was observed (Figure 9.1A) whereas the retention times of the two batches of detection anti-IDs were the same (Figure 9.1B). Further investigation using cation exchange chromatography (CEX) revealed that the small-scale and

Table 9.2 Assay performance differences between the small-scale and large-scale anti-idiotypic antibodies (anti-IDs).

Parameter	Small-scale anti-IDs	Large-scale anti-IDs
ECL count (a) of 10 ng/mL mAb drug in human serum without target	16,004	66,766
ECL count (b) of 10 ng/mL mAb drug in human serum with 25 μg/mL soluble target	16,312	6148
Percent (%) inhibition of soluble target*	-1.9	90.8

* Percent (%) inhibition = [(a–b)/a] *100%.

Abbreviations: anti-IDs, anti-idiotypic antibodies; ECL, electrochemiluminescence; mAb, monoclonal antibody.

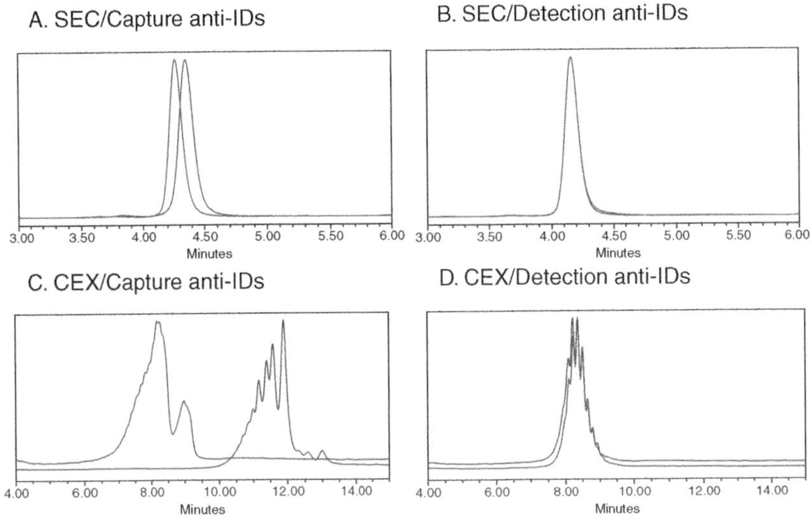

Figure 9.1 (A) and (B) Overlaid size-exclusion chromatography (SEC) chromatograms of small-scale (in red) and large-scale (in blue) reagents. (C) and (D) Overlaid cation exchange chromatography (CEX) chromatograms of small-scale (in red) and large-scale (in blue) reagents, for capture and detection anti-idiotypic antibodies (anti-IDs), respectively. *Reprinted with permission from ACS Pharmacol Transl Sci.* (2020). *28*(3(6)), 1310–1317. *Copyright (2020) American Chemical Society.*

the large-scale capture anti-IDs had different CEX chromatograms (Figure 9.1C). LC-MS studies also showed that the two capture anti-IDs had different molecular weights. In contrast, the two detection anti-IDs had similar charge profiles (Figure 9.1D) and the same molecular weight.

Taken together, these results suggest that the two batches of capture anti-IDs are distinct from each other. A new lot of large-scale capture anti-ID was regenerated and proved to exhibit similar analytical and performance characteristics as the initial small-scale capture anti-ID. This case study demonstrated that SEC and CEX can be used to identify lot-to-lot reagent differences while MS can be used to confirm those differences.

Case Study #2

This next case study demonstrates the broad utility of icIEF for reagent characterization. Typically, in therapeutic protein characterization, icIEF methods are extensively optimized; however, a generic icIEF method with minor alterations is often adequate for critical reagent characterization. The icIEF charge electropherograms of four different antibodies (with distinct amino acid sequences) used in clinical PK/ADA/NAb assays were generated using a generic icIEF pH 3–10 method (Figure 9.2A–D). Addition of 2 M urea in the master mix of the generic icIEF pH 3–10 method significantly improved the resolution of mAb1 (Figure 9.2E), but not for mAb2, mAb3, and mAb4 (Figure 9.2F–H). These data show that

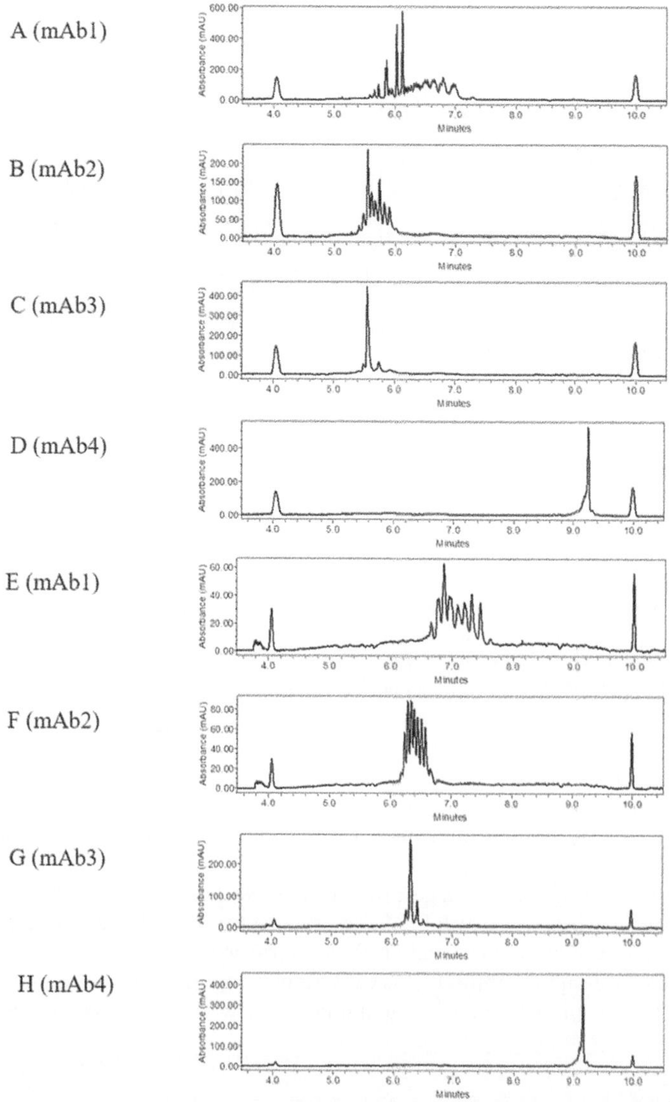

Figure 9.2 Representative charge profile electropherograms obtained from four different antibodies by using a generic imaging capillary isoelectric focusing (icIEF) method (A–D) and an optimized method in the presence of 2 M urea (E–H). *Adapted from ACS Pharmacol Transl Sci.* (2020). *28*(3(6)), 1310–1317.

a generic icIEF with minimum method development can distinguish mAbs with different amino acid sequences.

The critical reagent production process (hybridoma vs. recombinant, cell culture conditions and purification procedures, etc.) may change from batch to batch, which could impact PTM with altered critical reagent charge profile, and thus result in different charge electropherograms. icIEF can be used as one of several methods to identify batch-to-batch inconsistences due to production changes. It is important to note that if the modified residues are not involved in the epitope/paratope interactions, these alterations may not affect the performance of LBA assays, and thus there is no need to fail the batch of the reagents unless assay performance differences are observed.

Tier 2: BLI/SPR

Case Study #3

This case study illustrates the use of BLI to assess the affinity of positive control (PC) reagents during ADA method development. The 2019 US Food and Drug Administration (FDA) guidance on immunogenicity recommends using a polyclonal antibody (pAb) PC to assess assay sensitivity, while the low PC level should be at least 100 ng/mL (Plotkin, 2010; Zhou et al., 2013) with adequate drug tolerability. For one biotherapeutic program, this goal was not achieved with a pAb PC (Table 9.3) despite extensive assay development efforts, while the signal was too low for pAb PC at 100 ng/mL in the presence of 10 μg/mL drug in comparison to negative control. However, when the pAb PC was substituted with a mixture of three mAbs identified through anti-IDs screening, the sensitivity of this ADA assay increased five-fold (Table 9.3).

Using the Tier 1 analytical SEC characterization method, aggregation in the pAb PC was eliminated as a possible cause of this difference. A Tier 2 assessment using BLI demonstrated that the binding affinity to the drug was

Table 9.3 Different antidrug antibody sensitivities and drug tolerability using a polyclonal antibody (pAb) or monoclonal antibody (mAb) as a positive control (PC).

	Electrochemiluminescence (ECL) signal		
PC tested	**Negative control**	**PC at 100 ng/mL plus 10 μg/mL drug**	**PC at 100 ng/mL plus 100 μg/mL drug**
pAb PC	100	127	102
mAb PC	100	641	254

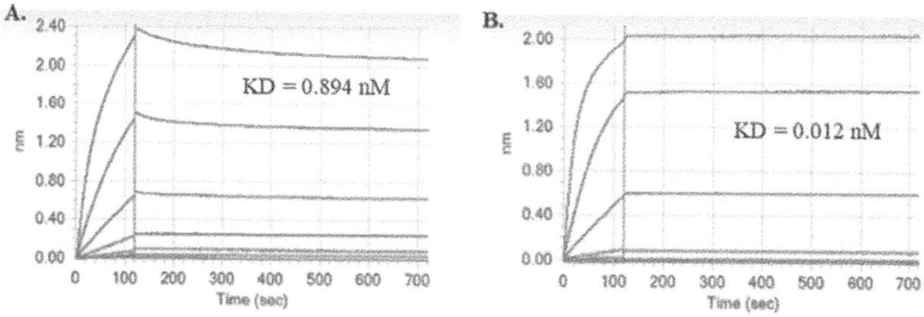

Figure 9.3 Representative affinity sensorgrams for binding between a therapeutic monoclonal antibody (mAb) drug and (A) polyclonal antibody (pAb) positive control (PC) or (B) mAb PC, respectively. *Reprinted with permission from ACS Pharmacol Transl Sci.* (2020). *28*(3(6)), 1310–1317. *Copyright (2020) American Chemical Society.*

lower for the pAb PC compared to that for the mAb PC (Figure 9.3A vs. 9.3B), which suggested that the affinity of the PC to the therapeutic drug (used as capture/detection reagents) was a main driver of sensitivity in this ADA assay. This case study illustrates the utility of binding assessments for troubleshooting in the LBA-based method development.

Tier 3: Peptide mapping by HRMS

Finally, in situations where Tier 1 and Tier 2 approaches are insufficient to determine the root cause of assay performance differences, peptide mapping by HRMS can be implemented as a Tier 3 approach to elucidate modifications at the amino acid level. To demonstrate such an approach, 100 μg of reduced and deglycosylated mAb was digested by trypsin and LysC, injected onto a ultra-performance liquid chromatography (UPLC) system coupled to an HRMS system, and subjected to analysis (Figure 9.4A, 9.4B). Peptide mapping revealed several PTMs, including an oxidation at heavy chain M254 (demonstrated by a 16 Da mass difference in b4 and y5 ions between unmodified vs. oxidized peptide). Further analysis revealed that only 1.52% of the mAb underwent oxidation at this residue, demonstrating the power of a Tier 3 approach to identify subtle reagent changes during troubleshooting.

9.6 Summary

In this chapter, the authors have laid out a comprehensive reagent strategy and defined the fundamentals of reagent identity that are critical to bioanalytical assay performance. The techniques for determination of reagent identity and

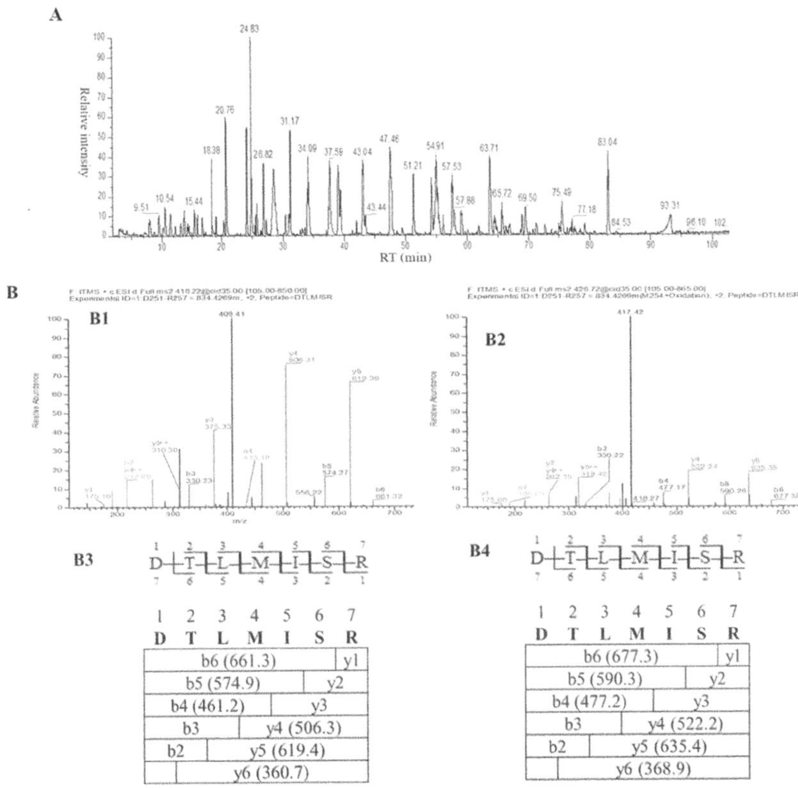

Figure 9.4 (A) Representative total ion chromatogram (TIC) of liquid chromatography-mass spectrometry (LC-MS) peptide mapping of a monoclonal antibody (mAb) after deglycosylation, reduction, and trypsin/Lys-C digestion. (B) MS/MS spectra (B1 and B2) and assigned fragment ions of the tryptic peptide DTLMISR (B3 and B4) for the unmodified peptide (B1/B3) and oxidized peptide (B2/B4), respectively. Oxidation at M254 (corresponding to position 4 in the peptide above) was identified. *Reprinted with permission from ACS Pharmacol Transl Sci.* (2020). *28*(3(6)), 1310–1317. *Copyright (2020) American Chemical Society.*

quality are suggested. The principles and considerations behind applying these tools are discussed. In addition, key elements of a reagent development plan, including reagent generation, reagent selection, and optimization during method development, and consideration for reagents' life cycle management are elaborated. Case studies for reagent characterization in assay development and troubleshooting are provided. We hope the contents described here are helpful for understanding the important aspects of reagent generation, characterization, and management, and thus beneficial to regulated bioanalysis in support of PK, ADA, and NAb assessments during drug development of biotherapeutics.

References

Adamson, H., Nicholl, A., Tiede, C., Tang, A. A., Davidson, A., Curd, H., Wignall, A., Ford, R., Nuttall, J., Mcpherson, M. J., Johnson, M., & Tomlinson, D. C. (2019). Affimers as anti-idiotypic affinity reagents for pharmacokinetic analysis of biotherapeutics. *Biotechniques*, 67(6), 261–269. doi:10.2144/btn-2019-0089

An, Y., Zhang, Y., Mueller, H. M., Shameem, M., & Chen, X. (2014). A new tool for monoclonal antibody analysis: Application of IdeS proteolysis in IgG domain-specific characterization. *MAbs*, 6(4), 879–893. doi:10.4161/mabs.28762

Andersen, T., Pepaj, M., Trones, R., Lundanes, E., & Greibrokk, T. (2004). Isoelectric point separation of proteins by capillary pH-gradient ion-exchange chromatography. *Journal of Chromatography A*, 1025(2), 217–226. doi:10.1016/j.chroma.2003.10.093

Bautista, A. C., Salimi-Moosavi, H., & Jawa, V. (2012). Universal immunoassay applied during early development of large molecules to understand impact of immunogenicity on biotherapeutic exposure. *The AAPS Journal*, 14(4), 843–849. doi:10.1208/s12248-012-9403-0

Bjellqvist, B., Basse, B., Olsen, E., & Celis, J. E. (1994). Reference points for comparisons of two-dimensional maps of proteins from different human cell types defined in a pH scale where isoelectric points correlate with polypeptide compositions. *Electrophoresis*, 15(3–4), 529–539. doi:10.1002/elps.1150150171

Bjellqvist, B., Hughes, G. J., Pasquali, C., Paquet, N., Ravier, F., Sanchez, J. C., Frutiger, S., & Hochstrasser, D. (1993). The focusing positions of polypeptides in immobilized pH gradients can be predicted from their amino acid sequences. *Electrophoresis*, 14(10), 1023–1031. doi:10.1002/elps.11501401163

Bradford, M. M. (1976). A rapid and sensitive method for the quantitation of microgram quantities of protein utilizing the principle of protein-dye binding. *Analytical Biochemistry*, 72(1–2), 248–254. doi:10.1006/abio.1976.9999

Brady, P. N., & Macnaughtan, M. A. (2015). Evaluation of colorimetric assays for analyzing reductively methylated proteins: Biases and mechanistic insights. *Analytical Biochemistry*, 491, 43–51. doi:10.1016/j.ab.2015.08.027

Chin, S. E., Ferraro, F., Groves, M., Liang, M., Vaughan, T. J., & Dobson, C. L. (2015). Isolation of high-affinity, neutralizing anti-idiotype antibodies by phage and ribosome display for application in immunogenicity and pharmacokinetic analyses. *Journal of Immunological Methods*, 416, 49–58. doi:10.1016/j.jim.2014.10.013

Duo, J., Bruno, J., Kozhich, A., David-Brown, D., Luo, L., Kwok, S., Santockyte, R., Haulenbeek, J., Liu, R., Hamuro, L., Peterson, J. E., Piccoli, S., Desilva, B., Pillutla, R., & Zhang, Y. J. (2018). Surface plasmon resonance as a tool for ligand-binding assay reagent characterization in bioanalysis of biotherapeutics. *Bioanalysis*, 10(8), 559–576. doi:10.4155/bio-2017-0271

Duo, J., Bruno, J., Piccoli, S., Desilva, B., & Zhang, Y. J. (2017). Characterization of labeled reagents in ligand-binding assays by a surface plasmon resonance biosensor. *Bioanalysis*, 9(2), 193–207. doi:10.4155/bio-2016-0204

European Medicines Agency, Committee for Medicinal Products for Human Use (CHMP). (2011). EMEA/CHMP/EWP/192217/2009 Guideline on Bioanalytical Method Validation. https://www.ema.europa.eu/en/documents/scientific-guideline/guideline-bioanalytical-method-validation_en.pdf

Fandozzi, C., Evans, C., Wilson, A., Su, D., Anderson, M., Clausen, V., Dillen, L., Garofolo, F., Holliman, C., Nickbarg, E., Olah, T., Ramanathan, R., Zhang, H., Kaur, S., Pillutla, R., Yu, H., Bateman, K., Donato, L. D., Hengel, S., Jian, W., ... Zeng, J. (2019). 2019 White paper on recent issues in bioanalysis: chromatographic assays (Part 1 – Innovation in small molecules and oligonucleotides & mass spectrometric method development strategies for large molecule bioanalysis). *Bioanalysis*, 11(22), 2029–2048. doi:10.4155/bio-2019-0260

Felten, C., Salas-Solano, O., & Michels, D. A. (2011). Imaged capillary isoelectric focusing for charge-variant analysis of biopharmaceuticals. *BioProcess International*, 9(10), 48–54. https://bioprocessintl.com/download-PDF/?article=33290&post=imaged-capillary-isoelectric-focusing-for-charge-variant-analysis-of-biopharmaceuticals-323451&spon=false&partner=&login=false

Food and Drug Administration. (2018). Guidance for Industry, Bioanalytical Method Validation. https://www.fda.gov/media/70858/download

ForteBio. (2020). *BLI Technology* [Online]. Available: https://www.fortebio.com/applications/bli-technology (Accessed 14 May 2020)

Geist, B. J., Egan, A. C., Yang, T. Y., Dong, Y., & Shankar, G. (2013). Characterization of critical reagents in ligand-binding assays: Enabling robust bioanalytical methods and lifecycle management. *Bioanalysis*, 5(2), 227–244. doi:10.4155/bio.12.304

Giorgetti, J., Beck, A., Leize-wagner, E., & François, Y.-N. (2020). Combination of intact, middle-up and bottom-up levels to characterize 7 therapeutic monoclonal antibodies by capillary electrophoresis – Mass spectrometry. *Journal of Pharmaceutical and Biomedical Analysis*, 182, 113107. doi:10.1016/j.jpba.2020.113107

Goldring, J. P. D. (2019). Measuring protein concentration with absorbance, Lowry, Bradford Coomassie Blue, or the Smith bicinchoninic acid assay before electrophoresis. In B. T. Kurien & R. H. Scofield (Eds.), Electrophoretic separation of proteins: Methods and protocols. Springer New York.

Harth, S., Ten Haaf, A., Loew, C., Frisch, C., & Knappik, A. (2019). Generation by phage display and characterization of drug-target complex-specific antibodies for pharmacokinetic analysis of biotherapeutics. *MAbs*, 11(1), 178–190. doi:10.1080/19420862.2018.1538723

Haulenbeek, J., & Beaver, C. J. (2021). The impact of ligand binding based assays critical reagent characterization and storage. *Bioanalysis*, 13(10), 797–805. doi:10.4155/bio-2020-0288

Haulenbeek, J., & Piccoli, S. P. (2014). Conjugated critical reagent characterization for ligand-binding assays: Using MALDI-TOF-MS as an orthogonal tool to assess assay performance. *Bioanalysis*, 6(7), 983–992. doi:10.4155/bio.14.65

Huttenhain, R., & Hess, S. (2010). A combined top-down and bottom-up MS approach for the characterization of hemoglobin variants in Rhesus monkeys. *Proteomics*, 10(20), 3657–3668. doi:10.1002/pmic.201000161

Jani, D., Nowak, J., Chen, Y., Boni, J., & Gorovits, B. (2018). Assessment of clinical immunogenicity of inotuzumab ozogamicin in patients with non-Hodgkin lymphoma and acute lymphoblastic leukemia. *AAPS Open*, 4(1), 1. doi:10.1186/s41120-018-0021-5

Japanese Ministry of Health, Labor and Welfare. (2014). Guideline on Bioanalytical Method (Ligand Binding Assay) Validation in Pharmaceutical Development. Ministry of Health, Labour and Welfare. http://www.nihs.go.jp/drug/BMV/260530_LBA-GL_E.pdf

Kelley, M., Ahene, A. B., Gorovits, B., Kamerud, J., King, L. E., McIntosh, T., & Yang, J. (2013). Theoretical considerations and practical approaches to address the effect of anti-drug antibody (ADA) on quantification of biotherapeutics in circulation. *The AAPS Journal*, 15(3), 646–658. doi:10.1208/s12248-013-9468-4

King, L. E., Farley, E., Imazato, M., Keefe, J., Khan, M., Ma, M., Pihl, K. S., & Sriraman, P. (2014). Ligand binding assay critical reagents and their stability: Recommendations and best practices from the Global Bioanalysis Consortium Harmonization Team. *The AAPS Journal*, 16(3), 504–515. doi:10.1208/s12248-014-9583-x

Kubiak, R. J., Lee, N., Zhu, Y., Franch, W. R., Levitskaya, S. V., Krishnan, S. R., Abraham, V., Akufongwe, P. F., Larkin, C. J., & White, W. I. (2016). Storage conditions of conjugated reagents can impact results of immunogenicity assays. *Journal of Immunology Research*, 2016, 1485615. doi:10.1155/2016/1485615

Kueltzo, L. A., Wang, W., Randolph, T. W., & Carpenter, J. F. (2008). Effects of solution conditions, processing parameters, and container materials on aggregation of a monoclonal antibody during freeze-thawing. *Journal of Pharmaceutical Sciences*, 97(5), 1801–1812. doi:10.1002/jps.21110

Lehermayr, C., Mahler, H. C., Mader, K., & Fischer, S. (2011). Assessment of net charge and protein-protein interactions of different monoclonal antibodies. *Journal of Pharmaceutical Sciences*, 100(7), 2551–2562. doi:10.1002/jps.22506

Li, J., Schantz, A., Schwegler, M., & Shankar, G. (2011). Detection of low-affinity anti-drug antibodies and improved drug tolerance in immunogenicity testing by Octet(®) biolayer interferometry. *Journal of Pharmaceutical and Biomedical Analysis*, 54(2), 286–294. doi:10.1016/j.jpba.2010.08.022

Mack, S., & Darling, S. (2014). Mastering charge heterogeneity analysis of therapeutic proteins. *Genetic Engineering & Biotechnology News*, 34(7). https://www.genengnews.com/magazine/221/mastering-charge-heterogeneity-analysis-of-therapeutic-proteins

Mahler, H. C., Friess, W., Grauschopf, U., & Kiese, S. (2009). Protein aggregation: Pathways, induction factors and analysis. *Journal of Pharmaceutical Sciences*, 98(9), 2909–2934. doi:10.1002/jps.21566

Mayer, A. P., & Hottenstein, C. S. (2016). Ligand-binding assay development: What do you want to measure versus what you are measuring? *The AAPS Journal*, 18(2), 287–289. doi:10.1208/s12248-015-9855-0

Mikulskis, A., Yeung, D., Subramanyam, M., & Amaravadi, L. (2011). Solution ELISA as a platform of choice for development of robust, drug tolerant immunogenicity assays in support of drug development. *Journal of Immunological Methods*, 365(1–2), 38–49. doi:10.1016/j.jim.2010.11.011

Mora, J. R., Obenauer-Kutner, L., & Patel, V. I. M. A. L. (2010). Application of the Gyrolab platform to ligand-binding assays: A user's perspective. *Bioanalysis*, 2(10), 1711–1715. doi:10.4155/bio.10.122

Murphy, B. M., Swarts, S., Mueller, B. M., Van Der Geer, P., Manning, M. C., & Fitchmun, M. I. (2013). Protein instability following transport or storage on dry ice. *Nature Methods*, 10(4), 278–279. doi:10.1038/nmeth.2409

Nencini, F., Pratesi, S., Petroni, G., Matucci, A., Maggi, E., & Vultaggio, A. (2014). Assays and strategies for immunogenicity assessment of biological agents. *Drug Development Research*, 75(S1), S4–6. doi:10.1002/ddr.21184

O'Hara, D. M., & Theobald, V. (2013). Life cycle management of critical ligand-binding reagents. *Bioanalysis*, 5(21), 2679–2696. doi:10.4155/bio.13.241

O'Hara, D. M., Theobald, V., Egan, A. C., Usansky, J., Krishna, M., Terwee, J., Maia, M., Spriggs, F. P., Kenney, J., Safavi, A., & Keefe, J. (2012). Ligand binding assays in the 21st century laboratory: Recommendations for characterization and supply of critical reagents. *The AAPS Journal*, 14(2), 316–328. doi:10.1208/s12248-012-9334-9

Parray, H. A., Shukla, S., Samal, S., Shrivastava, T., Ahmed, S., Sharma, C., & Kumar, R. (2020). Hybridoma technology a versatile method for isolation of monoclonal antibodies, its applicability across species, limitations, advancement and future perspectives. *International Immunopharmacology*, 85, 106639. doi:10.1016/j.intimp.2020.106639

Petersen, R. L. (2017). Strategies using bio-layer interferometry biosensor technology for vaccine research and development. *Biosensors (Basel)*, 7(4). doi:10.3390/bios7040049

Piccoli, S., Mehta, D., Vitaliti, A., Allinson, J., Amur, S., Eck, S., Green, C., Hedrick, M., Hopper, S., Ji, A., Joyce, A., Litwin, V., Maher, K., Mathews, J., Peng, K., Safavi, A., Wang, Y. M., Zhang, Y., Amaravadi, L., Palackal, N., ... Yang, T. Y. (2019). 2019 White paper on recent issues in bioanalysis: FDA Immunogenicity guidance, gene therapy, critical reagents, biomarkers and flow cytometry validation (Part 3 – Recommendations on 2019 FDA immunogenicity guidance, gene therapy bioanalytical challenges, strategies for critical reagent management, biomarker assay validation, flow cytometry validation & CLSI H62). *Bioanalysis*, 11(24), 2207–2244. doi:10.4155/bio-2019-0271

Pihl, S., Van Der Strate, B. W., Golob, M., Ryding, J., Vermet, L., Jaitner, B., Goodman, J., & Timmerman, P. (2019). EBF recommendation on practical management of critical reagents for antidrug antibody ligand-binding assays. *Bioanalysis*, 11(19), 1787–1798. doi:10.4155/bio-2019-0248

Pihl, S., Van Der Strate, B. W., Golob, M., Vermet, L., Jaitner, B., Goodman, J., Fjording, M. S., & Timmerman, P. (2018). EBF recommendation on practical management of critical reagents for PK ligand-binding assays. *Bioanalysis*, 10(19), 1557–1565. doi:10.4155/bio-2018-0230

Plotkin, S. A. (2010). Correlates of protection induced by vaccination. *Clinical and Vaccine Immunology*, 17(7), 1055–1065. doi:10.1128/cvi.00131-10

Rocha, A. G., Krynski, K., Mancino, A., Sciscione, M., & Beaver, C. J. (2019). New insights on critical reagent optimization for antidrug antibody assays. *Bioanalysis*, 11(9), 815–823. doi:10.4155/bio-2019-0040

Rocha, A. G., Tufano, E., Krynski, K., & Beaver, C. J. (2020). Purification of small batches of biotinylated antibodies by immobilized metal affinity chromatography for ligand binding method development. *Biomedical Journal of Scientific & Technical Research*, 27(3), 20760–20763. doi:10.26717/BJSTR.2020.27.004498

Rup, B., & O'Hara, D. (2007). Critical ligand binding reagent preparation/selection: When specificity depends on reagents. *The AAPS Journal*, 9(2), E148–55. doi:10.1208/aapsj0902016

Salimi-Moosavi, H., Winters, A., Abbott, C., Patel, J., Hager, T., Patel, V., Shih, J., Zhuang, Y., & Ma, M. (2015). A multifactorial screening strategy to identify anti-idiotypic reagents for bioanalytical support of antibody therapeutics. *Analytical Biochemistry*, 470, 52–60. doi:10.1016/j.ab.2014.10.007

Santockyte, R., Zeng, J., & Zheng, N. (2019). Critical reagent screening and characterization: Benefits and approaches for protein biomarker assays by hybrid LC-MS. *Bioanalysis*, 11(8), 785–795. doi:10.4155/bio-2018-0277

Sickert, D., Kroeger, K., Zickler, C., Chokote, E., Winkler, B., Grenet, J. M., Legay, F., & Zaar, A. (2008). Improvement of drug tolerance in immunogenicity testing by acid treatment on Biacore. *Journal of Immunological Methods*, 334(1–2), 29–36. doi:10.1016/j.jim.2008.01.010

Simonian, M. H., & Smith, J. A. (2006). Spectrophotometric and colorimetric determination of protein concentration. *Current Protocols in Molecular Biology*. Chapter 10, Unit 10 1A. doi:10.1002/0471142727.mb1001as76

Soderstrom, C. I., Spriggs, F. P., Song, W., & Burrell, S. (2011). Comparison of four distinct detection platforms using multiple ligand binding assay formats. *Journal of Immunological Methods*, 371(1–2), 106–113. doi:10.1016/j.jim.2011.06.020

Sosic, Z., Houde, D., Blum, A., Carlage, T., & Lyubarskaya, Y. (2008). Application of imaging capillary IEF for characterization and quantitative analysis of recombinant protein charge heterogeneity. *Electrophoresis*, 29(21), 4368–4376. doi:10.1002/elps.200800157

Spriggs, F. P., Zhong, Z. D., Safavi, A., Jani, D., Dontha, N., Kant, A., Ly, J., Brilando, L., Osterlund, K., Rouleau, N., Fischer, S. K., Boissonneault, M., & Ray, C. (2012). Ligand binding assays in the 21st century laboratory: Platforms. *The AAPS Journal*, 14(1), 113–118. doi:10.1208/s12248-012-9321-1

Staack, R. F., Stracke, J. O., Stubenrauch, K., Vogel, R., Schleypen, J., & Papadimitriou, A. (2011). Quality requirements for critical assay reagents used in bioanalysis of therapeutic proteins: What bioanalysts should know about their reagents. *Bioanalysis*, 3(5), 523–534. doi:10.4155/bio.11.16

Sultana, A., & Lee, J. E. (2015). Measuring protein-protein and protein-nucleic Acid interactions by biolayer interferometry. *Current Protocols in Protein Science*, 79, 19 25 1–19 25 26. doi:10.1002/0471140864.ps1925s79

Takkar, R., Yadav, A., Kumaraswamy, S., Liao-Chan, S., & Theunissen, J. W. 2020. *Cross-competition or epitope binning assays on the Octet HTX system*. FortéBio application note. [Online]. https://www.fortebio.com/sites/default/files/en/assets/app-note/cross-competition-or-epitope-binning-assays-on-octet-htx-system.pdf (Accessed)

Talbot, J. J., Calamba, D., Pai, M., Ma, M., & Thway, T. M. (2015). Measurement of free versus total therapeutic monoclonal antibody in pharmacokinetic assessment is modulated by affinity, incubation time, and bioanalytical platform. *The AAPS Journal*, 17(6), 1446–1454. doi:10.1208/s12248-015-9807-8

Thway, T. M., Magana, I., Bautista, A., Jawa, V., Gu, W., & Ma, M. (2013). Impact of anti-drug antibodies in preclinical pharmacokinetic assessment. *The AAPS Journal*, 15(3),856–863. doi:10.1208/s12248-013-9484-4

Tornetta, M., Fisher, D., O'Neil, K., Geng, D., Schantz, A., Brigham-Burke, M., Lombardo, D., Fink, D., Knight, D., Sweet, R., & Tsui, P. (2007). Isolation of human anti-idiotypic antibodies by phage display for clinical immune response assays. *Journal of Immunological Methods*, 328(1–2), 34–44. doi:10.1016/j.jim.2007.08.008

Verch, T., Chilewski, S., Bouaraphan, S., Yarovoi, H., Yin, K. C., Chen, D., & Washabaugh, M. W. (2010). Pharmacokinetic immunoassay methods in the presence of soluble target. *Journal of Immunological Methods*, 361(1–2), 75–81. doi:10.1016/j.jim.2010.07.014

Vira, S., Mekhedov, E., Humphrey, G., & Blank, P. S. (2010). Fluorescent-labeled antibodies: Balancing functionality and degree of labeling. *Analytical Biochemistry*, 402(2), 146–150. doi:10.1016/j.ab.2010.03.036

Wu, B., Schnarr, M., Devlin, J. L., Brown, S., & Yang, T. Y. (2019). Approaches to improve drug tolerance and target tolerance in the assessment of neutralizing anti-drug antibodies. *Bioanalysis*, 11(22), 2061–2074. doi:10.4155/bio-2019-0184

Yang, D., Singh, A., Wu, H., & Kroe-Barrett, R. (2017). Determination of high-affinity antibody-antigen binding kinetics using four biosensor platforms. *King et al.*, 2014(122). doi:10.3791/55659

Yang, K., Zhang, Y., Chou, R., Yeung, L., Letarte, S., Yang, R.-S., Li, X., Beaumont, M., Gunawan, R., Richardson, D., Dellatore, S., Woolf, E., & Xu, Y. (2020). A tiered approach for characterization to ensure quality, reproducibility, and long-term stability of critical reagents in regulated bioanalysis to support PK/ADA/NAb assays for biologics and vaccines programs. *ACS Pharmacology & Translational Science*, 3(6), 1310–1317. doi:10.1021/acsptsci.0c00135

Zaroff, S., & Tan, G. (2019). Hybridoma technology: The preferred method for monoclonal antibody generation for in vivo applications. *BioTechniques*, 67(3), 90–92. doi:10.2144/btn-2019-0054

Zbacnik, T. J., Holcomb, R. E., Katayama, D. S., Murphy, B. M., Payne, R. W., Coccaro, R. C., Evans, G. J., Matsuura, J. E., Henry, C. S., & Manning, M. C. (2017). Role of buffers in protein formulations. *Journal of Pharmaceutical Sciences*, 106(3), 713–733. doi:10.1016/j.xphs.2016.11.014

Zheng, J. Y., & Janis, L. J. (2006). Influence of pH, buffer species, and storage temperature on physicochemical stability of a humanized monoclonal antibody LA298. *International Journal of Pharmaceutics*, 308(1–2), 46–51. doi:10.1016/j.ijpharm.2005.10.024

Zhou, L., Hoofring, S. A., Wu, Y., Vu, T., Ma, P., Swanson, S. J., Chirmule, N., & Starcevic, M. (2013). Stratification of antibody-positive subjects by antibody level reveals an impact of immunogenicity on pharmacokinetics. *The AAPS Journal*, 15(1), 30–40. doi:10.1208/s12248-012-9408-8

10

Statistical Aspects of Bioanalytical Testing

Arkady M. Gershteyn[a] *and Mark Ma*[b]

[a]*Manufacturing Analytical Development, Lantheus Medical Imaging Inc. (corresponding author)*
[b]*Translational Clinical Science, Rallybio*

10.1 In Vitro Diagnostic Assays

10.1.1 In Vitro Diagnostic Assay Types

Bioanalytical testing is a wide field that can be separated into three classes:

1) **Qualitative**: Identifying the presence of some molecule (e.g., gas chromatography-mass spectrometry [GC-MS]), protein (e.g., enzyme-linked immunosorbent assay [ELISA] binding assay), or peptide (chromatography). In this case, the test is qualitative as it yields an affirmative or rejection answer to the question: "Is said substance present in the sample?"
2) **Quantitative:** Quantifying the concentration of some protein, peptide, or antibodies in the sample. In this situation, the test is quantitative as it yields a numeric output that answers the question: "How much of said substance is present in the sample?"
3) **Semiquantitative:** There is also a hybrid class called "semiquantitative" assays that use a number to quantify the amount and apply a threshold to it. If the obtained number equals or exceeds the threshold then the test yields a "Yes" response, and otherwise it yields a "No" response to the question: "Is said substance present in the sample?". However, the number is not reported to physicians since its accuracy and/or precision parameters may not warrant it being used on its own to quantify the sample concentration.

Bioanalytical Aspects in Biological Therapeutics, First Edition. Edited by Xiaohui Xu and Weifeng Xu.
© 2022 John Wiley & Sons, Inc. Published 2022 by John Wiley & Sons, Inc.

10.1.2 Qualitative In Vitro Diagnostic Assays

A. Qualitative

Implicitly, there is a certain minimum amount of substance that will register as positive on the platform/reagent test method. It is called the Limit of Detection (LOD).

- Limit of Detection (LOD) – the lowest amount of analyte in a sample that can be detected with stated probability, although perhaps not quantified as an exact value.
- Limit of Blank (LOB) – the highest measurement result that is likely to be observed with a stated probability for a blank sample.

There are two ways to determine the LOD: one based on the LOB and the second based on obtaining a satisfactory level of precision for repeat measurements of a sample with known concentration, usually obtained by dilution of a reference standard (Armbruster & Pry, 2008; CLSI, 2008; FDA, 2007).

The first method for determining LOD is obtained by taking the 95th percentile from individual measurements (replicate-level) of blank samples assuming normal distribution by applying the formulas below:

$$\text{LOB} = \text{mean blank} + 1.645(\text{SD blank})$$
$$\text{LOD} = \text{mean LOB} + 1.645(\text{SD LOB})$$

The LOB is the upper limit of the measuring system noise, so this is a "bottom-up" approach.

The second method for determining LOD is to predefine a maximum level of imprecision, for example, 20%. Then for replicates of known concentration tested multiple times, 1.96 x standard deviation (SD) (replicates)/mean (replicates) needs to be less than 15% to pass the criteria. A standard with the known concentration of the analyte is taken and diluted, say by 5, 10, 20, 30, 40, 50. As the tested concentration of the sample becomes smaller, the imprecision [the coefficient of variation (CV) = SD (replicates)/mean (replicates)] will typically rise. The smallest concentration level at which the sample still passes the imprecision criteria is considered to be the LOD. It is a "top-down" approach since the LOD is estimated by sequentially reducing the known concentration of the test sample.

Paradoxically both methods require that the test provides a numeric readout. Hence any bioanalytical test that is officially qualitative has "under the hood," a quantitative measuring instrument in operation. The second method requires that a standard of known concentration of the analyte be obtainable, which may not always be feasible, especially for truly novel analytes (e.g., chemicals or antibodies that have never been measured previously).

The cutoff for a qualitative test is chosen such that it is above the LOD, and assures an adequate level of sensitivity and specificity for the intended application. The receiver operator curve (ROC) is useful for evaluating various potential cutoffs. Clinical judgment needs to factor into an evaluation of the relative utility of balancing false-positives (FP) and false-negatives (FN). For example, in a breast cancer confirmatory assay, a false-positive may result in needless surgical elimination of a breast. In contrast, a false-negative may result in cancer severity progression, and potentially cancer spreading and metastasis. The cost function to minimize when selecting an optimal cutoff from the ROC may be a weighted sum of false-positives and false-negatives, where the weighting coefficients are predetermined by anticipated costs consequent to these respective errors. For establishing a ROC for an in vitro diagnostic product, the Clinical and Laboratory Standards Institute (CLSI) advises that one uses an equal number of positive and negative samples (Berk, 2010; CLSI, 2011; Maxim et al., 2014). Formal definitions are as follows:

- Cutoff-for a qualitative test – the threshold above which the result is reported as positive and below which the result is reported as negative.
- C50 – the analyte concentration near the cutoff that yields 50% positive results and 50% negative results when many replicates of a single sample at that concentration are tested.

For in vitro diagnostic assays applied to human samples:

- Sensitivity – what percentage of people with the disease are classified as such by the test [100 x TP/(TP + FN)].
- Specificity – what percentage of healthy people are classified as such by the test [100 x TN/(TN + FP)].
- Positive predictive value (PPV) – what percentage of people given a positive diagnosis by a test have said disease [100 x TP/(TP + FP)].
- Negative predictive value (NPV) – what percentage of people given a negative diagnosis by a test do not have said disease [100 x TN/(TN + FN)].
- Diagnostic accuracy – what percentage of people are given the correct diagnosis [(TP + TN)/(TP + FP + TN + FN)].
- ROC – a graphical plot that illustrates the diagnostic ability of a binary classifier system as its discrimination threshold is varied. On the ordinate (y-axis), the true-positive (TP) rate (sensitivity) is plotted, and the abscissa (x-axis) the false-positive (FP) rate (1−specificity). Consult Table 10.1 for definition of true-positive and false-positive cases.

The area under the curve (AUC) for a ROC can be interpreted as the probability that the classifier will assign a higher score to a randomly chosen positive example than to a randomly chosen negative example. This metric is useful for

Table 10.1 2 × 2 Contingency table for candidate test versus predicate test: diagnostic accuracy criteria.

		Predicate test (FDA approved)		
Candidate test		Positive	Negative	Total
	Positive	True-positive (TP)	False-positive (FP)	TP + FP (e.g., positive by candidate test)
	Negative	False-negative (FN)	True-negative (TN)	FN + TN (e.g., negative by candidate test)
	Total	TP + FN (e.g., "diseased")	FP + TN (e.g., "healthy")	

comparing two or more qualitative tests if both are in the development stage. See example in Graph 1 where the authors select a clinical efficacy biomarker for advanced renal cell carcinoma (English et al., 2016). If the thresholds have been chosen and fixed for the qualitative tests, then metrics such as sensitivity, specificity, PPV, NPV, and diagnostic accuracy would be more applicable for their comparison.

Graph 1: Receiver operator curve (ROC)

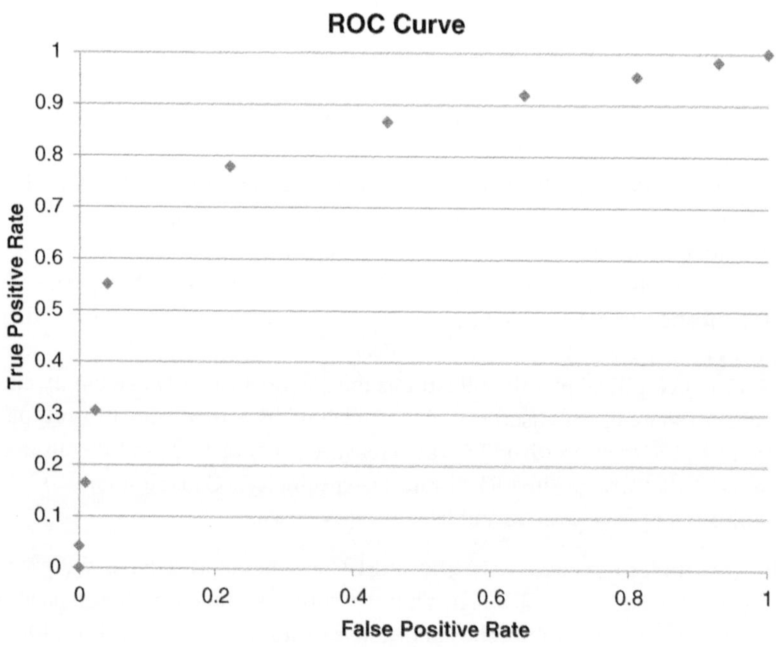

10.1.3 Quantitative In Vitro Diagnostic Assays

B. Quantitative

Statistical criteria:

- Accuracy (degree of bias) – the closeness of agreement of a measured quantity value and true quantity value of the measurand. Operationally this is defined as the average of multiple measurements of sample replicates by candidate assay/instrument minus the sample's known value (when using a known reference standard, or same sample previously measured on a "gold standard" measurement platform).
- Precision (run-to-run variability, operator-to-operator variability, day-to-day variability, site-to-site variability) – the closeness of agreement between independent measurement results obtained by stipulated conditions. Operationally this is obtained from the multiple-factor ANOVA model, and the most important component is within-device imprecision (includes between-run, and sometimes between-day and/or between-operator) estimate.
- Total analytical error/total allowable error (TAE) = bias + 2 x SD(precision). The maximum permitted TAE needs to be prespecified for an in vitro diagnostic (IVD) assay/instrument at each medical decision level (MDL). Bias is taken as a regression residual at specific MDL. Precision SD estimate, in this context, stands for within-device SD (estimated from sample concentration at or close to MDL under consideration).
- Medical decision level (MDL) – a concentration of the measurand that is used as a threshold for making a clinical statement.
- Method comparison – evaluates accuracy. Uses regression (ordinary least squares [OLS], weighted linear regression, Deming regression, weighted Deming regression, or Passing–Bablok regression) and has criteria to slope, intercept, and percent (%) bias at MDLs. Appropriate regression approach is determined by the shape of the Bland–Altman difference plot (scatter at same level-constant SD, scatter increases as x increases – constant CV, or scatter not uniform but does not increase consistently as x increases) (CLSI, 2013).

Method comparison – Step 1 – Difference plot

- Difference plot (e.g., Bland–Altman) used to evaluate bias and profile of variance of difference as a function of the x-axis.
- In Bland–Altman difference plot x-axis is an average of measurements done by two different devices (i.e., new device and comparator device).

- Bland–Altman applicable when neither device is "gold standard".
- The mean of two devices' measurements of the same sample is considered more accurate (closer to the true value) than either measurement taken separately.

Method comparison – Bland–Altman difference plot evaluation

Question 1: Is the bias increasing linearly with the x-axis?

- If Yes, then nonlinear recalibration required before proceeding with regression.
- Testing is done by applying linear regression to the difference data and estimating the slope, and its 95% confidence interval (95% CI).
- If 95% slope includes zero then no evidence for linear increasing (with concentration) bias.
- If Slope 95% CI excludes zero then conclude that there is a linearly increasing bias.

Question 2: What is the SD of difference profile throughout the tested concentration range (x-axis)?

- Situation A: Constant throughout – "constant SD" > Deming regression

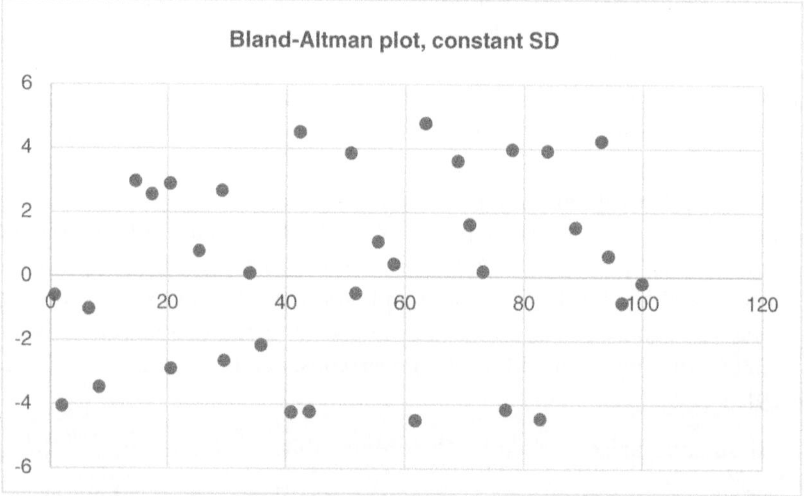

- Situation B: Increasing linearly with concentration – "constant %CV" > CV Deming regression/weighted linear regression

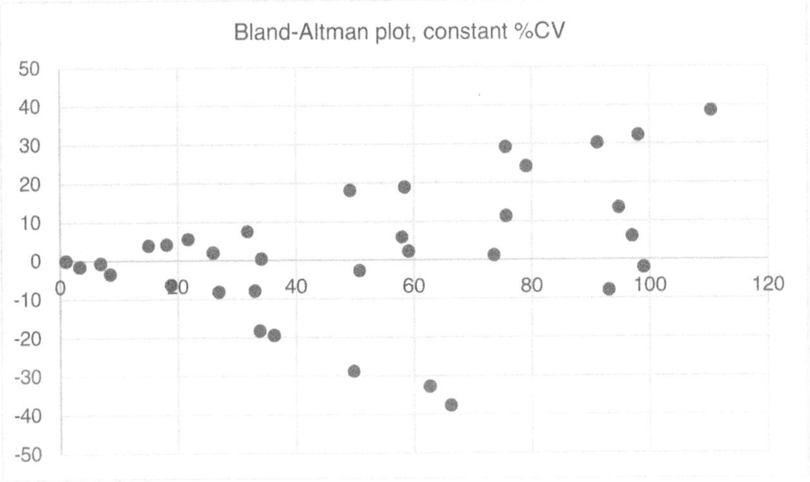

- Situation C: Other > Passing–Bablock regression – nonparametric method that does not require normality assumption nor constant SD/nor constant %CV.

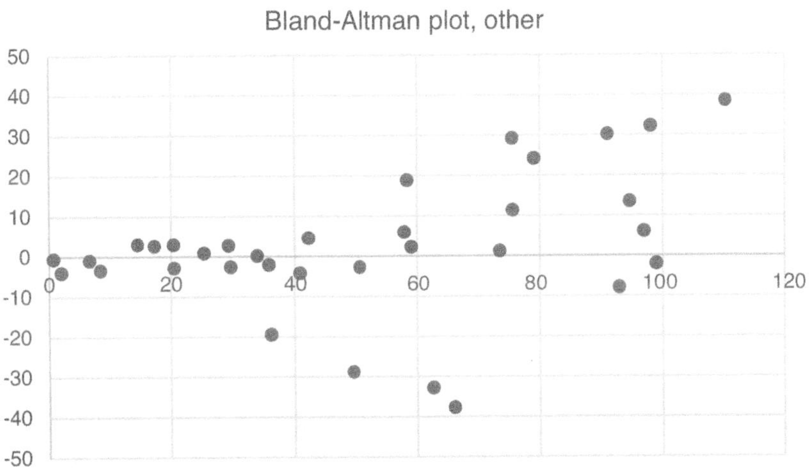

Step 2 – Regression parameters compare to criteria

Regression parameters that it is useful to have criteria on are slope (e.g., slope needs to be within 1 ± 0.2), intercept (e.g., ≤ 10 of mean of X), and R-squared (or R-squared adjusted) (e.g., R-squared adjusted ≥ 0.90).

Bland–Altman plot also provides a Limit of Agreement (LOA) defined as middle 95% of data for differences between candidate method versus predicate measurement method.

- Precision study – evaluates imprecision due to various sources, including repeatability (within-run) and reproducibility (two factors – run and operator, or three factors – run, operator, and site). It is important that the factors be balanced. Let us illustrate this with an example – if one is designing a precision study for a manual ELISA test then the operator may be significant. So, factors are day (1–4) and AM/PM. In the table below, A is assigned to AM twice, and B is assigned to AM twice, and the same is true for PM. Thus, the operator and time of day variables are indeed balanced.

Day	Time of day	Operator
1	AM	A
	PM	B
2	AM	B
	PM	A
3	AM	A
	PM	B
4	AM	B
	PM	A

Suppose that for each AM/PM, three runs are done, and each run consists of three replicates. Then using the replicate level data, one obtains the following table:

Day	AM/PM	Operator	Run (1–3)	Replicate (1–3)

When analyzing the data, one will use an ANOVA with four variables – run, operator, AM/PM, and day. The highest level variable is day, nested within it one has AM/PM and operator (these two are really at the same level), and then nested below these two is run. So, one needs to construct a nested ANOVA model. The results will yield estimates of repeatability (within-run variance), and reproducibility (sum of variance from between-operator, between-AM/PM, and between-day). Total imprecision variance would be a total of repeatability variance with reproducibility variance. Typically, imprecision is reported in terms of SD (square root of variance) and as %CV. The %CV equals 100 x SD/mean (e.g., what percent of the mean the SD represents).

Three common designs are:

1) Twenty days, two runs per day, with multiple reagent lots, operators, instruments, and calibration cycles. This closely resembles the running of the laboratory day-to-day.
2) More than two runs per day, use multiple instruments, at one location, and to be analyzed by nested ANOVA. This enables estimating within-device (aka within-laboratory precision).
3) More than two runs per day, use multiple instruments, at multiple locations and to be analyzed by nested ANOVA. This enables estimating both within-device (aka within-laboratory precision) and between-laboratory precision. This design enables one to estimate the imprecision of the sample's recovered value when it can be analyzed in one of several laboratories.

- Repeatability (of measurements) – the closeness of agreement between successive measurements of the same measurand carried out under the same conditions or measurements, also known as within-run precision.
- Reproducibility (of measurements) – the closeness of the agreement between results of measurements of the same measurand carried out under changed conditions of measurement (e.g., between-run, between-operator, between-site).
- Run – an interval within which the trueness and precision of a testing system are expected to be stable but cannot be greater than 24 hours, or less than the frequency recommended by the manufacturer.

LOB and LOD apply to qualitative assays too, and the definition is the same as above.

- Limit of Quantification (LOQ) – the lowest amount of analyte in a sample that can be quantitatively determined with stated precision and accuracy under stated experimental conditions.

Strictly speaking, the LOQ $\geq$ LOD, but usually LOQ > LOD. One needs to confirm experimentally that the concentration level above LOD does indeed meet the total allowable error (bias + 2 x SD precision). To establish LOQ, an LOD study can be used that includes a minimum of 40 replicates from 3–5 samples (at different concentrations), and with at least five runs. For example, let us assume the requirement is that the total allowable error is $\leq$ 25% of the sample true concentration value. Then by diluting the sample to five values and first identifying LOD, let us assume that there is one sample below LOD and three samples above LOD. Typically for some value above LOD, the criteria will start to be met. The lowest concentration tested at which the total allowable error criteria are met is chosen as the test's LOQ (CLSI, 2012).

- Measuring range – set of values of measurands for which the error of the measuring instrument is intended to lie within specified limits (CLSI EP-17-A2) (CLSI, 2012).
- Linearity – this is evaluated by taking a sample with known concentration and producing serial dilutions and/or spikes to ensure that the claimed linear range is covered. Developers of new methods (including both manufacturers and laboratories that modify existing methods) should use 7–11 concentrations across the anticipated measuring range. The number of replicates per level should be 2–4. Often a wider range is probed by the chosen concentrations than ultimately claimed as "linear". The samples can be individual samples, pools of multiple individual samples, standards of known concentration (calibrator), patient samples, and patient pools, and calibrators may be supplemented with the analyte and likewise can be diluted by water to contrive other concentration levels. The guidance adopts the polynomial evaluation of linearity approach, which consists of fitting three regression models to the replicate means for each concentration level:

Order	Polynomial	Regression df (Rdf)
First	$y = b_0 + b_1x$	2
Second	$y = b_0 + b_1x + b_2x^2$	3
Third	$y = b_0 + b_1x + b_2x^2 + b_3x^3$	4

The y values are known and true concentrations, and the x values are means of multireplicate (2–4) measurements by a novel test method for each sample.

The first step is to look at the 2nd and 3rd order regression equations coefficients for nonlinear terms (x^2 and x^3). Apply the t-test for regression coefficients, t = coefficient/standard error, and test at a two-sided alpha level of 5%. If none of the nonlinear terms are statistically significant (p value > 0.05), then linearity is demonstrated. If, however, this is not the case, then the model (1st, 2nd, 3rd order) with the smallest standard error of regression (Syx) is chosen as the best fit model. For each concentration level, the degree of nonlinearity is defined as the predicted value from the best fit model and the predicted value from the 1st order (i.e., linear) model. This error can be examined simply in concentration units or relatively (by dividing by the known concentration, yi). There must be predefined acceptance criteria for the maximum allowable level of nonlinearity at each level (e.g., < 10% for all seven levels). The claimed linear range must satisfy the error criteria at all levels encompassed within the said range. Note that precision information can also be captured from this study design due to the multiple replicates done at each level (CLSI, 2003).

The guidance uses standard error (SE) of regression as a criterion of best fit. The SE of regression provides the absolute measure of the typical distance that the data points fall from the regression line. Given we are comparing models of growing complexity, it would be natural to expect that a more complex model will more closely match the data; and will yield a smaller SE of regression. This more complex model may not be predictive for future similar linearity study data, better than a simpler model, and result in overfitting. I would suggest modifying the guidance to use Akaike Information Criteria (AIC) instead of SE of regression as the criteria for model fit, as it penalizes the fit by model complexity.

- Matrix effect evaluation

Matrix for in vitro diagnostic (IVD) diagnostic purposes typically relate to the temperature (frozen vs. fresh) or blood filtration process (e.g., whole blood is centrifuged to filter out the cell material to yield plasma, or whole blood is centrifuged even more to filter out clotting factors and yields serum). Matrix comparison studies, therefore, are inherently a paired test design where the processed sample is compared to raw patient samples (before processing). Typical comparisons include frozen 1st cycle vs. fresh, frozen 2nd cycle vs. fresh, plasma vs. whole blood, serum vs. whole blood. The recommended protocol requires that at least 20 patient samples be used, that they are measured on the assay/reagent or detection instrument (either manually or automatically), and processed (e.g., by extracting out all the cells for a plasma vs. whole blood comparison). The processed samples need to be randomly interspersed between the unprocessed samples so that from the test order one cannot infer their status (processed or unprocessed). Further several replicates for each sample are evaluated on the measurement device, and the mean of these replicates (e.g., triplicates) is used (CLSI, 2014).

On the x-axis, one places the unprocessed sample's mean replicate value; and on the y-axis, the processed sample's mean replicate value for at least 20 patient samples. Then regression analysis proceeds as described in the Method Comparison section. An exception is that instead of the Bland–Altman plot which uses as x the average of x and y, now the x-axis is considered the truth and there is no need to modify it. Another exception is that Demming regression would not be appropriate as it assumes an uncertainly in the x-values which is absent here. So the first one would draw a difference plot, where y = y−x and x = x, and then evaluate which of the situations applies – (a) constant SD scatter, (b) constant %CV scatter, or (c) other.

If (a) then apply OLS regression. If (b) then apply weighted OLS regression. If (c) then one can apply Passing–Bablok regression (CLSI, 2014; NCSS Statistical Software). Then calculate the 95% prediction interval for y at each x point; and see if the observed y values fall within this interval. If they fall within this interval, then there is no matrix effect. However, if for some x, there is a y that falls outside the 95% prediction interval then there is evidence of a matrix effect, and the processed samples are considered not equal to (or "commutable") to the unprocessed samples.

10.1.4 Semiquantitative In Vitro Diagnostic Assays

C. "Semiquantitative"

These are treated just like qualitative tests.

General approaches:

To validate an assay there are some general approaches that often are applicable to test accuracy and/or linearity, these include:

- The known quantity used for testing
- The gold standard test used for method comparison
- Food and Drug Administration (FDA)-approved predicate used for method comparison
- Subtract out background (e.g., chromatographic column with antibody-specific binding antigens front-loaded. Then the front of columns produces signal + noise, and the end of column yields only noise. Subtract the sum of response from the front of the column minus sum of response from the back of the column to yield signal.)
- Spiking the sample with a known quantity of analyte or dilute high sample by various factors. Useful for covering the entire dynamic range and for evaluating linearity and potential hook effect.

10.2 Biosimilars (351 K): Pharmacokinetics (PK)/ Pharmacodynamics (PD) Metrics and Their Comparison Including Acceptance Criteria; Equivalence Test (TOST); Sample Size for TOST; Effect Size (Cohen's d) for Binary and Continuous Metrics

10.2.1 Biosimilar Definition and Regulatory Pathway

- Biosimilar – biological product is highly similar to the reference product notwithstanding minor differences in clinically inactive components and there are no clinically meaningful differences between the biological product and the reference product in terms of the safety, purity, and potency of the product (Section 351(i) of the Public Health Service [PHS] Act). This regulatory pathway for biotechnology-derived products became open in 1996 with the publication "FDA Guidance Concerning Demonstration of Comparability of Human Biological Products, Including Therapeutic Biotechnology Products." Previously sponsors would need to submit a brand new New Drug Application (NDA)/Biologics License Application (BLA) and conduct a full set of clinical trials, even for highly similar protein-based therapeutic products (Christl; EMA, 2005, 2014; FDA, 1997, 2012, 2015a, 2015b, 2015c, 2016, 2017; International Council for Harmonisation of Technical Requirements for Pharmaceuticals for Human Use [ICH], 1995).

FDA approval pathways:

Drugs: small molecules:Approved via Federal Food Drug and Cosmetic Act (FDCA)		**Biologics:Approved via Public Health Service Act (PHSA)**	
New Drug Application (NDA) 505 (b) (1)	Abbreviated New Drug Application (ANDA) 505 (b) (2)	Biologics License Application (BLA) 351 (a)	Biosimilars Biologics License Application 351 (k)
Safety and efficacy must be demonstrated	Bioequivalence must be demonstrated	Safety and efficacy must be demonstrated	Must demonstrate that it is highly similar to 351 (a) reference. Interchangeable biosimilars require more data

10.2.2 Components of Biosimilars Biologics License Application 351(K) (Christl; EMA, 2005, 2014; FDA, 1996, 1997, 2012, 2015a, 2015b, 2015c, 2016, 2017; ICH, 1995)

- Manufacturing information – same manufacturer? Same manufacturing process? Describe raw materials, equipment, process, process controls, and acceptance criteria. Typically need to provide information on multiple lots to show new process/equipment passes the same acceptance criteria.
- Analytical studies (multiple) – structural analysis, analyze finished dosage form, and functional assays evaluated in vitro assays (biological assay, binding assay, enzyme kinetics).

In vivo example: cell-based bioactivity assays may be used to detect the potential for induced cytokine release syndrome in vivo.

- The animal study, including toxicity and immunogenicity and PK and PD secondary endpoints. Contingent on there being a human-comparable animal model for an indication of interest. If no adequate animal model exists, then using human cells or tissue may be a viable solution.
- At least one human pharmacokinetics (PK) study.
- At least one human pharmacodynamics (PD) (optional).

10.2.3 Biosimilars Metrics by Study Type (Christl; EMA, 2005, 2014; FDA, 1996, 1997, 2012, 2015a, 2015b, 2015c, 2016, 2017; ICH, 1995)

- Functional assay – percent (%) activity.
- PD study (in vitro or animal studies, human studies optional; measure clinical outcome variable or its biological proxy [clinical correlate] vs. drug concentration):
  - Potency: EC_{50} – concentration of drugs when 50% of its maximum effect is achieved.
  - Tolerance: dose–response curve shows at what point small increase in dose brings in diminished returns in terms of efficacy.
- PK-study (animal and/or human studies; measure drug concentration over time): See Table 10.2.5 for an example.
  - AUC – total drug exposure over time.
  - C_{max} – maximum (or peak) serum concentration that a drug achieves in a specified compartment or test area of the body after the drug has been administrated and before the administration of a second dose.
- Immunogenicity – The risk to patients of mounting an immune response to a therapeutic protein product should be monitored by using a validated anti-drug antibody (ADA) assay:
  - Assay types: ADA screening assay, confirmatory ADA assay, characterization assay, neutralization assay.
  - One-assay approach: biosimilar reagent used to detect both biosimilar and reference product ADAs.
  - Two-assay approach: biosimilar reagent used to detect biosimilar ADAs and reference product reagent used to detect reference product ADAs.

10.2.4 Biosimilars PK Metric in Depth

- PK:
  - For intravenous studies, AUC (0–infinity) is the primary endpoint.
  - For subcutaneous studies, C_{max} and AUC are coprimary endpoints.
  - For single-dose studies, AUC is calculated as the area under the biological product concentration–time curve from start to infinity. AUC (0–infinity) = AUC(0–t) + Ct/kel, where Ct, concentration at last time point, kel, elimination rate constant.
  - C_{max} determined from data without extrapolation.
  - For multiple-dose studies, AUC is calculated from time zero to end of dosing interval at steady state: AUC (0–tau) – primary endpoint.
  - For multiple-dose studies, concentration prior to the next dose (C_{trough}) and C_{max} are both secondary endpoints.

10.2.5 Biosimilars – Example of PK Graph

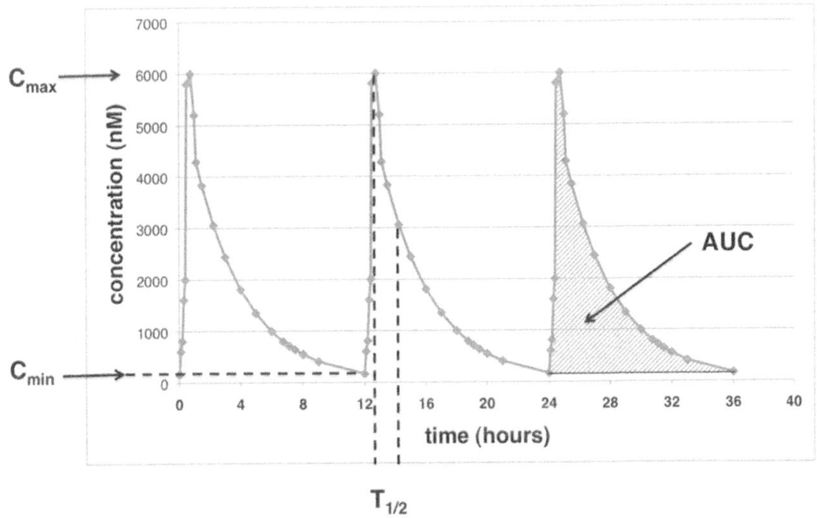

Illustration of the time course of plasma drug concentrations over 36 hours, following administration of a drug every 12 hours. Resource: https://commons.wikimedia.org/wiki/File:Pharmacokinetic_parameters.png (public domain).

10.2.6 Biosimilars Example of PK Table

Table 10.2 Pharmacokinetic parameters following a single subcutaneous administration of test or comparator drug (hypothetical data).

Parameter	Test (n = 36)	Comparator (n = 36)	GMR (90% CI)
T_{max} (h)	10.0 (4.0–34.0)	8.0 (4.0–14.0)	-
C_{max} (mlU/mL)	66.2 ± 24.1 (36.2)	69.1 ± 19.8 (28.7)	0.958 (0.871–0.947)
$AUC_{0\text{-}144h}$ (h*mlU/mL)	3031.8 ± 672.5 (22.2)	3040.7 ± 665.9 (21.9)	0.997 (0.961–1.038)
$AUC_{0\text{-}inf}$ (h*mlU/mL)	4433.2 ± 1169.3 (26.4)	4192.4 ± 982.5 (23.4)	1.057 (0.989–1.121)
$t_{1/2}$ (h)	89.5 ± 51.3 (57.3)	76.3 ± 41.3 (54.1)	-
CL/F (mL/h)	943.9 ± 220.7 (23.4)	985.1 ± 217.3 (22.1)	-
MRT	52.6 ± 5.9 (11.2)	48.9 ± 6.2 (12.7)	-

Notes: Data are presented as mean ± SD (coefficient of variation, %), except for T_{max} where data are presented as median (minimum–maximum).
Abbreviations:
T_{max} (h), time of C_{max}.
C_{max} (mlU/mL), maximum observed serum erythropoietin concentration.
$AUC_{0\text{-}144h}$ (h*mlU/mL), area under the curve from time zero to the time of the last observation.
$AUC_{0\text{-}inf}$ (h*mlU/mL), area under the curve extrapolated to infinity.
$t_{1/2}$ (h), terminal half-life.
CI, confidence interval; CL/F, total clearance; GMR, geometric mean ratio; MRT, mean residence time; SD, standard deviation.

10.2.7 Biosimilars PD Metric in Depth

- PD:
  - PD biomarker needs to reflect the drug mechanism of action and have a wide dynamic range over the range of drug concentrations achieved during the PK study.
  - Safety and immunogenicity need to be evaluated in addition to efficacy.
  - Key metric is the area under the effect curve (AUC_{EC}) for PD studies. See Table 10.3 for full list of metrics.
  - Concentration dynamics may depend on organ or tissue (e.g., brain, kidney, muscle, plasma).
  - Repeated doses will provide periodic curve for plasma concentration versus time.

10.2.8 Biosimilars PD Example of Graph

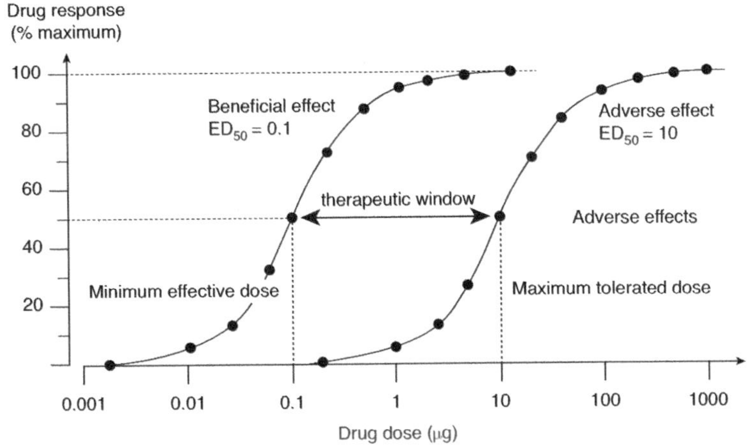

10.2.9 Biosimilars PD Example of Table

Table 10.3 Pharmacodynamic parameters for the reticulocyte count, hemoglobin, hematocrit, and red blood cell count following a single subcutaneous administration of the test and comparator drug (hypothetical data).

Parameters	Test (n = 36)	Comparator (n = 36)	Mean difference (90% CI)	P value (ANOVA)
Reticulocyte count				
T_{max} (h)	132.8 (0.0–214.7)	136.1 (94.8–317.1)	–	–
ΔE_{max} (%)	0.75 ± 0.25 (33.33)	0.71 ± 0.23 (32.39)	0.04 (–0.04–0.10)	0.431
ΔAUEC (h*%)	126.91 ± 54.28 (42.77)	130.74 ± 58.19 (44.51)	−3.83 (−21.65−14.27)	0.732
Hemoglobin				
T_{max} (h)	93.0 (0.0–310.7)	93.0 (0.0–312.5)	–	–
ΔE_{max} (%)	8.2 ± 5.1(62.2)	8.2 ± 5.3 (67.9)	0.04 (−0.04−0.12)	0.813
ΔAUEC (h*%)	981.2 ± 849.1 (86.6)	985.8 ± 973.5 (98.8)	−4.61 (−22.32−13.20)	0.641
Hematocrit				
T_{max} (h)	93.0 (0.0–310.7)	93.0 (0.0–312.5)	–	–

(Continued)

Table 10.3 (Continued)

Parameters	Test (n = 36)	Comparator (n = 36)	Mean difference (90% CI)	P value (ANOVA)
ΔE_{max} (%)	2.5 ± 1.1 (44.0)	2.5 ± 1.3 (52.1)	−0.03 (−0.23–0.14)	0.952
ΔAUEC (h*%)	319.5 ± 254.9 (79.8)	304.0 ± 262.7 (86.4)	15.53 (−73.92–118.74)	0.817
Red blood cell count				
T_{max} (h)	93.0 (0.0–310.7)	93.0 (0.0–312.5)	–	–
ΔE_{max} (%)	0.32 ± 0.15 (46.88)	0.33 ± 0.19 (57.58)	−0.01 (−0.08–0.07)	0.897
ΔAUEC (h*%)	31.71 ± 27.37 (86.31)	31.25 ± 29.59 (94.69)	0.46 (−10.05–12.31)	0.941

Notes: Data are presented as mean ± SD (coefficient of variation, %), except for T_{max} where data are presented as median (minimum–maximum).
Abbreviations:
T_{max} (h), time of E_{max} (maximum effect).
ΔE_{max} (%), maximum effect change.
ΔAUEC (h*%), area under the baseline-adjusted effect curve.
CI, confidence interval; SD, standard deviation.

10.2.10 Statistical Comparison of PK/PD Results (Christl; EMA, 2005, 2014; FDA, 1997, 2012, 2015a, 2015b, 2015c, 2016, 2017; ICH, 1995)

- Log transform exposure measure prior to analysis.
- Average equivalence approach to compare PK/PD parameters.
  - Applies to AUC and C_{max} metrics.
  - Based on two one-sided tests to determine whether the average values for the PK/PD measures determined after administration of test and reference products are comparable.
  - Involved calculating 90% CI for log-transformed averages of the measures for test and reference products.
- 90% CI for the ratio between geometric means of parameters of proposed biosimilar product and reference product.
- The calculated CI needs to fall within the limit, typically 80% to 125%, as per FDA guidance.

10.2.11 Analytical Similarity Assessment (Christl; EMA, 2005, 2014; FDA, 1996, 1997, 2012, 2015a, 2015b, 2015c, 2016, 2017; ICH, 1995)

- FDA recommends using at least 10 reference product lots, and at least 10 biosimilars (candidate) lots.
- FDA suggests that the reference product lots should have expiration time spanning the entire product shelf life range (i.e., include both "fresh" lots and those close to expiration).
- Analytical similarity acceptance criteria need to be derived from an analysis of US licensed products (reference).
- If biosimilar lots were manufactured by different processes then additional data showing their comparability needs to be provided prior to pooling them. Simpler if all biosimilar lots made by the same process.
- Factors for consideration: age of lot produced at testing, multiple testing, assay performance, differences in attributes that will be considered acceptable (no clinical impact).

10.2.12 Ranking Metrics Based on Risk and Determining Comparison Method (Christl; EMA, 2005, 2014; FDA, 1996, 1997, 2012, 2015a, 2015b, 2015c, 2016, 2017; ICH, 1995)

- Rank reference product's quality attributes based on potential impact on the clinical performance.
- Factors to rank: product's activity, PK/PD, safety, immunogenicity profiles.
- Apply more stringent statistical methods to higher risk factors.
- If a factor has more uncertainty associated with its estimate then it may be considered higher risk.
- Statistical method: equivalence testing. Limit equivalence testing to 2–3 factors with the highest risk.
- Tier 1 – high-risk factors – equivalence testing.
- Tier 2 – moderate-risk factors – quality ranges approach.
- Tier 3 – low-risk factors – visual comparison approach.
- Statistical Analysis Plan (SAP) needs to be developed for inclusion in the analytical similarity assessment plan. However, allowed to obtain preliminary data first to estimate the variability of the reference product's quality metrics/ select an adequate assay.

10.2.13 Analytical Similarity Assessment Plan (Christl; EMA, 2005, 2014; FDA, 1997, 2012, 2015a, 2015b, 2015c, 2016, 2017; ICH, 1995)

- Risk ranking of attributes, tiers for each attribute.
- Statistical Analysis Plan (SAP).
- Number of reference and biosimilar lots.
- Discuss with FDA early on and submit the final version prior to starting associated studies.

10.2.14 Tier 1 – Equivalence Test (FDA, 2017; Isakov, 2016; Little, 2015)

- Null hypothesis: $\mu_T - \mu_R \leq$ -delta or $\mu_T - \mu_R \geq$ delta.
- Alternative hypothesis: -delta $< \mu_T - \mu_R <$ delta.
- Delta is a positive number denoting the largest acceptable difference between biosimilar and reference product that has no clinical impact ("equivalence margin").
- Control type I error at 5%, by requiring 90% CI on mean difference to lie within (–delta, delta) interval. If both one-sided tests are simultaneously rejected, this CI will be contained in the ±delta interval.
- Advice on how to set equivalence margin (delta) when there is no consensus on what is a clinically significant difference:
  - Use multiple of reference product's variability (e.g., 1.5 x standard deviation of metric for the reference product).
  - Note: SD of metric needs to include both within-lot and between-lot sources of variation.
  - Example: equivalence test for AUC after one dose with predefined limits of 80%–125%.

10.2.15 Sample Size for Two One-Sided T-Tests (TOST) (FDA, 2017; Isakov, 2016; Little, 2015; Rapidmicro Biosystems & The Rapid Blog, 2010)

- Determine delta based on what is a clinically significant difference for a two-sided equivalence test.
- Assume that one observes no shift in mean for metric of interest (e.g., AUC, C_{max}, EC_{50}, activity level, etc.).

- Find sample size such that power for the one-sided t-test is at least XX% power and alpha level is at 5%. Note: use of the historical value of SD.
- Type 1 error controlled because to accept alternative hypothesis both null hypothesis $\mu_T - \mu_R \leq$ -delta and $\mu_T - \mu_R \geq$ delta needs to be rejected.
- Alternatively, one can calculate the two-sided 90% CI on the mean difference. If this CI falls within (–delta, +delta) then equivalence is established.

10.2.16 TOST Figure

Data series	Upper bound	Lower bound	Mean	Margin of equivalence	T-test	TOST
A	2	−4	−1	(−10, 10)	Fail	Pass
B	14	−12	1	(−10, 10)	Fail	Fail
C	12	−2	5	(−10, 10)	Fail	Fail
D	5	3	4	(−10, 10)	Pass	Pass
E	13	5	9	(−10, 10)	Pass	Fail

Margin of equivalence is determined ahead of time, based on regulatory requirements.

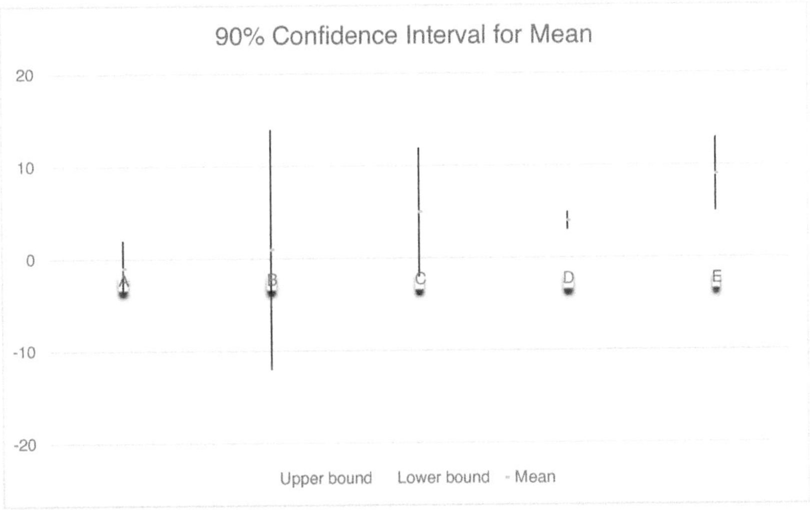

If 90% CI of mean difference falls within bounds for margin of equivalence, then TOST passes (establish that difference is within margin of equivalence).

If 90% CI of mean difference does not intersect with zero (0) then t-test passes (establish difference exists).

10.2.17 Tier 2 – Quality Range Approach (Christl; EMA, 2005, 2014; FDA, 1996, 1997, 2012, 2015a, 2015b, 2015c, 2016, 2017; ICH, 1995)

- Range: [μ_R-X SD (reference), μ_R + X SD (reference)].
- Need range to encompass 90% of the data.
- X = 1.645 if X is normally distributed.
- Use 10 lots of each, same ones as used in equivalence testing.

10.2.18 Tier 3 – Visual Displays (Christl; EMA, 2005, 2014; FDA, 1996, 1997, 2012, 2015a, 2015b, 2015c, 2016, 2017; ICH, 1995)

- Applicable to either lowest risk attributes or to qualitative ones.
- Compare the distribution of values of biosimilar vs. reference products.
- In cases where limited lot-to-lot variation is expected one lot for each is acceptable.
- Provide descriptive statistics as well.

10.2.19 Advice on How to Select Metrics

Pharmacodynamics (PD) (Christl; EMA, 2005, 2014; FDA, 1996, 1997, 2012, 2015a, 2015b, 2015c, 2016, 2017; ICH, 1995)

- Identify PD biomarkers used to measure the response to the biological product.
  - This may be a single biomarker or composite of biomarkers (e.g., protein or mRNA microarray analysis).
  - Factors to consider when selecting biomarker:
    - Time of onset of change in PD biomarker relative to dosing and its return to baseline with discontinuation of dosing.
    - The dynamic range of PD biomarker overexposure range to the biological product.
    - The sensitivity of PD biomarker to differences between the proposed biosimilar product and the reference product.
    - The relevance of PD biomarker to the mechanism of action of the drug.
    - Analytical validity of PD biomarker.

If PD biomarker has large dynamic range but is not sensitive to the difference between the proposed biosimilar product and reference, then one may use PK parameters as a primary endpoint.

Pharmacodynamics (PD) and Pharmacokinetics (PK) study design (Christl; EMA, 2005, 2014; FDA, 1996, 1997, 2012, 2015a, 2015b, 2015c, 2016, 2017; ICH, 1995**)**

- Crossover design:
  - For PK similarity studies a single-dose, randomized, crossover design is generally preferred.
  - Recommended for products with short half-life (less than 5 days).
  - Recommended for product with rapid PD response (time of onset, maximal effect, and disappearance after drug exposure).
  - Most sensitive design for PK similarity studies and requires lowest number of patients.
  - For PD similarity studies where onset is not rapid, then use multiple-dose crossover design (e.g., Williams design).
  - Consider washout period when using crossover design.
- Parallel design:
  - Recommended for products that have long half-life and elicit immunogenic response.
  - When repeated exposure may yield higher immune response.
  - Optimal for diseases that have time-dependent changes associated with exposure to the drug.

Other PK/PD Study Design Considerations (Christl; EMA, 2005, 2014; FDA, 1996, 1997, 2012, 2015a, 2015b, 2015c, 2016, 2017; ICH, 1995**)**

- Heathy individual vs. patients:
  - Clinical PK/PD studies should be conducted in healthy subjects if the product can safely be administered to them. More sensitive than using patients for assessing similarity.
  - Sometimes not ethical to perform study on healthy patients, or PD biomarker only expressed in patients. Then conduct study on patient population for which drug is indicated.
- Route of administration:
  - Same as reference product (e.g., intravenous, subcutaneous).
  - If multiple approved routes of administration for reference product, then select subcutaneous or other extravascular routes of administration. These are most sensitive for potential PK differences during absorption phase in addition to distribution and elimination phases.
- Dose selection:
  - If study done in patients, then use approved dose for reference product.
  - If study performed on healthy individuals, lower dose is fine.

Analytical Characterization (Christl; EMA, 2005, 2014; FDA, 1996, 1997, 2012, 2015a, 2015b, 2015c, 2016, 2017; ICH, 1995)

- Extensive and robust comparative structural and functional studies need to be performed on biosimilar and reference.
- Examples: bioassay, binding assays, studies of enzyme kinetics.
- Factors to evaluate include:
  - Molecular weight.
  - Higher-order and posttranslational modifications.
  - Heterogeneity.
  - Functional properties.
  - Impurity profiles.
  - Degradation profiles denoting stability of protein.

Specific Assay and Manufacturing Process (Christl; EMA, 2005, 2014; FDA, 1996, 1997, 2012, 2015a, 2015b, 2015c, 2016, 2017; ICH, 1995)

- Specific assay types:
  - Ligand binding – concentration of most biological products is measured using ligand binding assays.
  - Concentration and activity assays.
  - PD assays.
- Manufacturing process:
  - If there are changes, then 351(k) may be required.
  - Stability testing in real time to support shelf life claim.
  - Stability under challenging conditions needs to be explored including – high temperature, freeze–thaw, light exposure, agitation.

Immunogenicity Studies (Christl; EMA, 2005, 2014; FDA, 1996, 1997, 2012, 2015a, 2015b, 2015c, 2016, 2017; ICH, 1995)

- Neutralizing antibodies.
- Immune-mediated toxicity.
- Include relevant patient populations that are not immunocompromised.
- Consider data from reference product with regard to safety and immunogenicity.

Pediatric Assessment (Christl; EMA, 2005, 2014; FDA, 1996, 1997, 2012, 2015a, 2015b, 2015c, 2016, 2017; ICH, 1995)

- Under Pediatric Research Equity Act (PREA) all applicants for:
  - New active ingredients.
  - New indications.

- New dosage forms.
- New dosing regimens.
- Or new routes of administration

are required to contain pediatric assessment to support dosing, safety, and effectiveness of the product unless this requirement is waived.

10.2.20 Binary Metric (Example – Responders) Sample Size Calculation

- Power calculation for TOST (two one-sided t-tests) (i.e., equivalence test):
  - Given reference product has 30% responders' rate.
  - Given "margin of equivalence" is 8%.
  - Then as long as biosimilar candidate product has responder rate 90% CI within (0.22, 0.38) we will conclude that it is equivalent to reference.
  - Find sample size such that power is 80% or 90% to establish equivalence with type I error at 0.05.

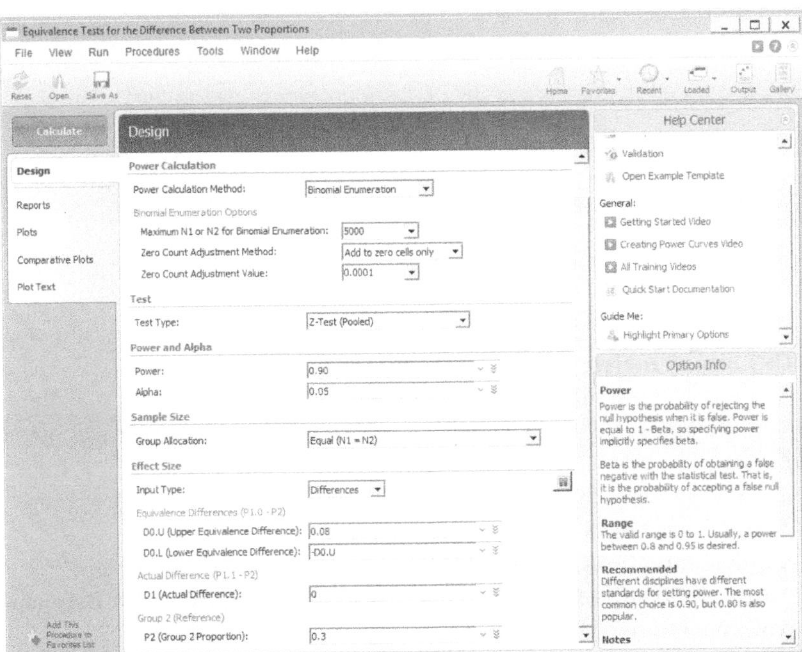

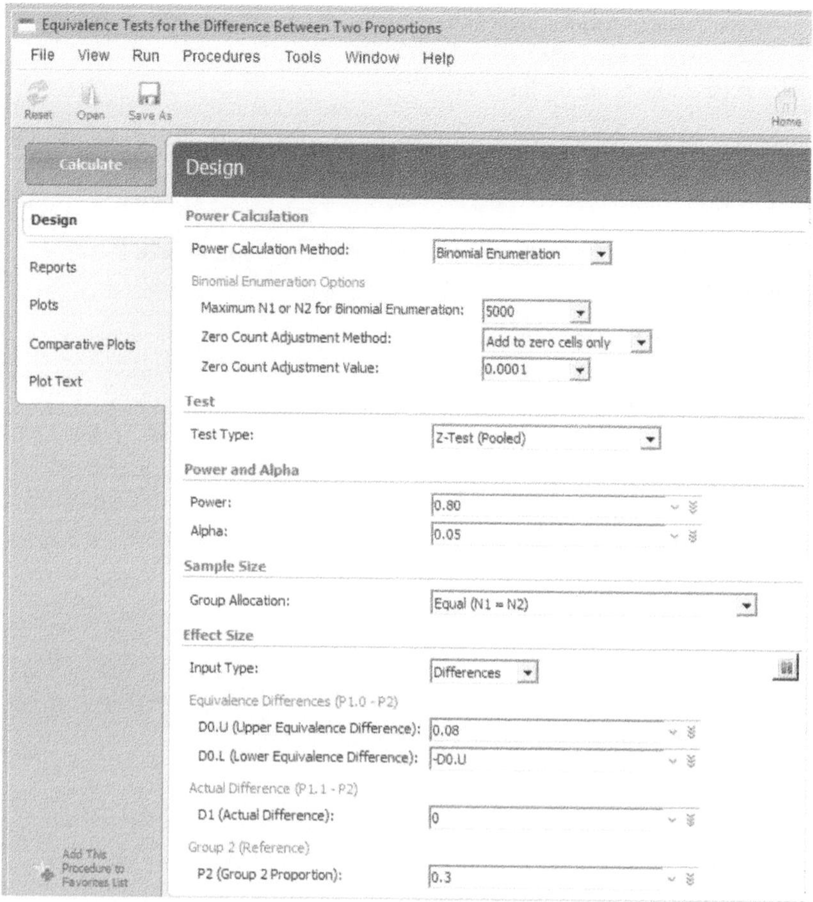

TOST binary metric sample size – PASS

Pref = 0.3, Pbiosimiliar = 0.3, Power = 80% or Power = 90%.

Following sample sizes based on a binomial enumeration via PASS software.

For 80% power need 563 per group and for 90% power need 710 per group.

Sample size: PASS results for 80% power

Equivalence Tests for the Difference Between Two Proportions

Numeric Results for Equivalence Tests for the Difference Between Two Proportions

Test statistic: Z-test with pooled variance

H0: P1 − P2 ≤ D0.L or P1 − P2 ≥ D0.U. H1: D0.L < P1 − P2 = D1 < D0.U.

Target Power	Actual Power*	N1	N2	N	Ref. P2	P1.0 L	P1.0 U	D0.L	D0.U	D1	Target Alpha	Actual Alpha*†
0.80	0.8004	563	563	1126	0.300	0.220	0.380	−0.080	0.080	0.000	0.0500	0.0497

*Power and actual alpha were computed using binomial enumeration of all possible outcomes.
†Warning: when solving for sample size with power computed using binomial enumeration, the target alpha level is not guaranteed. Actual alpha may be greater than target alpha in some cases. We suggest that one investigates sample sizes near the solution to find designs with an actual alpha that is tolerable.

Sample size: PASS results for 90% power

Equivalence Tests for the Difference Between Two Proportions

Numeric Results for Equivalence Tests for the Difference Between Two Proportions

Test statistic: Z-test with pooled variance

H0: P1 − P2 ≤ D0.L or P1 − P2 ≥ D0.U. H1: D0.L < P1 − P2 = D1 < D0.U.

Target Power	Actual Power*	N1	N2	N	P2	Ref. P1.0 L	P1.0 U	D0.L	D0.U	D1	Target Alpha	Actual Alpha*†
0.90	0.9002	710	710	1420	0.300	0.220	0.380	−0.080	0.080	0.000	0.0500	0.0498

*Power and actual alpha were computed using binomial enumeration of all possible outcomes.
†Warning: when solving for sample size with power computed using binomial enumeration, the target alpha level is not guaranteed. Actual alpha may be greater than target alpha in some cases. We suggest that one investigates sample sizes near the solution to find designs with an actual alpha that is tolerable.

10.2.21 Continuous Metric (Example – Disease Severity/Activity Score)

- Power calculation for TOST (two one-sided t-tests) (i.e., equivalence test):
  - FDA guidance on noninferiority trials suggests that a margin for noninferiority preserves at least 50% efficacy from historical data.
  - If reference product has efficacy value estimate of 1.00 (treatment mean–placebo mean), then 50% of that would be 0.5. So then the "margin of equivalence" is the amount remaining 1–0.5 = 0.5.
  - Setting delta (aka "margin for equivalence") to (−0.5, + 0.5).

   - Next assume the true difference in efficacy is zero.
   - Find sample size such that power is 80% or 90% to establish equivalence with type I error at 0.05.

10.2.22 Continuous Metric, Sample Size Example for Biosimilar

Biosimilar versus Solaris (Reference) for Paroxysmal Nocturnal Hemoglobinuria (PNH) Indication

- Proxy for transfusion may be hemoglobin level.
- Note: transfusion must occur when patient's hemoglobin value was 9.0 g/dL or lower with symptoms or when patient's hemoglobin value is 7.0 g/dL or lower without symptoms.
- 7% change in hemoglobin is considered clinically significant.

Hence, if efficacy biosimilar study has patients with transfusion-dependent hemolytic PNH, then it is reasonable to expect mean baseline hemoglobin value to be 7.7 (placebo, N = 44, SD = 0.75) or 7.8 (eculizumab, N = 43, SD = 0.79).

Let us assume mean will be 7.7, then 7% of that is 0.54.

If we apply equivalence test with acceptable range on difference in mean change in hemoglobin between groups at (−0.54, 0.54) then we can calculate power.

TOST sample size (per group) for 80% and 90% power, continuous metric.

nQuery Advisor - [MTE1tg-1]

File Edit View Options Assistants Randomize Plot Window Help

Two-group t-tests (TOST) of equivalence in means (equal n's)

	1	2
Test significance levels, α (one-sided)	0.050	0.050
Lower equivalence limit for $\mu_T - \mu_S$, Δ_L	-0.540	-0.540
Upper equivalence limit for $\mu_T - \mu_S$, Δ_U	0.540	0.540
Expected difference, $\mu_T - \mu_S$	0.000	0.000
Common standard deviation, σ	0.800	0.800
Power (%)	80	90
n per group	39	49

- When the sample size in each group is 39, a two-group design will have 80% power to reject both the null hypothesis that the test mean minus the standard mean is below –0.540 and the null hypothesis that the test mean minus the standard mean is above 0.540 (i.e., that the test and standard are not equivalent), in favor of the alternative hypothesis that the means of the two groups are equivalent, assuming that the expected difference in means is 0.000, the common SD is 0.800 and that each test is made at the 5.0% level.
- When the sample size in each group is 49, a two-group design will have 90% power to reject both the null hypothesis that the test mean minus the standard mean is below –0.540 and the null hypothesis that the test mean minus the standard mean is above 0.540 (i.e., that the test and standard are not equivalent), in favor of the alternative hypothesis that the means of the two groups are equivalent, assuming that the expected difference in means is 0.000, the common SD is 0.800 and that each test is made at the 5.0% level.

10.2.23 Effect Size – Continuous and Binary Endpoints

- Cohen's d (for two-sample t-test): delta/SD; delta, difference actual mean from expected mean (reference product) (Rosnow & Rosenthal, 2003).

Size of effect	Cohen's d
Small	0.2
Medium	0.5
Large	0.8

- Cohen's h (for two-sample proportion test): h = arcsin(sqrt(p1)) – arcsin(sqrt(p2)):
  - P1 – proportion of responders in proposed bioequivalent product assuming "equivalence margin" increase from reference product's responders' proportion.
  - P2 – proportion of responders in reference product.
- Consider whether "equivalence margin" constitutes small effect size for continuous variable.
- If "equivalence margin" translates to "medium" or "large" Cohen's d then may be problematic.

10.2.24 Interchangeable Biosimilars (Christl; EMA, 2005, 2014; FDA, 1996, 1997, 2012, 2015a, 2015b, 2015c, 2016, 2017; ICH, 1995; Isakov, 2016; Little, 2015; Brennan, 2017)

- A protein-based therapeutic product that is interchangeable with a reference product is:
  - Biosimiliar to reference product for all indications stated in reference product, AND
  - Can be expected to produce the same clinical result as the reference product in any given patient.
- Requires switching study (e.g., patients start on reference product and then switch to proposed interchangeable product).
- Idea is that variation in clinical effect from one dose of reference to another dose of reference is same as from one dose of reference to one dose of interchangeable product for same patient.
- "A switching study should evaluate changes in treatment that result in two or more alternating exposures (switch intervals) to the proposed interchangeable product and to the reference product."
- Is more difficult to prove than simple biosimilar claim.
- If product gains "interchangeable" status then FDA allows patients to switch from reference to interchangeable product without consulting physician/healthcare provider.
- May require postmarketing data commitment if fingerprint-like analytical similarity comparison does not show "same" for all factors.

10.3 Cutpoint Analysis of Neutralizing Drug Antibodies and ADA Screening Assays

The determination of a screening cutoff point is one of the most important elements in neutralizing antibody (NAb) and ADA assay validation (FDA, 2016; Shankar et al., 2008; Zhang et al., 2013).

There are three types of cutpoints:

- *Fixed cutpoint*: a cutpoint that is determined in prestudy validation and the same value is used for the in-study phase. *One should demonstrate similar means and variances across assay runs during prestudy validation* (Shankar et al., 2008).
- *Floating cutpoint*: a cutpoint calculated by multiplying a specific normalization factor, determined from the prestudy validation data, to the biological background obtained during the in-study phase. The biological background

may be represented by the negative control (pool of matrix from subjects that are negative for ADA), the assay diluent, or the predose subject sample (subject-specific cutpoint). A floating cutpoint may be plate-specific, unique for each run (common for a number of plates within a run), or unique for each subject (using the subject's pretreatment/"baseline" sample result). *One should demonstrate homogeneity of sample variance across runs* (Shankar et al., 2008).

- *Dynamic cutpoint*: a cutpoint that changes between the plates in a run, between runs in a study, or between studies, and *does not use the variation estimates from prestudy validation*. This approach is necessary only when the variability between sample results from each run is significantly different between assay runs. A practically limiting factor is that a significant number of samples are needed in each plate/run to compute this cutpoint (Shankar et al., 2008).

10.3.1 NAb Assay

10.3.1.1 Floating Cutpoint Analysis – NAb Assay

- Obtain 50 drug-naïve samples, conduct three runs with NAb assay, measuring in duplicate and computing a run-specific, sample-specific mean for each duplicate.
- Compute percent inhibition as follows: 100–100 x (sample RLU readout/ negative control average relative light unit (RLU) on plate).
- Four negative controls were run on each plate. This varies by assay experiment, sometimes investigators run six negative controls, for example.
- Note: higher percent inhibition corresponds to a higher concentration of NAb to drug ALXN1210 present in the sample.

10.3.1.2 Floating Cutpoint Analysis – NAb Assay – Outlier Analysis

- Analytical and biological outliers were identified and removed before the screening cutoff point analysis.
- Grubbs' test was employed to detect analytical outliers.
- Tukey's boxplot with 3*IQR is employed to detect biological outliers (see sample table below):

Q1	Q3	IQR	Lower fence (Q1 – 3 x IQR)	Upper fence (Q3 + 3 x IQR)
0.9333	5.1667	4.2333	–11.7667	17.8667

Tukey's boxplot method: any biological sample with a mean inhibition percent value ≥ 17.8667 or ≤ −11.7667 was considered a biological outlier. If a sample was deemed a biological outlier, all three replicates were removed.

10.3.1.3 Floating Cutpoint Analysis – NAb Assay, Cutpoint Analysis

Below are formulas for calculating the parametric and robust parametric cutpoint values:

- Parametric = mean + 1.645 x total SD. The total SD from a one-way ANOVA (PI by run) is calculated as the square root of the total sums of squares divided by the degrees of freedom.
- Robust parametric = median + 1.645 x 1.483 x MAD (median absolute deviation).
- Lower 90% CI for 95% percentile was computed via JMP 13 software's custom quantile tool.

Floating point requires variance be the same throughout runs – satisfied in this case

- Means and variances between runs were compared to determine if the fixed cutoff point, the floating cutoff point, or the dynamic cutoff point should be selected for this analysis.

Level	Number	Mean	SD	SEM	Lower 95%	Upper 95%
Run_1	48	0.73333	4.64553	0.67053	−0.616	2.0823
Run_2	48	5.16250	4.19385	0.60533	3.945	6.3803
Run_3	47	3.80213	4.36067	0.63607	2.522	5.0825

- Dispersion comparisons by Levene's test on equality of variance ($p = 0.5443$) and non-parametric Brown–Forsythe test on median absolute deviations ($p = 0.6158$) show that dispersion is comparable across runs.
- Given we are dealing with a situation where means differ by run but the variance is consistent across runs, the floating-point approach is valid.

Compare means to see if simple fixed-point method would do

- ANOVA analysis shows that between-run factor is significant at $p < 0.0001$, therefore the global null hypothesis that means are equal across the three runs is rejected.
- Likewise, the non-parametric Kruskal–Wallis test for equality of medians across the three runs yields $p < 0.0001$, clearly indicating that the medians are not all equal.

- The pairwise comparison of means across runs yielded significant differences by both the Tukey–Kramer honest significance difference (HSD) test and the nonparametric Steel–Dwass method. Run 1 (mean = 0.73) is statistically significantly different from Run 2 (mean = 5.16) and Run 3 (mean = 3.80).
- Means are different across runs, so fixed-point approach is not an option whereas floating-point remains viable.

Are data normally distributed within each run? Yes, the p value for Shapiro–Wilks Normality Test yields a p value of above 0.05 for each run (see Figure 10.1).

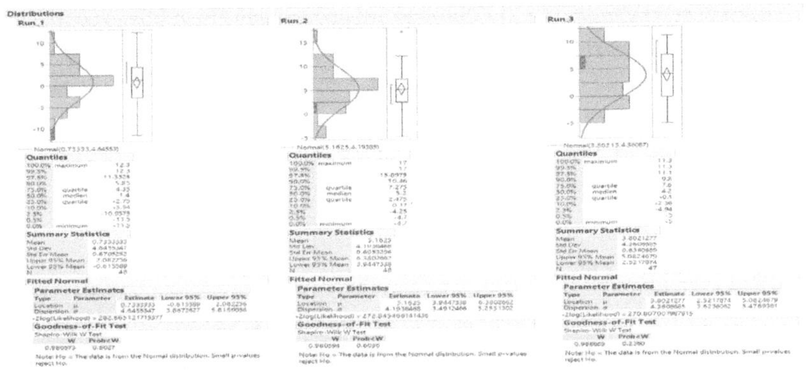

Figure 10.1 Frequency histograms of percent inhibitions in data with normal fit and Shapiro–Wilk normality test by run (outliers removed).

Summary of cutoff points (example):

Method	Cutoff point	Reactivity rate (%)
95% percentile	10.38	4.90
90% one-sided lower bound of 95% percentile	9.7	9.79
Parametric	11.05	3.50
Robust parametric	10.04	7.69

10.3.1.4 NAb Findings Summary (Example)

- For this analysis, the robust parametric method is preferred because (1) the data are normally distributed, (2) the cutoff point provided a false-positive rate of at least 5%; and (3) the false-positive rate is at an acceptable level of ~ 7.7%.

- The data were normally distributed after removal of the analytical outliers. ANOVA results provided evidence that means were different across runs. Since the homogeneity of variance tests were not statistically significant, a floating cutoff point is appropriate.
- The cutoff point derived using the parametric estimate was relatively close to the robust parametric method (11.05 vs. 10.04, i.e., within 10.1%). This makes sense since percent inhibition was normally distributed.
- The robust parametric method and 90% one-sided lower bound of the 95% percentile method yielded almost the same results (9.7 and 10.04, respectively). The corresponding reactivity rates were 7.7% and 9.8%. The parametric method yielded a cutpoint value of 11.05 and a reactivity rate of 3.5%.
- This report concludes that the cutoff point provided by the robust parametric method is appropriate for the following reasons: (1) the data are normally distributed, (2) the cutoff point provided a false-positive rate of at least 5%, and (3) the false-positive rate is at an acceptable level of ~ 7.7%.

10.3.2 ADA

10.3.2.1 ADA Assay Analysis

- The only difference compared to NAb assay is that a different metric is used.
- Instead of percent inhibition, use signal-to-noise ratio (SNR) and log SNR (or another transform that may equalize variance across runs).
- Outlier removal methods remain the same.
- Means and variances compared across runs.
- Floating cutpoint method selected in this example.
- Four analysis methods used and the reactivity calculated for each method (see sample table below):

10.3.2.2 Example of antidrug antibodies (ADA) summary table

Method	Original scale		Log scale*	
	Cutoff point	Reactivity rate (%)	Cutoff point	Reactivity rate (%)
95% percentile	1.444	4.8	1.444	4.8%
90% one-sided lower bound of 95% percentile	1.236	8.1	1.236	8.1
Parametric	1.391	5.6	1.285	6.5
Robust parametric	0.999	12.9	1.039	11.3

10.3.2.3 ADA Findings (Example)

- The obtained cutoff points from the original and log scales are similar to each other. The same biological outliers were identified in each case.
- Although the 95% percentile is the least conservative method, the reactivity rate was close to the 5% level. A false-positive rate of 5% is generally considered desirable for screening ADA assays.
- The 90% one-sided lower bound of the 95% percentile method is more conservative than the 95% percentile method.
- The robust parametric method was the most conservative method, with reactivity rates > 10% in both the original and log scales.
- For this analysis, the 90% one-sided lower bound of 95% percentile method was preferred since (1) it required no distributional assumptions, (2) it provided a prespecified confidence level on the cutoff point, and (3) a reactivity rate close to 5% was achieved.

Acknowledgments

Arkady Gershteyn thanks Jim Richards of Instrumentation Laboratory for insightful discussions regarding in vitro diagnostic statistics.

References

Armbruster, D. A., & Pry, T. (2008, August). Limit of Blank, Limit of Detection and Limit of Quantitation. *Clinical Biochemist Reviews*, 29 (Suppl (i)). https://www.ncbi.nlm.nih.gov/pmc/articles/PMC2556583/pdf/cbr29_s_pgs49.pdf

Berk, R.(2010). *Balancing the Costs of Forecasting Errors in Parole Decisions*. http://www.albanylawreview.org/Articles/Vol74_3/74.3.1071%20Berk.pdf

Clinical and Laboratory Standards Institute (CLSI). (2003, April). *EP06-A Evaluation of the Linearity of Quantitative Measurement Procedures: A Statistical Approach 1st Edition: Approved Guideline*. https://clsi.org/standards/products/method-evaluation/documents/ep06

Clinical and Laboratory Standards Institute (CLSI). (2008). *EP12-A2: User Protocol for Evaluation of Qualitative Test Performance; Approved Guideline-2nd Edition*. https://clsi.org/media/1433/ep12a2_sample.pdf

Clinical and Laboratory Standards Institute (CLSI). (2011). *EP24-A2 Assessment of the Diagnostic Accuracy of Laboratory Tests Using Receiver Operator Curves; Approved Guideline, 2nd Edition*. https://clsi.org/media/1425/ep24a2_sample.pdf

Clinical and Laboratory Standards Institute (CLSI). (2012, June). *EP17-A2 Protocols for Determination of Limits of Detection and Limits of Quantification: Approved Guideline 2nd Edition.* https://clsi.org/media/1430/ep17a2_sample.pdf

Clinical and Laboratory Standards Institute (CLSI). (2013, August). *EP09-A3 Measurement Procedure Comparison and Bias Estimation Using Patient Samples: Approved Guideline.* https://clsi.org/media/1435/ep09a3_sample.pdf

Clinical and Laboratory Standards Institute (CLSI). (2014, August). *EP14-A3 Evaluation of Commutability of Processed Samples; Approved Guideline – 3rd Edition.* https://clsi.org/media/1432/ep14a3_sample.pdf

English, P. A., Williams, J. A., Martini, J.-F., Motzer, J.A., Valota, O., & Buller, R.E. (2016). A case for the use of receiver operating characteristic analysis of potential clinical efficacy biomarkers in advanced renal cell carcinoma. *Future Oncology*. 12(2), 175–182. https://www.ncbi.nlm.nih.gov/pmc/articles/PMC5549778/pdf/fon-12-175.pdf

Food and Drug Administration (FDA). (2007). *Statistical Guidance on Reporting Results from Studies Evaluating Diagnostic Tests: Guidance for Industry and FDA Staff.* https://www.fda.gov/downloads/MedicalDevices/DeviceRegulationandGuidance/GuidanceDocuments/ucm071287.pdf

Maxim, L. D., Niebo, R., & Utell, M. J. (2014). Screening tests: A review with examples. *Inhalation Toxicology*, 26(13): 811–828. https://www.ncbi.nlm.nih.gov/pmc/articles/PMC4389712/pdf/uiht26_811.pdf

NCSS Statistical Software. *Passing-Bablok Regression for Method Comparison.* https://www.ncss.com/wp-content/themes/ncss/pdf/Procedures/NCSS/Passing-Bablok_Regression_for_Method_Comparison.pdf

Biosimiliars Section References

Brennan, Z. (2017). *FDA: Interchangeable Biosimilar Approvals Expected Within 2 Years.* Regulatory Affairs Professional Society https://www.raps.org/news-articles/news-articles/2017/6/fda-interchangeable-biosimilar-approvals-expected-within-2-years

Christl, L. *FDA's Overview of the Regulatory Guidance for the Development and Approval of Biosimilar Products in the US.* https://www.fda.gov/downloads/drugs/developmentapprovalprocess/howdrugsaredevelopedandapproved/approvalapplications/therapeuticbiologicapplications/biosimilars/ucm428732.pdf

European Medicines Agency (EMA). (2005, June). *Guidance for Industry Q5E Comparability of Biotechnological/Biological Products Subject to Changes in Their Manufacturing Process.* http://www.ema.europa.eu/docs/en_GB/document_library/Scientific_guideline/2009/09/WC500002805.pdf

European Medicines Agency (EMA). (2014). *EMA Guidance Document: Guidelines on Similar Biological Medicinal Products*. http://www.ema.europa.eu/docs/en_GB/document_library/Scientific_guideline/2014/10/WC500176768.pdf

Food and Drug Administration (FDA). (1996, April). *FDA Guidance Concerning Demonstration of Comparability of Human Biological Products, Including Therapeutic Biotechnology Products*. https://www.fda.gov/Drugs/GuidanceComplianceRegulatoryInformation/Guidances/ucm122879.htm

Food and Drug Administration (FDA). (1997, July). *S6 Preclinical Safety Evaluation of Biotechnology-Derived Pharmaceuticals*. https://www.fda.gov/downloads/Drugs/GuidanceComplianceRegulatoryInformation/Guidances/UCM074957.pdf

Food and Drug Administration (FDA). (2007, January). *Considerations in Demonstrating Interchangeability with a Reference Product*. https://www.fda.gov/downloads/Drugs/GuidanceComplianceRegulatoryInformation/Guidances/UCM537135.pdf

Food and Drug Administration (FDA) (2012, May). *S6 Addendum to Preclinical Safety Evaluation of Biotechnology-Derived Pharmaceuticals*. https://www.fda.gov/downloads/Drugs/GuidanceComplianceRegulatoryInformation/GuidancesCM194490.pdf

Food and Drug Administration (FDA). (2015a, April). *Scientific Considerations in Demonstrating Biosimiliarity to a Reference Product*. https://www.fda.gov/downloads/drugs/guidances/ucm291128.pdf

Food and Drug Administration (FDA). (2015b, April). *Quality Consideration in Demonstrating Biosimiliarity of a Therapeutic Protein Product to a Reference Product*. https://www.fda.gov/downloads/drugs/guidances/ucm291134.pdf

Food and Drug Administration (FDA). (2015c, April). *Biosimiliar: Questions and Answers Regarding Implementation of the Biologics Price Competition and Innovation Act of 2009*. https://www.fda.gov/downloads/drugs/guidances/ucm444661.pdf

Food and Drug Administration. (2016, December). *Clinical Pharmacology Data to Support a Demonstration of Biosimilarity to a Reference Product*. https://www.fda.gov/downloads/drugs/guidancecomplianceregulatoryinformation/guidances/ucm397017.pdf

Food and Drug Administration (FDA). (2017, September). *Statistical Approaches to Evaluate Analytical Similarity*. https://www.fda.gov/downloads/drugs/guidancecomplianceregulatoryinformation/guidances/ucm576786.pdf

International Council for Harmonisation of Technical Requirements for Pharmaceuticals for Human Use (ICH). (1995). *ICH Harmonized Tripartite Guideline: Quality of Biotechnological Products: Stability Testing of Biotechnological/Biological Products Q5C*, November 30. https://www.ich.org/

fileadmin/Public_Web_Site/ICH_Products/Guidelines/Quality/Q5C/Step4/Q5C_Guideline.pdf

Isakov, L., Jin, B., & Jacobs, I.A. (2016). Statistical Primer on Biosimilar Clinical Development. *American Journal of Therapeutics*. https://www.ncbi.nlm.nih.gov/pmc/articles/PMC5102275/pdf/ajt-23-e1903.pdf

Little, T. A. (2015). *Equivalence Testing for Comparability*. BioPharm International. http://thomasalittleconsulting.com/publications/articles/Equivalence_Testing_for_Comparability_2015.pdf

RapidMicro Biosystems, & The Rapid Blog. (2010). *Equivalency Tests & Rapid Method Validation or "Lies, Damn Lies, and Statistics"*. RapidMicro Biosystems. Posted August 13, 2010. http://www.rapidmicrobio.com/blog/bid/49445/Equivalency-Tests-Rapid-Method-Validation-or-Lies-damn-lies-and-statistics

Rosnow, R. L., & Rosenthal, R. (2003). Effect Sizes for Experimenting Psychologists. *Revue Canadienne de Psychologie Expérimentale*, 57(3), 237. http://www2.psych.ubc.ca/~schaller/528Readings/RosnowRosenthal2003.pdf

NAb and ADA Section References

Food and Drug Administration (FDA). (2016, April). *Draft Guidance for Industry: Assay Development and Validation for Immunogenicity Testing of Therapeutic Protein Products*.

Shankar, G., Devanarayan, V., Amaravadi, L., Barrett, Y.C., Bowsher, R., Finco-Kent, D., Fiscella, M., Gorovits, B., Kirschner, S., Moxness, M., Parish, T., Quarmby, V., Smith, H., Smith W., Zuckerman, L.A., & Koren, E. (2008). Recommendations for the validation of immunoassays used for detection of host antibodies against biotechnology products. *Journal of Pharmaceutical and Biomedical Analysis*, 48(5), 1267–1281. https://pubmed.ncbi.nlm.nih.gov/18993008

Zhang, L., Zhang, J., Kubiak, R. J., & Yang, H. (2013). Statistical methods and tool for cut point analysis in immunogenicity assays. *Journal of Immunological Methods*, 389(1–2), 79–87. https://pubmed.ncbi.nlm.nih.gov/23305975

11

Bioanalytical Aspects in Biological Therapeutics: Biotransformation

Wenying Jian[a], Cong Wei[b] and Jinping Gan[c]

[a]Janssen Research & Development, Johnson & Johnson
[b]Drug Metabolism and Pharmacokinetics, Biogen Inc.
[c]HiFiBiO Therapeutics (corresponding author)

11.1 Introduction

Biotransformation is a biological process in vivo that results in the chemical modification of xenobiotics. Biotransformation typically occurs for all biologics as the primary means of clearance from the body. The rate of this process determines the pharmacokinetics (PK) of therapeutic proteins therefore affecting the exposure over time, which is key to the therapeutic effect of biologic drugs. The biotransformation process results in the generation of metabolites, which may have similar or different biological activity to the parent entity and, in some cases, may possess biological properties that are toxic. Moreover, the metabolites may have different PK properties than the parent, resulting in accumulation in plasma or specific organs, leading to unintended consequences.

Fortunately, well-behaved protein therapeutics go through proteolysis by various proteases throughout the body as the primary means of biotransformation. The proteolytic process generates amino acids/small peptide fragments that are typically not of concern except for the potentially immunogenic activities of peptides. There are, however, many biotherapeutics that are subjected

Bioanalytical Aspects in Biological Therapeutics, First Edition. Edited by Xiaohui Xu and Weifeng Xu.
© 2022 John Wiley & Sons, Inc. Published 2022 by John Wiley & Sons, Inc.

to other types of biotransformation, resulting in relatively stable metabolites that circulate in the body. The characterization of these metabolites is important in explaining unanticipated PK or pharmacodynamic (PD) effects. In addition to profiling the serum or plasma samples from subjects dosed with the protein therapeutics, sometimes it is also important to investigate their disposition in the tissues of interest. Moreover, the investigation into the identity and amount of these metabolites across species may help explain differences in drug clearance, adverse effects, or pharmacological activities.

Mass spectrometry (MS) is an important tool in the characterization of these metabolites. Recent advances in high-resolution MS have enabled qualitative and quantitative analysis of proteins of different sizes and formats in complex biological matrices such as plasma and tissue homogenates. Because of the complex nature of biological matrices, it is often critical to have robust sample preparation techniques to enrich the analytes of interest while getting rid of most of the proteins and lipids from the matrices. This is often accomplished by specific antibody-enabled affinity capturing, but in some cases other methods can be used based on the physicochemical properties of the analytes. Depending on the molecular nature of the therapeutic protein and its anticipated metabolites, different methods of sample processing are often used to obtain either the intact analytes of interest or the peptides generated by tryptic digestion. In some cases, specific chemical or enzyme cleavages are used to yield medium-sized protein fragments. The resulting products are subjected to different liquid chromatography (LC)-MS-based analysis such as intact analysis or peptide mapping, depending on the nature of the samples and the anticipated metabolites.

In this chapter we will introduce the concept of biotransformation of biologics, and its importance in the drug discovery and development process. In addition, we will also introduce different MS techniques that are utilized in the investigation of biologics biotransformation.

11.2 Biotransformation of Biologics

The sections that follow will go into detail about the biotransformation of different types of protein therapeutics. Each section will provide a general strategy, workflow, and technical highlights for the protein type of interest with examples from the literature.

11.2.1 Biotransformation of Monoclonal Antibodies

Therapeutic monoclonal antibodies (mAbs) are among the most promising and fast-growing classes of drug molecules due to several advantages such as high specificity, low off-target effect, and low attrition rate in drug discovery and development. In comparison to traditional small molecule drugs, mAbs exhibit unique ADME (absorption, distribution, metabolism, and elimination) properties because of their large molecular size (~150 kDa), complex structure (multidomain protein made of two light chains and two heavy chains conjugated by multiple disulfide bonds), and complicated interaction with other molecules in the body (Lee, 2013; Tibbitts et al., 2016). In general, mAbs are considered to be intrinsically stable in the body and their biotransformation products, usually peptides and amino acids, are biocompatible with no pharmacological or toxic effects. Therefore, extensive biotransformation information for mAbs is generally not required for regulatory submission. However, it has been recognized that certain biotransformations may impact the stability, potency, PK, and immunogenicity of the antibody and therefore warrant investigation (Ezan et al., 2014; Hall, 2014).

11.2.1.1 MS "Bottom-Up" Approach

Ligand binding assays (LBAs), such as enzyme-linked immunosorbent assay (ELISA), have been the gold standard for analyzing biotherapeutics such as mAbs in biological samples. They have the advantage of being highly sensitive and efficient but lack the capability to elucidate biotransformation events. Conversely, LC-MS has long been the primary tool to analyze and characterize small molecule therapeutics for investigation of their ADME properties. In recent years, with the advancement of sample preparation and instrumentation, LC-MS has emerged as an important complementary analytical platform for large molecules for understanding their plasma/tissue exposure, immunogenicity, target engagement, and biotransformation (Kellie & Karlinsey, 2018; Van De Merbel, 2019). Due to its intrinsic ability to elucidate structure information, LC-MS provides a unique advantage in revealing biotransformation events of biotherapeutics in the body.

It is particularly challenging to analyze large proteins such as mAbs directly by MS due to their large size. As molecular weight increases, ionization efficiency decreases, and number of charge states increases, both severely impacting the signal intensity that can be obtained for certain molecules in MS detection. Therefore, the most commonly used approach to analyze proteins by LC-MS is a "bottom-up" one, by enzymatically digesting the protein into peptides that are more compatible for LC-MS analysis, a process also known as peptide mapping. By means of a "bottom-up" approach, biotransformations at the amino acid level can be detected, such as deamidation, Asp isomerization, oxidation, pyroglutamate formation, C-terminal lysine clipping, glycosylation changes, and so on. These are among the critical quality attributes (CQAs) that define the characteristics of biotherapeutic products and are routinely monitored during the manufacturing process. In recent years, there has been increased interest in the detection of changes in these CQAs of therapeutic mAbs from in vivo samples, which provide insight into the PK effects of different modifications and make it possible to establish meaningful manufacturing ranges of CQAs that are supported by clinical evidence (Li et al., 2016a).

The LC-MS methodology for detecting in vivo biotransformation of mAbs usually involves reversed-phase LC separation of the peptides generated from trypsin digestion followed by full scan triggered tandem MS (MS^2) detection on hybrid high-resolution MS (HRMS) such as the triple-TOF or QExactive instrument (Figure 11.1) (Li et al., 2016b).-The HRMS spectra of the parent ions and their *b*-ion and *y*-ion fragments generated under collision-induced dissociation (CID) can be matched to a database that contains not only the predicted *m/z* of unmodified peptides but those with predefined modifications. Recently, Yao et al. reported a workflow of untargeted determination of universal biotransformation profile of biologics using LC-MS-based methodology with the assistance of bioinformatics tools. Differential analysis can filter away the modifications that are shared by control and treated samples and therefore improve the specificity of the detection (Yao et al., 2018). Alternatively, there have been reports on monitoring known modifications such as deamidation by multiple reaction monitoring (MRM) on triple quadrupole MS (Bults et al., 2016; Mehl et al., 2016). The highly specific and sensitive detection with MRM provided targeted quantitation of known modifications.

11.2.1.2 MS "Top-Down" Approach

The "bottom-up" approach provides sensitive and specific detection of biotransformation at the amino acid level. However, digesting the protein into a mixture of peptides increases the complexity of the samples and eliminates

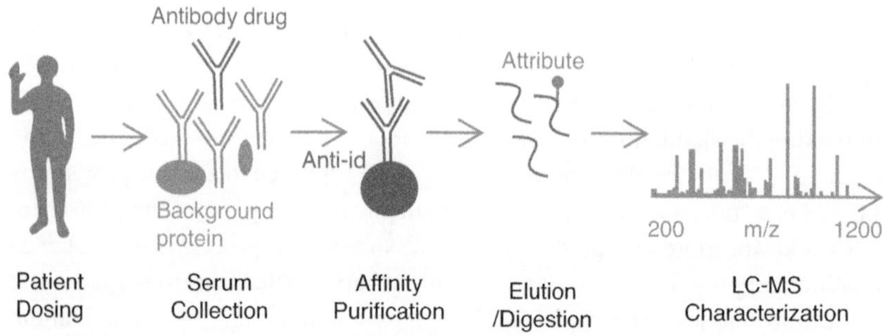

Figure 11.1 Scheme of affinity-purification coupled with mass spectrometric analysis for profiling quality attributes of antibody drug from patient samples (Li et al., 2016b) (open access).

high-level sequence and structure information. Particularly, for biotransformations involving proteolysis, peptides derived from proteolytic catabolites may not be differentiable from peptides generated by enzymatic digestion, or not suitable for LC-MS detection. Therefore a "top-down"-based intact protein approach is a necessary complementary platform for revealing protein biotransformation.

In a typical "top-down" workflow, after sample purification, the proteins are directly subjected to reversed-phase LC separation followed by full-scan high-resolution MS analysis (Van Den Broek & Van Dongen, 2015). Under electrospray ionization, the raw spectra of an mAb usually show multiple charged peaks (typical charge state 40–70) in multiple glycosylation forms (if a deglycosylation procedure is not done), which can be further complicated by the existence of biotransformation products. The raw spectra need to be deconvoluted by the software to reconstruct the spectra of zero or singly charged peaks for data interpretation and to reveal biotransformation products (Figure 11.2). There have also been reports of using off-line MALDI-TOF for intact protein analysis to investigate protein biotransformation. Under MALDI conditions, the analyte is primarily singly charged and therefore does not require deconvolution. "Top-down" analysis can readily detect high-level structure information

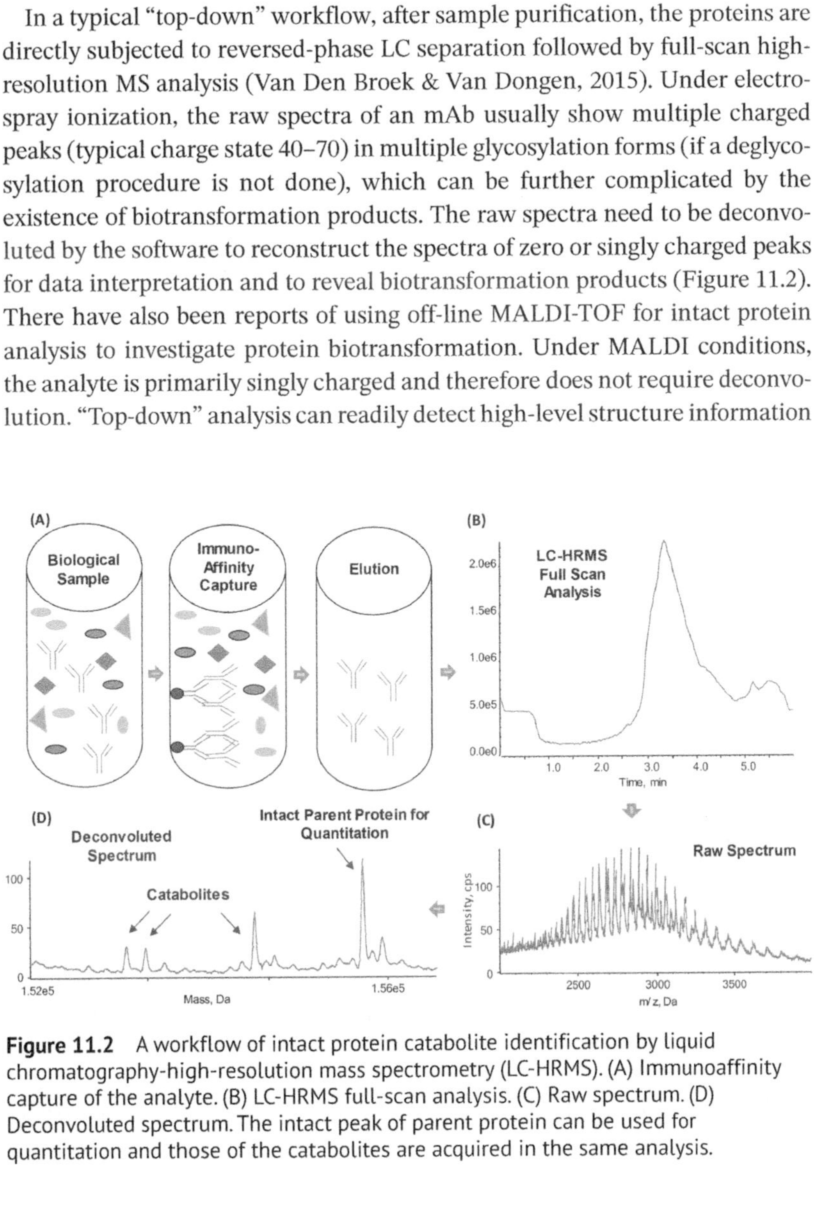

Figure 11.2 A workflow of intact protein catabolite identification by liquid chromatography-high-resolution mass spectrometry (LC-HRMS). (A) Immunoaffinity capture of the analyte. (B) LC-HRMS full-scan analysis. (C) Raw spectrum. (D) Deconvoluted spectrum. The intact peak of parent protein can be used for quantitation and those of the catabolites are acquired in the same analysis.

such as existence of multiple subunits that are conjugated by disulfide bonds. Modifications that have large mass shift, such as C-terminal lysine clipping and glycosylation change, can also be visualized directly in deconvoluted intact MS spectra.

One major drawback of the "top-down" approach is its limited sensitivity and resolution for large proteins, because of their large size, low ionization efficiency, and dilution of signal among multiply charged states. Low-abundant biotransformation species may be buried as a shoulder peaks off main peaks and therefore cannot be detected or their abundance reliably estimated. An effective alternative approach to improve sensitivity and peak resolution is "middle-down" analysis in which the molecules are broken down into relatively smaller "intact" pieces. For mAbs this can be achieved by a reduction using DTT (dithiothreitol) or TCEP (tris(2-carboxyethyl)phosphine) to break the disulfide bonds, resulting in two copies of light chain (~25 kDa) and two copies of heavy chain (~50 kDa) protein that are more compatible with LC-MS analysis in comparison to the 150 kDa whole protein. Furthermore, mAbs can be specifically cleaved around the hinge region by IgG protease such as IdeS (immunoglobulin-degrading enzyme from *Streptococcus pyogenes*), SpeB, and Kpg, to reduce the size of the heavy chain. IdeS is the most commonly used IgG protease that cleaves below the hinge region to generate F(ab′)2 and Fc/2 fragments, which upon reduction produce three types of fragments approximately ~ 25 kDa each, amenable to LC-MS analysis. Kellie et al. used this approach to quantify mAbs in plasma samples at the intact level and, more importantly, to monitor different drug variants such as glycoforms, which is a critical aspect of CQA analysis and of general understanding of biotransformation (Kellie et al., 2016).

11.2.1.3 Sample Preparation

It should be noted that LC-MS-based analysis including both peptide mapping and intact analysis are mature techniques for characterization of purified materials and there is copious literature on their applications. However, their utilization for investigation of in vivo protein biotransformation has just emerged in recent years thanks to advancement in sample preparation techniques that effectively clean up the biological samples for detection of targeted analytes and their catabolites. Generally speaking, different techniques can be used for sample preparation for LC-MS-based quantitation of proteins in biological samples, ranging from protein precipitation/pellet digestion, solid phase extraction, abundant protein depletion (serum albumin depletion), to immunoaffinity capture (IAC). Among them, the most effective clean-up method is IAC and it is the only one that can provide samples clean enough for biotransformation investigation. In a typical immunoaffinity workflow, capture antibody immobilized on magnetic beads or a solid phase (such as resin) is used to pull-down the target mAb while the other components such as proteins, peptides, salts, and lipids are washed away. The bound analyte antibody can be then eluted for trypsin digestion for peptide mapping or direct intact LC-MS analysis. For investigation of human mAbs in preclinical species, a general anti-human Fc can serve the purpose and effectively concentrate the human antibody without pulling down the IgGs of the dosed animals. For human plasma samples, however, an anti-idiotype (anti-ID) antibody is required to specifically pull down the target analyte without interference from other IgGs in human plasma. Alternatively, a target protein that is recognized and bound by the analyte mAb can be used as the capture reagent.

For the purposes of biotransformation investigation, the capture reagent should be "nonspecific" enough to enable capture and detection of the catabolites. For example, if the major biotransformation is expected to take place in the complementarity-determining region (CDR) region of an mAb, a capture antibody recognizing the Fc region would be expected to bind to both unchanged drug and its catabolites. Conversely, an antibody targeting the CDR may be blocked from binding to the catabolites due to biotransformation in that region, as has been observed by Bults et al. for deamidation at the CDR of trastuzumab (Bults et al., 2016). In this sense, a polyclonal antibody that recognizes multiple epitopes may be more effective in detecting biotransformation events.

Monoclonal antibodies are glycosylated and these sugar chains may play an important role in their interaction with other molecules, their PK behavior, and functions. On the other hand, glycosylation introduces heterogeneity to the molecules, particularly when it comes to LC-MS analysis. The MS signal can be diluted among different glycoforms and the closely distributed peaks could complicate spectra interpretation. If the focus of the biotransformation investigation is not on glycosylation, *N*-linked glycans, which are the major glycosylation ??? in mAb biotherapeutics, can be removed enzymatically by PNGase F. The resulting mass spectra can be simplified with a signal concentrated on the deglycosylated form. However, care should be taken to avoid incomplete deglycosylation or introducing additional artifacts, both of which increase the complexity of the spectra.

11.2.1.4 Common Biotransformation of mAbs

11.2.1.4.1 Deamidation and Isomerization

Spontaneous deamidation could take place on asparagine (Asn) or glutamine (Gln) residues in proteins during manufacture, formulation, storage, as well as in vivo in circulation as a biotransformation event. Deamidation of Asn involves formation of a succinimide intermediate which quickly converts to aspartic acid (Asp) or isoaspartic acid (isoAsp), and Asp can undergo the same intermediate transition to convert to isoAsp, and vice versa (Figure 11.3). It has been shown that deamidation and isomerization at certain regions of an mAb could impact its target binding affinity, potency, efficacy, safety, and PK (Vlasak et al., 2009). Therefore, it is important to characterize in vivo deamidation and isomerization events and in some cases quantify their relative abundances to gain a better understanding of the PK–PD relationship. The rate of deamidation is dependent upon multiple factors such as pH, temperature, ionic strength, primary sequence, local polypeptide conformation, and tertiary structure. Regarding primary sequence, adjacent Gly and Ser are the most destabilizing residues, while Ser, Thr, and Lys also favor deamidation when they are located on the N-side of Asn (Chelius et al., 2005). Even though both Asn and Gln can undergo deamidation, the rate of deamidation of Gln is much slower and therefore Asn has been of primary interest for in vivo biotransformation investigation. Characterization of in vivo deamidation and isomerization involves IAC followed by proteolytic enzyme (typically trypsin) digestion and LC-MS detection of tryptic peptides. HRMS has been the standard platform when characterizing unknown biotransformations. In the case when the modifications have been identified, tryptic peptides containing the modifications can be monitored by MRM on a triple quadrupole MS instrument in a targeted fashion to afford better sensitivity and more reliable quantitation. The molecular weight of Asn and its deamidated product Asp differ by only one mass unit (+1 Da from Asn to Asp) while there is no mass difference between Asp and isoAsp. It is important to have chromatographic separation for peptides containing these three distinct residues in order to have unambiguous identification/quantification. Typically, in vitro thermally stressed stability

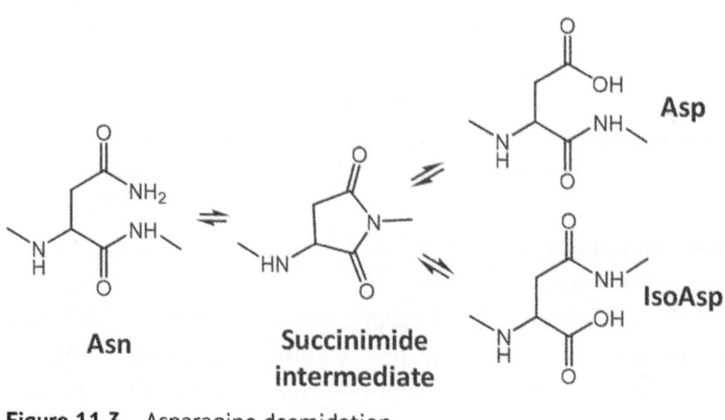

Figure 11.3 Asparagine deamidation.

samples that have been confirmed to have undergone deamidation and isomerization can serve as a reference for verification of the separation.

Huang et al. investigated in vivo deamidation on Asn55 located in the CDR region of a humanized mAb, which is of particular interest since deamidation at this site greatly decreases binding activity (Huang et al., 2005). The half-life of deamidation at Asn55 in vivo in cynomolgus monkeys after intravenous (IV) or subcutaneous dose were determined to be ~ 277 hours. The ratio of the isoAsp55/Asp55 remains constant at 2:1 during the time course studied, suggesting the clearance rate of the two derivatives are similar. Oqullette et al. studied deamidation of Asn in an Asn-Gly motif in the heavy chain binding region of an mAb and revealed the presence of a succinimide intermediate, which was stable in vitro but unstable in vivo in cynomolgus monkeys and rapidly converted to Asp/isoAsp (Ouellette et al., 2013). By day 50 after dosing, complete deamidation to Asp/isoAsp was observed. Even though the succinimide intermediate had much lower binding affinity to the antigen than the unchanged form, the deamidation product showed little loss of binding affinity. This study demonstrated the importance of evaluating modifications to mAbs in vivo for a better understanding of their biological relevance. Mehl et al. investigated in vivo Asp isomerization and Asn deamidation in cynomolgus monkeys on CDR of two different mAbs, respectively (Mehl et al., 2016). It was shown that the isomerization took place relatively slowly and isoAsp comprised less than 10% of the dosed drug 6 weeks after IV dosing. In comparison, deamidation was more extensive and 80% of the dosed mAb was converted to the Asp or isoAsp form by day 42 after IV dosing. Li et al. reported rapid Asn deamidation in Fc domains of mAbs both in cynomolgus monkeys and in humans, which suggested that patient exposure to deamidated species is primarily dependent on in vivo modification rather than those generated during drug production and storage (Li et al., 2016a, 2016b), Bults et al. developed a bioanalytical method to quantify deamidation products and the succinimide intermediate of Asn55 in CDR of trastuzumab in cancer patients. After 365 days of treatment with trastuzumab, a total of 24% of the mAb was deamidated mainly into the isoAsp55 form and much less to Asp55 (Bults et al., 2016). The same study also showed that deamidation leads to a loss of binding affinity to the antibody used in the ELISA assay, reflected by a lower concentration measured by ELISA than those measured by using a signature peptide from a stable part of the mAb by LC-MS/MS. The impact of deamidation/isomerization on bioanalytical assays, particularly those involving affinity capture by antibody such as hybrid ligand-binding LC-MS/MS assay or LBA assay, was investigated in depth in a recent study and it was proposed that binding conditions should be carefully evaluated to avoid generating misleading concentration data (Liu et al., 2018).

11.2.1.4.2 Oxidation

Met (methionine) oxidation is a very common modification that can impact the bioactivity and PK behavior of the mAb and potentially induce an immunogenic response (Mo et al., 2016). In IgG mAb, the sequences in the constant region involved in FcRn binding are highly conservative and contain Met residues. It has been reported that Fc Met oxidation can result in reduction in FcRn binding affinity, and accordingly a decrease in serum circulation half-life (Wang et al., 2011). There are also mAbs with Met in CDR and oxidation could impact antigen binding. Li et al. assessed multiple CQAs of an mAb during circulation in cynomolgus monkey, including oxidation of Met249 in the heavy chain which is involved in FcRn binding (Li et al., 2016a). It was found that the oxidation remained relatively constant at 2.5% and then decreased to below 1% beyond 10 days, which is consistent with potentially increased clearance of oxMet249 via decreased FcRn binding. Another investigation in clinical samples showed that oxMet on the Fc of an mAb decreased slightly from ~ 8.5% to 7% over the course of 6 weeks, consistent with slight accelerated clearance of Fc-oxidized mAb (Li et al., 2016b). Overall, this suggests that blood is not an oxidation environment and, as a result, oxidation of mAbs is not increased during circulation.

11.2.1.4.3 N-Terminal Glutamate to Pyroglutamate

Both glutamine and glutamate at the N-termini of an mAb can cyclize to form the cyclic amino acid pyroglutamate (pE) (Figure 11.4). Because the N-terminal residues of both the light chains and the heavy chains are in the variable region near the CDR, pE formation can potentially impact target binding. Liu et al. investigated pE formation of different mAbs in human in vivo (Liu et al., 2011). It was found that the rate of pE formation is variable among different antibodies and the different sequence context within he same mAb. For one mAb tested, the pE level on its heavy chain increased linearly from less than 2%

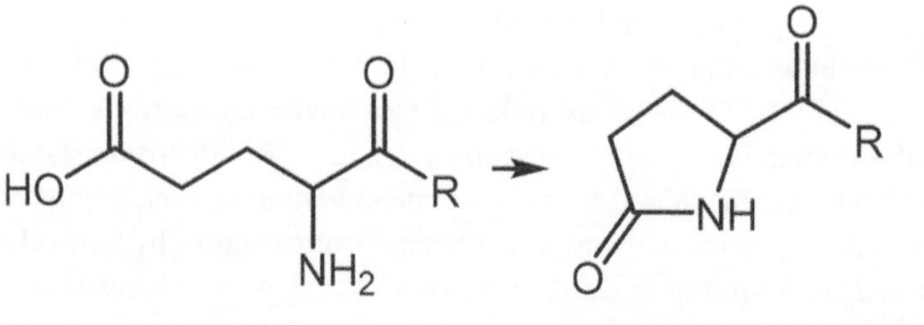

Figure 11.4 Formation of pyroglutamate from glutamate.

to ~ 12% over 34 days after dosing, while that on the light chain showed little change. The changes and their rates could be reproduced in phosphate-bufferd sline (PBS) incubation in vitro, suggesting that the changes observed in vivo were due to chemical conversion but not enzymatic reaction or differential clearance. In another clinical study, Li et al. observed a complete conversion of N-terminal glutamine of an mAb into pE from 96% to 100% in humans within 2 weeks of dosing (Li et al., 2016b).

11.2.1.4.4 C-Terminal Lysine Processing

The heavy chain gene of all the human IgG subtypes encodes a C-terminal Lys, while this residue is mostly absent from the endogenous antibodies isolated from human serum. Cai et al. studied the rate of C-terminal Lys clipping of a therapeutic mAb injected into humans (Cai et al., 2011). It was found that the C-terminal Lys was rapidly lost in vivo with a half-life of about an hour (62 minutes). In vitro incubation in conditions that inactivate enzyme activity such as inhibitors or heat stress significantly slowed down the process, indicating that C-terminal Lys clipping is an enzymatic process (Cai et al., 2011; Yin et al., 2013). No known function has been attributed to C-terminal Lys and this variability is rapidly lost upon injection.

11.2.1.4.5 Glycosylation and Glycation Change

Glycosylation on the CH2 domain of the Fc region of IgG is essential for specific effector functions and interaction with FcRn, and therefore plays an important role in determining the efficacy and in vivo half-life of an mAb. Goetze reported that the relative level of high mannose-containing glycoform (Man5) decreased as a function of circulation time in vivo in humans, while other glycans remained constant (Goetze et al., 2011). This indicates that the Man5 glycoform is cleared more rapidly in humans than other glycan forms. Li et al. made similar observations in both the cynomolgus monkey and humans (Li et al., 2016a, 2016b). In the monkey study, the relative percentage of major glycoforms (G0F, G0F-GlcNac, G1F) of an mAb remained constant in circulation 7 weeks after administration of the mAb. In comparison, the percentage of Man5 glycoform was reduced from ~ 4.5% to below detection level within 7 days. In the human study, the percentage of Man5 decreased from ~ 3% to ~ 1% within 2 weeks. Therefore, Man5 is considered a critical quality attribute which may impact the PK profile of the mAb therapeutics if present at a high level in the drug product.

Glycation is a type of nonenzymatic glycosylation in which a protein is modified by a reducing sugar by means of a reaction between a primary amine on the protein and the carbonyl carbon of the sugar. Goetze et al. studied glycation of an IgG in vivo and found that the glycation levels slowly increased over time at 0.00092 and 0.0021 glucose additions per chain per day for light chain and heavy chain, respectively (Goetze et al., 2012). Levels of glycation found on the Fc portion of endogenous IgG isolated from humans showed an average of 0.045 glucose additions per fragment. Highly glycated IgGs showed no change in binding to FcRn or protein A in comparison with less glycated ones. It was concluded that glycation is a naturally occurring process which modifies therapeutic mAbs and endogenous antibodies alike, and has no impact on Fc function.

11.2.1.4.6 Disulfide Bond Modifications

IgG contain multiple intra- and inter-chain disulfide bonds which are important for assembling and retaining the correct structure of the protein. Due to their relatively low dissociation energy, disulfide bonds can be subjected to multiple modifications such as disulfide bond reduction, thioether bond formation, disulfide scrambling, addition of Cys/glutathione/free light chain, and so on, some of which have been investigated in vivo(Moritz & Stracke, 2017). Liu et al. studied different subclasses of human IgG2 antibody that differ in disulfide linkage and showed that these disulfide linkages interconvert while circulating in humans, which may alter the activity of the antibody over its lifetime (Liu et al., 2008). Another modification to disulfide bonds is formation of thioether, also called lanthionine, as the result of elimination of one of the sulfur atoms in the bond, producing a carbon-sulfur-carbon linkage (-C-S-C-). Zhang et al. found that thioether at LC214-HC220 disulfide linkage in an IgG1 is formed at a rate of 0.1%/day while circulating in human blood (Zhang et al., 2013). The same modifications were found in the same position in endogenous antibodies isolated from human subjects, at levels consistent with this conversion rate. Overall, the impact of disulfide bond modifications on the function and safety of the mAb should be considered in the context of their in vivo occurrence in endogenous antibodies and in vivo repair mechanisms.

11.2.1.4.7 Challenges of In Vitro–In Vivo Correlation

Unlike small molecule drugs, for which there are mature systems and tools to investigate their biotransformation properties in vitro and in vivo and to establish their correlation, there is a significant knowledge gap for large molecules. For therapeutic mAbs, an in vitro buffer system (such as PBS) or plasma incubation under stressed conditions have been used to investigate their potential biotransformation. However, it was found that the in vitro system does not fully represent in vivo situations in terms of rates and types of modifications (Yin et al., 2013). For example, an investigation conducted by Yin et al. showed that C-terminal lysine processing, deamidation, pyroglutamate formation, and thioether bond formation are enhanced for the tested mAbs in plasma incubation and in vivo circulation in comparison to PBS incubation. In contrast, PBS showed no C-terminal lysine processing, and slower rates of the other three modifications. Other modifications such as oxidation, Asp isomerization, glycation, and aggregation do not show significant differences among the three systems. This work demonstrated the challenges of using an in vitro system to predict in vivo biotransformation and the importance of investigating using relevant matrix and biological systems. In general, comparison across the in vitro, preclinical, and clinical spaces might provide valuable information regarding biotransformation properties.

11.2.2 Biotransformation of Fusion and Recombinant Proteins

11.2.2.1 Biotransformation Investigation of Fusion and Recombinant Proteins

Many pharmacologically active peptides or proteins exhibit a short in vivo half-life due to rapid proteolytic degradation and/or fast urinary clearance. Newer generation of biotherapeutics have been developed to improve the PK characteristics of these biologics so that they can be developed into medications with the desired dose level and dose frequency. Examples of the strategies to extend serum half-life include fusion/conjugation/binding with Fc or serum albumin, conjugation to polymers, fusion to polypeptides, and so on (Kontermann, 2016). Increasing the size and adding a large moiety to a molecule can protect it from proteolytic degradation and urinary clearance. In addition, a molecule containing Fc or serum albumin moieties can go through the FcRn-mediated recycling pathway to take advantage of the natural half-life extension mechanism (Wu & Sun, 2014). For the novel engineered molecules, there has been significant interest in investigating their biotransformation in order to understand its impact on stability and to develop molecules with the desired PK profile (Pearson & Rock, 2015).

Half-life extended fusion molecules such as peptide-Fc fusion protein (also known as Fc fusion protein or peptibody) typically contain a naturally unstable part (the peptide) fused to the more stable scaffold (the Fc). The main type of biotransformation is proteolytic degradation of the peptide which determines the serum half-life of its biological activity, while other changes typical for Fc may take place but have been of less interest for biotransformation investigation. Biotransformation can be revealed by differential LBAs utilizing different reagents recognizing specific parts of the analytes (Hager et al., 2013; Murphy et al., 2010). For example, for a GLP1-mAb conjugated protein, quantitation was done with two different LBAs that both used anti-ID capture of the mAb but an anti-human IgG for detection of mAb only (total assay), or anti-GLP1

(N-terminal specificity) for detection of intact GLP-1 (N-terminal assay) (Murphy et al., 2010). The drug concentration measured by the two assays showed that N-terminal concentrations are significantly lower than that of total mAb concentration, indicating that the GLP-1 portion was quickly degraded in vivo. However, to determine the exact site of the proteolytic degradation, MS studies are warranted.

11.2.2.2 MS "Top-Down" Approach

For investigation of proteolytic degradation, "top-down" intact analysis has been the method of choice as it retains the high-level structure information of the analyte. In comparison, in the "bottom-up" peptide mapping approach, peptide derived from proteolytic catabolite may not be detectable under LC-MS conditions or not be differentiated from peptide derived from enzymatic digestion used for sample preparation. It is highly challenging to analyze large proteins in biological matrices at an intact level due to their large size and complex structure. Recent advancement in sample preparation techniques and instrumentation has made it possible, which usually involves IAC of the target protein and its catabolites from biological samples, followed by high-resolution MS analysis. An early study on Decoy receptor 3 (DcRs) used tissue homogenate to evaluate stability and catabolite profile of the recombinant DcRs protein (Wroblewski et al., 2003). IAC capture followed by RP-LC fractionation and MALDI-TOF analysis revealed a major cleavage site at R218, which was engineered into a glutamine residue. The resulting DcR3 (R218Q) protein demonstrated improved stability in vitro and in vivo. For the GLP1-mAb that demonstrated a discrepancy between total and N-terminal assays discussed earlier, online two-dimensional LC-MS analysis that utilized an anti-ID affinity column followed by RPLC and QTOF MS analysis revealed the major catabolite with mass loss of 210 Da from the intact conjugate, corresponding to the loss of the first two amino acids from the N-terminus of GLP1 peptide (Murphy et al., 2010). Another biotransformation study on Fc fusion protein of peptides that target thrombopoietin receptor using anti-human Fc IAC and MALDI-TOF MS analysis revealed major proteolytic sites on three different constructs (Hall et al., 2010). Biotransformation study of Fc-FGF21 (fibroblast growth factor 21) fusion proteins utilizing anti-human Fc capture and MALDI-TOF MS analysis revealed a proteolytic site of Pro171 on FGF21, which enabled development of more stable drug candidates by mutating the Pro into Gly (Hecht et al., 2012). Similar observation of the catabolite profile of a wild-type Fc-FGF21 construct and Fc-FGF21 (P171G) was made by a recent work utilizing IAC followed by capillary electrophoresis-TOF MS for intact mass analysis (Han et al., 2017). In an in vivo biotransformation investigation of a novel recombinant protein of tetranectin and apolipoprotein (TN-ApoA1), the protein was purified from rabbit plasma by IAC using an antibody that binds to the fusion region, followed by monolithic RP capillary LC and TOF HRMS analysis (Zell et al., 2016). Two catabolites, the major one formed by truncation from the N-terminus and the minor one formed by truncation from both sides, were identified. The individual PK profile of the unchanged protein and the two catabolites were calculated by applying the total protein concentration measured by ELISA to the MS signal intensity at each sampling point, assuming equal MS responses.

Recently, a workflow for simultaneous intact quantitation and catabolite identification has been proposed in which immune-affinity captured analytes are analyzed by RPLC TOF mass analysis. The deconvoluted full MS spectra contain rich information not only of the intact parent protein that can be used for quantitation, but also that of catabolites for elucidation of proteolytic sites on the fused peptide. Combination of this intact workflow with a "bottom-up" approach is exemplified by a recent publication using dulaglutide, a GLP1-Fc fusion protein as the model molecule (Kang et al., 2017). In this work, dulaglutide along with its in vivo catabolites, were purified from mouse plasma using anti-Fc antibody, followed by LC-MS/MS analysis on a triple quadrupole instrument after digestion, or by intact LC-HRMS analysis with triple TOF MS without digestion. As demonstrated in Figure 11.5, the total Fc measured by using a surrogate peptide on Fc was

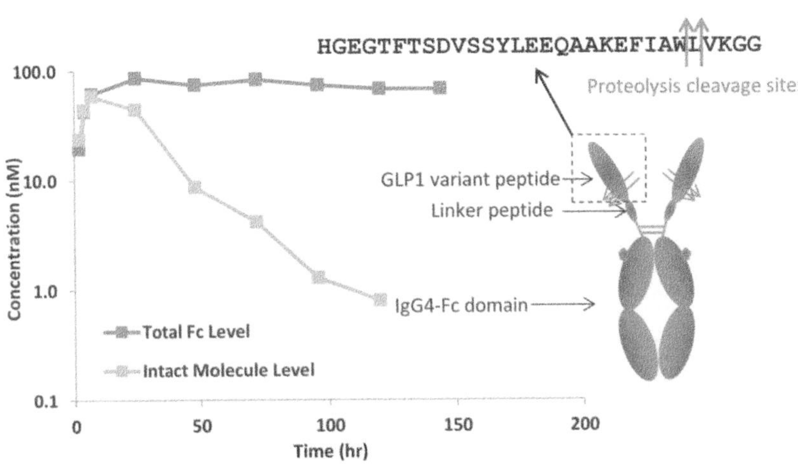

Figure 11.5 Simultaneous catabolite identification and quantitation of dulaglutide at intact level by immunoaffinity capture liquid chromatography high-resolution mass spectrometry. Reprinted with permission from Kang et al. (2017). Copyright (2017) American Chemical Society.

stable and maintained at a high level during circulation. Meanwhile, glucagon-like peptide 1 (GLP1) peptides fused to the N-terminus of Fc are subjected to proteolytic degradation, leading to loss of intact concentration, as revealed by simultaneous qualitative and quantitative intact MS analysis. The integrated LCMS approach provided an effective way to gain an in-depth understanding of the in vivo fate of the fusion proteins.

11.2.2.3 MS "Bottom-Up" Approach

As discussed earlier, it is uncommon to investigate proteolytic biotransformation using the "bottom-up" approach. However, there is a novel example of utilizing a tagging technique in combination with peptide mapping for characterization of certain catabolites (Kullolli et al., 2017). In this reported workflow, a neurotensin human Fc (NTs-huFc) fusion protein is purified from mouse serum using anti-human Fc antibody, followed by on-beads tagging with (2,4,6-trimethoxyphenyl N-terminus (TMPP-NTT), which specially reacts with alpha-amine at the N-terminus of the protein as well as its catabolites. After trypsin digestion, the samples were analyzed on HRMS for a data-dependent scan for peptide identification. The positively charged hydrophobic TMPP tag facilitates unambiguous sequence identification of all N-terminus peptides, including those of the two major catabolites due to truncation of the N-terminus. As a powerful orthogonal technique to intact LC-MS method, this workflow can be utilized to investigate all possible N-terminus proteolytic degradations for a protein of interest. A similar approach can be applied for C-terminus catabolites, if a suitable tagging reagent is available.

11.2.3 Biotransformation of Antibody–Drug Conjugates (ADCs)

11.2.3.1 Biotransformation Identified with Intact ADCs

To better understand ADC biotransformations, it is ideal to analyze ADC in the intact form with minimal treatment so that its structural information can be preserved as much as possible. Due to the limited sensitivity of the detection of intact ADC, there has been little investigation of biotransformations with intact ADC. A recent publication has shown that small mass changes on ADCs such as glycated and/or hydrolyzed species of ADC were identified using full-scan LC-HRMS analysis with the deconvolution of raw spectra (He et al., 2017). Further analysis showed that the hydrolysis/hydration observed in the ADC was attributed to the linker-drug rather than the antibody. When the structure of the linker-drug was examined, one potential hydrolysis site (i.e., the succinimide ring in the linker) and two potential hydration sites (i.e., the two pyrrolobenzodiazepine [PBD] imines in the payload) were identified. In addition, potential protein adducts derived from ADCs can also be identified using HRMS as discussed in the next section. When the linker of ADC is acid-labile, enzymatically cleavable, or chemically cleavable (e.g., desulfurization; Wang et al., 2019), the payload can be released for further characterization of biotransformation.

11.2.3.1.1 Maleimide Exchange/Hydrolysis of Succinimide on ADCs

Due to the reversible nature of maleimide chemistry, ADCs based on maleimide conjugation may release the payload. When the payload is deconjugated from ADCs in vivo, besides the payload in the unconjugated form, the released payload may also covalently bind to matrix proteins forming protein-adduct. Several studies have identified an albumin-linker-payload adduct forming in plasma as a major adduct during in vivo or ex vivo studies of maleimide-containing ADCs (Figure 11.6) (Shen et al., 2012b; Wei et al., 2016). A workflow for adduct identification and quantification has been proposed (Figure 11.7A) (Wei et al., 2016). To identify and characterize the payload-protein adduct, the ADCs would be removed/depleted from the matrix sample by immunoprecipitation using antigen or anti-Fc Ab, and the ADC-depleted sample is then further immunocaptured with anti-payload Ab followed by LC-full-scan HRMS analysis on the eluted samples. The deconvoluted full MS spectra would illustrate the major intact linker-payload-protein adduct MS spectra and potentially any other catabolites/metabolites (Figure 11.7B-F). To quantify how much of the payload forms the protein-adduct, the ADC-depleted sample would be further digested with proteases (e.g., cathepsin) to release the payload which is then followed by LC-MS/MS payload quantification, which could calculate back by molar ratios to achieve adduct quantification (Figure 11.8). Challenges remain to characterizing and quantifying the destiny of the released payload for which certain critical reagents (i.e., anti-payload Ab) may not be

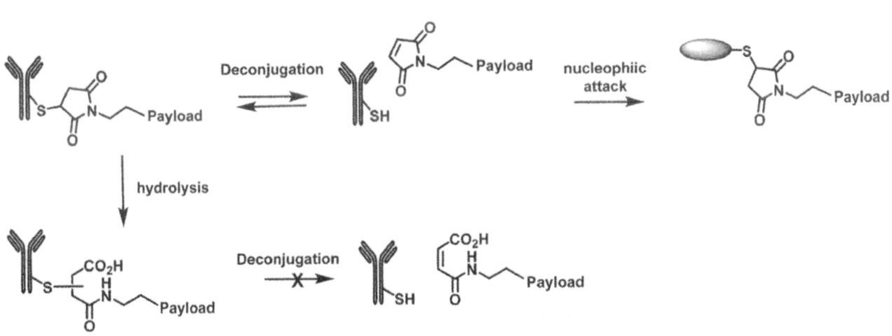

Figure 11.6 Scheme of the maleimide exchange from the antibody–drug conjugate (ADC). Copyright (2016) American Chemical Society.

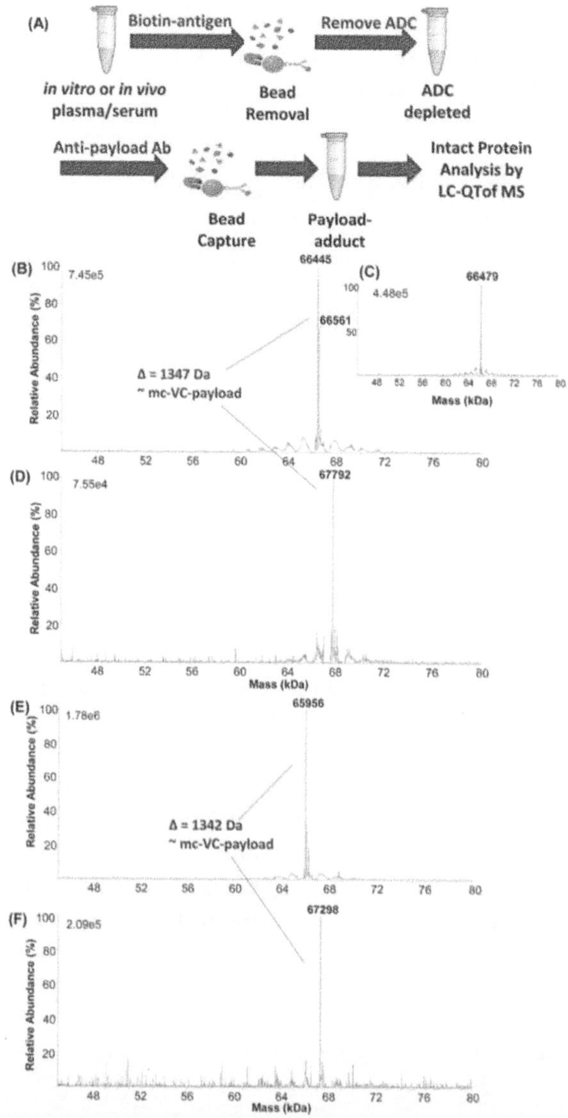

Figure 11.7 Identification of albumin-linker-payload adduct formation from in vitro human plasma incubation and from in vivo antibody–drug conjugate (ADC)-dosed cynomolgus monkey serum by intact liquid chromatography-mass spectrometry (LC-MS). (A) Scheme of the workflow to extract the protein-payload adduct from in vitro human plasma. (B) Deconvoluted mass spectrum of extracted native human albumin. (C) Deconvoluted mass spectrum of extracted native human albumin reduced with TCEP treatment. (D) Deconvoluted mass spectrum of payload-adduct from in vitro human plasma incubation for 144 hours. (E) Deconvoluted mass spectrum of extracted native monkey albumin. (F) Deconvoluted mass spectrum of payload-adduct from 5 mg/kg ADC post-dosed and ADC-depleted monkey serum at 168 hour. Copyright (2016) American Chemical Society.

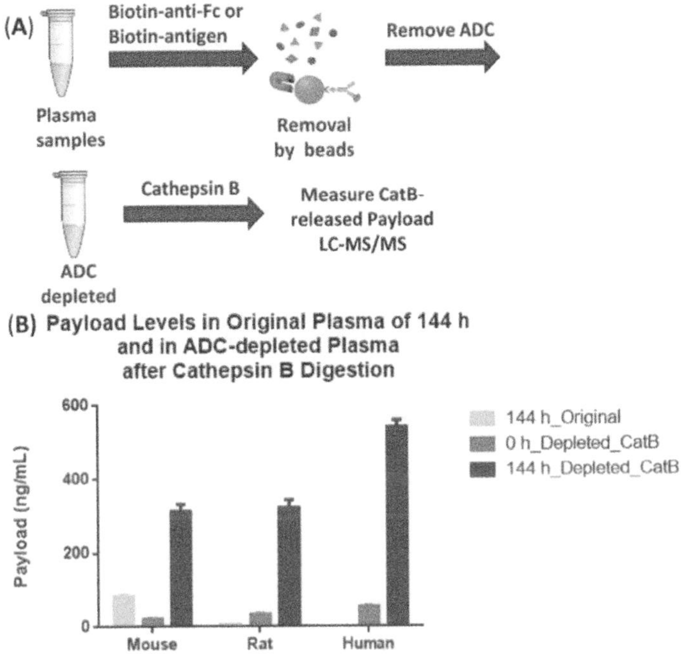

Figure 11.8 Recovery of the payload levels from the incubated samples after enzymatic cleavage of the antibody–drug conjugate (ADC)-depleted plasma. (A) Scheme of the workflow on payload quantification from the ADC-depleted plasma followed by cathepsin B (CatB) digestion. (B) Payload levels in the original incubated plasma at 144 hours and in the ADC-depleted plasma (0 hours and 144 hours) after digestion with cathepsin B. The original plasma samples (144 hours) were the samples that ADC was incubated in the plasma (mouse, rat, and human, respectively) for 144 hours, and the unconjugated payload concentrations were measured in those samples. Copyright (2016) American Chemical Society.

readily available. In this case, an excess amount of anti-albumin antibody could be used to pull down albumin-linker-payload adducts, and the quantification of the albumin-linker-payload adducts can be achieved by quantifying the payload after it is released by proteases. Another challenge is that when the linker is not cleavable, payload binding on the adduct may not be released for quantification analysis. When directly quantifying the intact payload-albumin adduct in the samples of ADC-depletion and anti-payload immunnocapture, limited sensitivity of LC-MS analysis for large molecule is a challenge. The succinimide ring resulting from maleimide chemistry can undergo hydrolysis to generate a ring-opened form, which in turn stabilizes the linker structure and also improves the stability, PK, and efficacy of the ADCs. A similar intact LC-MS approach can be applied to reveal the presence of the ring-opened form in biological samples (Shen et al., 2012b; Shi et al., 2018).

11.2.3.2 Biotransformation Identified with Reduced Size of ADC

Since it is still challenging to identify the detailed biotransformations in the intact ADC due to the large size of the molecule, IAC LC-MS F(ab')2 assay was developed to increase the detection sensitivity by cleaving the ADC into two smaller parts as a F(ab')2 fragment with linker-drug conjugated on it, and a Fc fragment (Su et al., 2016). The assay used a capture reagent against the Fab region of human IgGs such as extracellular domain (ECD) or anti-Fab antibody. IdeS digestion was employed to remove the glycan-containing Fc fragment and generate the linker-drug containing fragment F(ab')2 (~100 kDa) for analysis. The reduced size of the analytes results in improved sensitivity and resolution of mass spectra compared to the original intact ADC analytes (~150 kDa) with PNGaseF treatment for deglycosylation. Besides identifying the linker-payload deconjugation, this assay enabled the identification of ADC catabolites due to deacetylation (−43 Da) in the payload, which inactivated the payload and significantly affected the potency of the ADC (Su et al., 2016).

11.2.3.3 Biotransformation Identified with Deconjugated Linker-Payloads and Free Payloads

The linker-payload catabolites such as lysine-MCC-maytansinoid (Lys-MCC-DM1) and MCC-DM1 (MCC = maleimidomethyl cyclohexane-1-carboxylate) from the ADC with a noncleavable linker (MCC or N-succinimidyl-4-[N-maleimidomethyl] cyclohexane-1-carboxylate [SMCC]) have been identified as the major metabolites in bile and urine (Shen et al., 2012a; Singh et al., 2016). Previous in vivo biotransformation studies also revealed that from the ADC with a cleavable triglycyl peptide linker (CX, Gly-Gly-Gly), two major small molecule catabolites DM-CX1 and DM-CX2 that contain part of linker or the entire linker have been identified as the major circulating payload metabolites (Erickson & Lambert, 2012; Singh et al., 2016).

The in vitro and in vivo catabolism of SMCC and CX ADCs were often studied by the following four methods: (i) HPLC with radiometric detection (for [^{3}H]-payload-labeled ADCs), (ii) HPLC with UV detection (for non-radiolabeled ADCs), (iii) binding-competition ELISA (for non-radiolabeled ADCs), and (iv) LC-MS (for non-radiolabeled ADCs). Among them, LC-MS is the most sensitive method applied to non-radiolabeled ADCs for identification of catabolites of payload and linker-payload. The HPLC method with UV detection had low sensitivity and has only been demonstrated for application in one cell line with high antigen expression, which gave rise to a high level of antigen-mediated uptake of ADC and subsequent biotransformation. The more sensitive methods of radiometric HPLC detection and binding-competition ELISA were employed for evaluating catabolites in multiple cell lines with a range of antigen expression using [^{3}H]-labeled or non-radiolabeled ADCs, respectively. The ELISA methods allow for the measurement of conjugate concentrations (concentration of species containing at least one linked payload) as well as total antibody concentrations in plasma. The clearance profiles for a panel of ADCs were assessed using an ELISA method for the detection of conjugate (containing at least one linked payload) and found to correlate with their relative susceptibility to chemical cleavage via thiol-disulfide exchange of their linker moiety in vitro(Erickson & Lambert, 2012; Kellogg et al., 2011) giving an estimate of biotransformation occurring. For example, Ab-SPP-DM1 [SPP = *N*-succinimidyl-4-(2-pyridyldithio)pentanoate] conjugate can undergo reductive cleavage with dithiothreitol in vitro and was found to be cleared faster in mice than the uncleavable Ab-SMCC-DM1 conjugate. A similar relationship was observed between the clearance of T-DM1 with SMCC and the T-SPP-DM1 design (Erickson et al., 2012). The more sterically hindered disulfide of the AbSPDB-DM4 [SPDB = *N*-succinimidyl 4-(2-pyridyldithio)butanoate] conjugate was more resistant to reductive cleavage than the disulfide of Ab-SPP-DM1 conjugate in vitro (Kellogg et al., 2011) and cleared more slowly from circulation.

11.2.4 Biotransformation of PEGylated Proteins and Peptides

PEGylation is a commonly used strategy to prolong the circulating half-lives of protein therapeutics. PEGylation is achieved by conjugation of the protein/peptide with polyethylene glycol (PEG) through chemical modification of amino, hydroxyl, carbonyl, or sulfhydryl groups on the amino acid side chains (Baumann et al., 2014; Turecek et al., 2016). PEG, being a polymer, comes in different sizes and size distributions, ranging from a few hundred daltons to as high as 40 kDa. In addition to the size differences, there are also differences in the branching of PEG polymers. The different size and branching offer a wide selection of solubility, viscosity, and shielding of proteins with various conformations against proteolytic degradation.

The analysis of PEGylated proteins in biological fluids is typically accomplished by ELISA or other immunoassay formats with antibodies specifically targeting the protein of interest or a combination of protein and PEG to improve specificity, especially in the case of PEGylated endogenous proteins. However, there is a need for orthogonal analytical approaches such as LC-MS for PEGylated proteins in biological fluids when there is suspected clipping of the protein, or there is the need to analyze the composition of PEG-related material in tissues, especially ones where vacuoles were observed in animal toxicity studies (Baumann et al., 2014; Elliott et al., 2012; Gong et al., 2014). Tissue vacuolation is a common observation for PEGylated proteins, typically considered to be a result of macromolecule (such as PEG) engulfing by phagocytes, but sometimes target-specific uptake was suspected in the distribution of affected tissues (Ivens et al., 2015; Turecek et al., 2016; Webster et al., 2007). To further study clipping and tissue distribution, MS techniques are preferred due to their molecular specificity and versatility in dealing with different biological matrices.

The complexity of mass spectra of PEG and PEGylated proteins is highly dependent on the ionization technique. Conventional electrospray ionization (ESI) produces heavily charged ions, with overlaps of charge states due to the size distribution heterogeneity of PEG, and complicated with in-source fragmentation of PEG chains. All of these make deconvolution of ESI spectra of PEG very challenging (Forstenlehner et al., 2014; Gong et al., 2014; Huang et al., 2009; Lu et al., 2010). Conversely, MALDI ionization usually yields singly-charged ions, and it typically works better with PEGylated proteins than PEG itself due to better ionization of amino acid side chains. However, the resolution of MALDI-based techniques is typically not sufficient to account for small molecular weight changes in PEGylated proteins due to clipping or other biotransformation of single amino acids (Gong et al., 2014).

To overcome the PEG-related spectrum complexity in ESI mode, a variety of approaches have been introduced, namely dePEGylation, post-column charge reduction, and in-source PEG fragmentation. For PEGylated proteins with a cleavable linker such as an N-linked succinimide, base catalyzed hydrolysis results in proteins with modified side chain amines at the attachment site (Turecek et al., 2016; Veronese et al., 2001; Vestling et al., 1993). The resulting sample can then be subjected to LC-MS for intact protein analysis or tryptic digestion followed by LC-MS for conjugation site analysis. DePEGylation therefore is very powerful if clipping/modifications are suspected in the protein portion of the molecule. However, this approach is not applicable to non-cleavable PEGylation chemistry; furthermore, information regarding PEG size and size distribution is lost through the dePEGylation process.

Post-column charge reduction by T connection ("tee-in") infusion of triethylamine solution was shown to significantly reduce the charge states and to improve the charge state separations of PEGylated proteins, resulting in quality spectra suitable for deconvolution (Forstenlehner et al., 2014; Huang et al., 2009). Forstenlehner et al. demonstrated that they were able to separate and determine methionine oxidation states of an oxidative stressed sample of PEGylated rhG-CSF (average MW ~ 40 kDa) by the combination of HPLC, post-column charge reduction, and HRMS such as Q Exactive MS (Forstenlehner et al., 2014).

In-source fragmentation (ISF) of PEGylated proteins was shown to be achieved by elevating the source cone voltages or aperture voltage to induce

PEG fragmentation leading to proteins/peptides with a short PEG linker (Lu et al., 2010). In this example, 5 kDa and 20 kDa PEGylated glucagons, and 20 kDa PEGylated IgG4 light chain, were used to demonstrate the utility of ISF. ISF resulted in intact protein with a distribution of different sizes of small PEG fragments attached. Interestingly, 5 kDa and 20 kDa PEGylated glucogons gave similar fragment ion distribution, demonstrating that ISF efficiency is not PEG size-dependent. Further collision-induced dissociation (CID) fragmentation of the PEG-containing portion revealed the information of the PEG conjugation site. ISF followed by CID was also utilized in a triple quadrupole LC-MS/MS system to quantitatively determine the concentrations of PEG and PEGylated proteins in plasma and various tissue samples (Gong et al., 2014). Short PEG fragments of various sizes were subjected to further CID in the collision cell, and multiple reaction monitoring was used to quantitate the PEG and PEG proteins. Assay sensitivity was very good in the example with a lower limit of quantitation (LLOQ) of 0.05 μg/mL for a 40 kDa PEG molecule.

In summary, despite the molecular complexity of PEGylated proteins, a variety of tools and techniques are now available to address the biotransformation and disposition/distribution-related questions regarding PEGylated proteins, including dePEGylation through chemical pretreatment or in source PEG fragmentation, HPLC separation, post-HPLC column charge reduction, and finally HRMS (Figure 11.9).

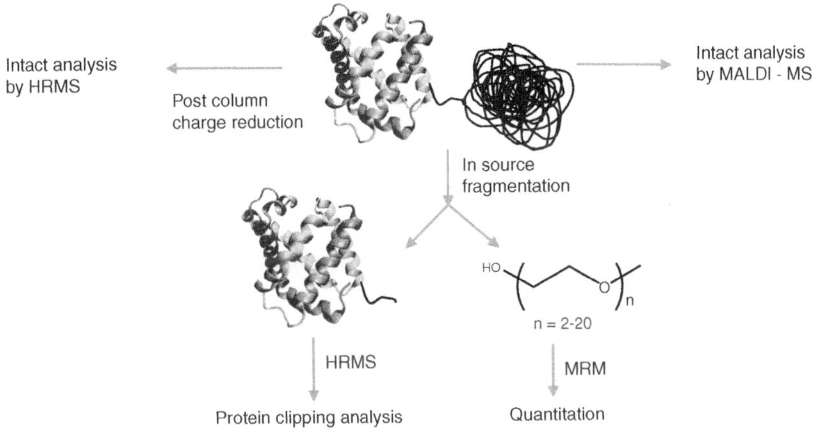

Figure 11.9 Schematic representation of different mass spectrometry approaches to study the disposition of PEGylated proteins. The protein illustrated is adapted from a three-dimensional structure of myoglobin by *Opabinia regalis*. HRMS, high-resolution mass spectrometry; MRM, multiple reaction monitoring. Self-created from PDB entry IMBA using the freely available visualization and analysis package VMD, CC BY-SA 3.0 (https://commons.wikimedia.org/w/index.php?curid = 1024254).

11.3 Conclusions and Future Perspectives

Biotransformation is one of the emerging aspects of ADME studies on biologics. Insufficient understanding of the nature of biotransformation and disposition of biologic therapeutics would pose challenges for the accurate depiction of its impact on PK and toxicity profiles. Some common biotransformations of mAb include deamidation, isomerization, oxidation, N-terminal glutamate to pyroglutamate, C-terminal lysine processing, and so on. Challenges remain with the limited sensitivity to characterize the biotransformation of intact mAb or protein therapeutics. The analytical challenges associated with ADCs are around characterizing and quantifying the dynamically changing mixture of ADC species in circulation due to catabolism of the antibody, linker, and payload. IAC and HRMS provide powerful tools for sample preparation and biotransformation detection of biologic drug candidates. Despite significant progress in the technologies pertaining to ADME studies of therapeutic biologics, a breakthrough on analytical methodology and an in-depth understanding about biotransformation in both nonclinical and clinical settings will be required to bring the ADME science of therapeutic biologics to a new horizon, such as disposition characterizations in tissues. The knowledge obtained about drug biotransformation properties will be instrumental in the optimization of disposition properties, and ultimately human efficacy and safety profiles of new therapeutic biologics.

References

Baumann, A., Tuerck, D., Prabhu, S., Dickmann, L., & Sims, J. (2014). Pharmacokinetics, metabolism and distribution of PEGs and PEGylated proteins: Quo vadis? *Drug Discovery Today*, 19(10), 1623–1631. https://doi.org/10.1016/j.drudis.2014.06.002

Bults, P., Bischoff, R., Bakker, H., Gietema, J. A., & Van De Merbel, N. C. (2016). LC-MS/MS-Based monitoring of in vivo protein biotransformation: Quantitative determination of trastuzumab and its deamidation products in human plasma. *Analytical Chemistry*, 88(3), 1871–1877. https://doi.org/10.1021/acs.analchem.5b04276

Cai, B., Pan, H., & Flynn, G. C. (2011). C-terminal lysine processing of human immunoglobulin G2 heavy chain in vivo. *Biotechnology and Bioengineering*, 108(2), 404–412. https://doi.org/10.1002/bit.22933

Chelius, D., Rehder, D. S., & Bondarenko, P. V. (2005). Identification and characterization of deamidation sites in the conserved regions of human immunoglobulin gamma antibodies. *Analytical Chemistry*, 77(18), 6004–6011. https://doi.org/10.1021/ac050672d

Elliott, V. L., Edge, G. T., Phelan, M. M., Lian, L. Y., Webster, R., Finn, R. F., Park, B. K., & Kitteringham, N. R. (2012). Evidence for metabolic cleavage of a PEGylated protein in vivo using multiple analytical methodologies. *Molecular Pharmaceutics*, 9(5), 1291–1301. https://doi.org/10.1021/mp200587m

Erickson, H. K., & Lambert, J. M. (2012). ADME of antibody-maytansinoid conjugates. *The AAPS Journal*, 14(4), 799–805. https://doi.org/10.1208/s12248-012-9386-x

Erickson, H. K., Lewis Phillips, G. D., Leipold, D. D., Provenzano, C. A., Mai, E., Johnson, H. A., Gunter, B., Audette, C. A., Gupta, M., Pinkas, J., & Tibbitts, J. (2012). The effect of different linkers on target cell catabolism and pharmacokinetics/pharmacodynamics of trastuzumab maytansinoid conjugates. *Molecular Cancer Therapeutics*, 11(5), 1133–1142. https://doi.org/10.1158/1535-7163.MCT-11-0727

Ezan, E., Becher, F., & Fenaille, F. (2014). Assessment of the metabolism of therapeutic proteins and antibodies. *Expert Opinion on Drug Metabolism & Toxicology*, 10(8), 1079–1091. https://doi.org/10.1517/17425255.2014.925878

Forstenlehner, I. C., Holzmann, J., Scheffler, K., Wieder, W., Toll, H., & Huber, C. G. (2014). A direct-infusion- and HPLC-ESI-Orbitrap-MS approach for the characterization of intact PEGylated proteins. *Analytical Chemistry*, 86(1), 826–834. https://doi.org/10.1021/ac403390y

Goetze, A. M., Liu, Y. D., Arroll, T., Chu, L., & Flynn, G. C. (2012). Rates and impact of human antibody glycation in vivo. *Glycobiology*, 22(2), 221–234. https://doi.org/10.1093/glycob/cwr141

Goetze, A. M., Liu, Y. D., Zhang, Z., Shah, B., Lee, E., Bondarenko, P. V., & Flynn, G. C. (2011). High-mannose glycans on the Fc region of therapeutic IgG antibodies increase serum clearance in humans. *Glycobiology*, 21(7), 949–959. https://doi.org/10.1093/glycob/cwr027

Gong, J., Gu, X., Achanzar, W. E., Chadwick, K. D., Gan, J., Brock, B. J., Kishnani, N. S., Humphreys, W. G., & Lyer, R. A. (2014). Quantitative analysis of polyethylene glycol (PEG) and PEGylated proteins in animal tissues by LC-MS/MS coupled with in-source CID. *Analytical Chemistry*, 86(15), 7642–7649. https://doi.org/10.1021/ac501507g

Hager, T., Spahr, C., Xu, J., Salimi-Moosavi, H., & Hall, M. (2013). Differential enzyme-linked immunosorbent assay and ligand-binding mass spectrometry for analysis of biotransformation of protein therapeutics: Application to various FGF21 modalities. *Analytical Chemistry*, 85(5), 2731–2738. https://doi.org/10.1021/ac303203y

Hall, M. P. (2014). Biotransformation and in vivo stability of protein biotherapeutics: Impact on candidate selection and pharmacokinetic profiling. *Drug Metabolism and Disposition*, 42(11), 1873–1880. https://doi.org/10.1124/dmd.114.058347

Hall, M. P., Gegg, C., Walker, K., Spahr, C., Ortiz, R., Patel, V., Yu, S., Zhang, L., Lu, H., DeSilva, B., & Lee, J. W. (2010). Ligand-binding mass spectrometry to study biotransformation of fusion protein drugs and guide immunoassay development: Strategic approach and application to peptibodies targeting the thrombopoietin receptor. *The AAPS Journal*, 12(4), 576–585. https://doi.org/10.1208/s12248-010-9218-9

Han, M., Pearson, J. T., Wang, Y., Winters, D., Soto, M., Rock, D. A., & Rock, B. M. (2017). Immunoaffinity capture coupled with capillary electrophoresis – Mass spectrometry to study therapeutic protein stability in vivo. *Analytical Biochemistry*, 539, 118–126. https://doi.org/10.1016/j.ab.2017.10.005

He, J., Su, D., Ng, C., Liu, L., Yu, S.-F., Pillow, T. H., Del Rosario, G., Darwish, M., Lee, B.-C., Ohri, R., Zhou, H., Wang, X., Lu, J., Kaur, S., & Xu, K. (2017). High-resolution accurate-mass mass spectrometry enabling in-depth characterization of in vivo biotransformations for intact antibody-drug conjugates. *Analytical Chemistry*, 89(10), 5476–5483. https://doi.org/10.1021/acs.analchem.7b00408

Hecht, R., Li, Y. S., Sun, J., Belouski, E., Hall, M., Hager, T., Yie, J., Wang, W., Winters, D., Smith, S., Spahr, C., Tam, L. T., Shen, Z., Stanislaus, S., Chinookoswong, N., Lau, Y., Sickmier, A., Michaels, M. L., Boone, T., Veniant, M. M., & Xu, J. (2012). Rationale-based engineering of a potent long-acting FGF21 analog for the treatment of type 2 diabetes. *PLoS One*, 7(11), e49345. https://doi.org/10.1371/journal.pone.0049345

Huang, L., Gough, P. C., & Defelippis, M. R. (2009). Characterization of poly(ethylene glycol) and PEGylated products by LC/MS with postcolumn addition of amines. *Analytical Chemistry*, 81(2), 567–577. https://doi.org/10.1021/ac801711u

Huang, L., Lu, J., Wroblewski, V. J., Beals, J. M., & Riggin, R. M. (2005). In vivo deamidation characterization of monoclonal antibody by LC/MS/MS. *Analytical Chemistry*, 77(5), 1432–1439. https://doi.org/10.1021/ac0494174

Ivens, I. A., Achanzar, W., Baumann, A., Brandli-Baiocco, A., Cavagnaro, J., Dempster, M., Depelchin, B. O., Rovira, A. R., Dill-Morton, L., Lane, J. H., Reipert, B. M., Salcedo, T., Schweighardt, B., Tsuruda, L. S., Turecek, P. L., & Sims, J. (2015). PEGylated biopharmaceuticals: Current experience and considerations for nonclinical development. *Toxicologic Pathology*, 43(7), 959–983. https://doi.org/10.1177/0192623315591171

Kang, L., Camacho, R. C., Li, W., D'Aquino, K., You, S., Chuo, V., Weng, N., & Jian, W. (2017). Simultaneous catabolite identification and quantitation of large therapeutic protein at the intact level by immunoaffinity capture liquid chromatography-high-resolution mass spectrometry. *Analytical Chemistry*, 89(11), 6065–6075. https://doi.org/10.1021/acs.analchem.7b00674

Kellie, J. F., & Karlinsey, M. Z. (2018). Review of approaches and examples for monitoring biotransformation in protein and peptide therapeutics by MS. *Bioanalysis*, 10(22), 1877–1890. https://doi.org/10.4155/bio-2018-0113

Kellie, J. F., Kehler, J. R., Mencken, T. J., Snell, R. J., & Hottenstein, C. S. (2016). A whole-molecule immunocapture LC-MS approach for the in vivo quantitation of biotherapeutics. *Bioanalysis*, 8(20), 2103–2114. https://doi.org/10.4155/bio-2016-0180

Kellogg, B. A., Garrett, L., Kovtun, Y., Lai, K. C., Leece, B., Miller, M., Payne, G., Steeves, R., Whiteman, K. R., Widdison, W., Xie, H., Singh, R., Chari, R. V., Lambert, J. M., & Lutz, R. J. (2011). Disulfide-linked antibody-maytansinoid conjugates: Optimization of in vivo activity by varying the steric hindrance at carbon atoms adjacent to the disulfide linkage. *Bioconjugate Chemistry*, 22(4), 717–727. https://doi.org/10.1021/bc100480a

Kontermann, R. E. (2016). Half-life extended biotherapeutics. *Expert Opinion on Biological Therapy*, 16(7), 903–915. https://doi.org/10.1517/14712598.2016.1165661

Kullolli, M., Rock, D. A., & Ma, J. (2017). Immuno-affinity capture followed by TMPP N-terminus tagging to study catabolism of therapeutic proteins. *Journal of Proteome Research*, 16(2), 911–919. https://doi.org/10.1021/acs.jproteome.6b00863

Lee, J. W. (2013). ADME of monoclonal antibody biotherapeutics: Knowledge gaps and emerging tools. *Bioanalysis*, 5(16), 2003–2014. https://doi.org/10.4155/bio.13.144

Li, Y., Huang, Y., Ferrant, J., Lyubarskaya, Y., Zhang, Y. E., Li, S. P., & Wu, S. L. (2016a). Assessing in vivo dynamics of multiple quality attributes from a therapeutic IgG4 monoclonal antibody circulating in cynomolgus monkey. *MAbs*, 8(5), 961–968. https://doi.org/10.1080/19420862.2016.1167298

Li, Y., Monine, M., Huang, Y., Swann, P., Nestorov, I., & Lyubarskaya, Y. (2016b). Quantitation and pharmacokinetic modeling of therapeutic antibody quality attributes in human studies. *MAbs*, 8(6), 1079–1087. https://doi.org/10.1080/19420862.2016.1186322

Liu, L., Xu, K., Li, J., Maia, M., Mathieu, M., Elliott, R., Yang, J., Nijem, I., & Kaur, S. (2018). Optimizing hybrid LC-MS/MS binding conditions is critical: Impact of biotransformation on quantification of trastuzumab. *Bioanalysis*, 10(22). https://doi.org/10.4155/bio-2018-0196

Liu, Y. D., Chen, X., Enk, J. Z., Plant, M., Dillon, T. M., & Flynn, G. C. (2008). Human IgG2 antibody disulfide rearrangement in vivo. *Journal of Biological Chemistry*, 283(43), 29266–29272. https://doi.org/10.1074/jbc.M804787200

Liu, Y. D., Goetze, A. M., Bass, R. B., & Flynn, G. C. (2011). N-Terminal glutamate to pyroglutamate conversion in vivo for human IgG2 antibodies. *Journal of Biological Chemistry*, 286(13), 11211–11217. https://doi.org/10.1074/jbc.M110.185041

Lu, X., Gough, P. C., Defelippis, M. R., & Huang, L. (2010). Elucidation of PEGylation site with a combined approach of in-source fragmentation and CID MS/MS. *Journal of the American Society for Mass Spectrometry*, 21(5), 810–818. https://doi.org/10.1016/j.jasms.2010.01.011

Mehl, J. T., Sleczka, B. G., Ciccimaro, E. F., Kozhich, A. T., Gilbertson, D. G., Vuppugalla, R., Huang, C. S., Stevens, B., Mo, J., Deyanova, E. G., Wang, Y., Huang, R. Y., Chen, G., & Olah, T. V. (2016). Quantification of in vivo site-specific Asp isomerization and Asn deamidation of mAbs in animal serum using IP-LC-MS. *Bioanalysis*, 8(15), 1611–1622. https://doi.org/10.4155/bio-2016-0035

Mo, J., Yan, Q., So, C. K., Soden, T., Lewis, M. J., & Hu, P. (2016). Understanding the impact of methionine oxidation on the biological functions of IgG1 antibodies using hydrogen/deuterium exchange mass spectrometry. *Analytical Chemistry*, 88(19), 9495–9502. https://doi.org/10.1021/acs.analchem.6b01958

Moritz, B., & Stracke, J. O. (2017). Assessment of disulfide and hinge modifications in monoclonal antibodies. *Electrophoresis*, 38(6), 769–785. https://doi.org/10.1002/elps.201600425

Murphy, R. E., Kinhikar, A. G., Shields, M. J., Del Rosario, J., Preston, R., Levin, N., & Ward, G. H. (2010). Combined use of immunoassay and two-dimensional liquid chromatography mass spectrometry for the detection and identification of metabolites from biotherapeutic pharmacokinetic samples. *Journal of Pharmaceutical and Biomedical Analysis*, 53(3), 221–227. https://doi.org/10.1016/j.jpba.2010.04.028

Ouellette, D., Chumsae, C., Clabbers, A., Radziejewski, C., & Correia, I. (2013). Comparison of the in vitro and in vivo stability of a succinimide intermediate observed on a therapeutic IgG1 molecule. *MAbs*, 5(3), 432–444. https://doi.org/10.4161/mabs.24458

Pearson, J. T., & Rock, D. A. (2015). Bioanalytical approaches to assess the proteolytic stability of therapeutic fusion proteins. *Bioanalysis*, 7(23), 3035–3051. https://doi.org/10.4155/bio.15.217

Shen, B. Q., Bumbaca, D., Saad, O., Yue, Q., Pastuskovas, C. V., Khojasteh, S. C., Tibbitts, J., Kaur, S., Wang, B., Chu, Y. W., Lorusso, P. M., & Girish, S. (2012a). Catabolic fate and pharmacokinetic characterization of trastuzumab emtansine (T-DM1): An emphasis on preclinical and clinical catabolism. *Current Drug Metabolism*, 13(7), 901–910. https://doi.org/10.2174/138920012802138598

Shen, B. Q., Xu, K., Liu, L., Raab, H., Bhakta, S., Kenrick, M., Parsons-Reponte, K. L., Tien, J., Yu, S. F., Mai, E., Li, D., Tibbitts, J., Baudys, J., Saad, O. M., Scales, S.

J., McDonald, P. J., Hass, P. E., Eigenbrot, C., Nguyen, T., Solis, W. A., ... Junutula, J. R. (2012b). Conjugation site modulates the in vivo stability and therapeutic activity of antibody-drug conjugates. *Nature Biotechnology*, 30(2), 184–189. https://doi.org/10.1038/nbt.2108

Shi, C., Goldberg, S., Lin, T., Dudkin, V., Widdison, W., Harris, L., Wilhelm, S., Jmeian, Y., Davis, D., O'Neil, K., Weng, N., & Jian, W. (2018). LC/MS/MS bioanalysis of protein-drug conjugates-The importance of incorporating succinimide hydrolysis products. *Analytical Chemistry*, 90(8), 5314–5321. https://doi.org/10.1021/acs.analchem.8b00411

Singh, A. P., Maass, K. F., Betts, A. M., Wittrup, K. D., Kulkarni, C., King, L. E., Khot, A., & Shah, D. K. (2016). Evolution of antibody-drug conjugate tumor disposition model to predict preclinical tumor pharmacokinetics of trastuzumab-emtansine (T-DM1). *The AAPS Journal*, 18(4), 861–875. https://doi.org/10.1208/s12248-016-9904-3

Su, D., Ng, C. K., Khosraviani, M., Yu, S. F., Cosino, E., Kaur, S., & Xu, K. (2016). Custom-designed affinity capture LC-MS F(ab')2 assay for biotransformation assessment of site-specific antibody drug conjugates. *Analytical Chemistry*, 88(23), 11340–11346. https://doi.org/10.1021/acs.analchem.6b03410

Tibbitts, J., Canter, D., Graff, R., Smith, A., & Khawli, L. A. (2016). Key factors influencing ADME properties of therapeutic proteins: A need for ADME characterization in drug discovery and development. *MAbs*, 8(2), 229–245. https://doi.org/10.1080/19420862.2015.1115937

Turecek, P. L., Bossard, M. J., Schoetens, F., & Ivens, I. A. (2016). PEGylation of biopharmaceuticals: A review of chemistry and nonclinical safety information of approved drugs. *Journal of Pharmaceutical Sciences*, 105(2), 460–475. https://doi.org/10.1016/j.xphs.2015.11.015

Van De Merbel, N. C. (2019). Protein quantification by LC-MS: A decade of progress through the pages of Bioanalysis. *Bioanalysis*, 11(7), 629–644. https://doi.org/10.4155/bio-2019-0032

Van Den Broek, I., & Van Dongen, W. D. (2015). LC-MS-based quantification of intact proteins: Perspective for clinical and bioanalytical applications. *Bioanalysis*, 7(15), 1943–1958. https://doi.org/10.4155/bio.15.113

Veronese, F. M., Sacca, B., Polverino De Laureto, P., Sergi, M., Caliceti, P., Schiavon, O., & Orsolini, P. (2001). New PEGs for peptide and protein modification, suitable for identification of the PEGylation site. *Bioconjugate Chemistry*, 12(1), 62–70. https://doi.org/10.1021/bc000061m

Vestling, M. M., Murphy, C. M., Keller, D. A., Fenselau, C., Dedinas, J., Ladd, D. L., & Olsen, M. A. (1993). A strategy for characterization of polyethylene glycol-derivatized proteins. A mass spectrometric analysis of the attachment sites in polyethylene glycol-derivatized superoxide dismutase. *Drug Metabolism and Disposition*, 21(5), 911–917. https://dmd.aspetjournals.org/content/21/5/911.long

Vlasak, J., Bussat, M. C., Wang, S., Wagner-Rousset, E., Schaefer, M., Klinguer-Hamour, C., Kirchmeier, M., Corvaia, N., Ionescu, R., & Beck, A. (2009). Identification and characterization of asparagine deamidation in the light chain CDR1 of a humanized IgG1 antibody. *Analytical Biochemistry*, 392(2), 145–154. https://doi.org/10.1016/j.ab.2009.05.043

Wang, J., Zhang, W., Salter, R., & Lim, H. K. (2019). Reductive desulfuration as an important tool in detection of small molecule modifications to payload of antibody drug conjugates. *Analytical Chemistry*, 91(3), 2368–2375. https://doi.org/10.1021/acs.analchem.8b05134

Wang, W., Vlasak, J., Li, Y., Pristatsky, P., Fang, Y., Pittman, T., Roman, J., Wang, Y., Prueksaritanont, T., & Ionescu, R. (2011). Impact of methionine oxidation in human IgG1 Fc on serum half-life of monoclonal antibodies. *Molecular Immunology*, 48(6–7), 860–866. https://doi.org/10.1016/j.molimm.2010.12.009

Webster, R., Didier, E., Harris, P., Siegel, N., Stadler, J., Tilbury, L., & Smith, D. (2007). PEGylated proteins: Evaluation of their safety in the absence of definitive metabolism studies. *Drug Metabolism and Disposition*, 35(1), 9–16. https://doi.org/10.1124/dmd.106.012419

Wei, C., Zhang, G., Clark, T., Barletta, F., Tumey, L. N., Rago, B., Hansel, S., & Han, X. (2016). Where did the linker-payload go? A quantitative investigation on the destination of the released linker-payload from an antibody-drug conjugate with a maleimide linker in plasma. *Analytical Chemistry*, 88(9), 4979–4986. https://doi.org/10.1021/acs.analchem.6b00976

Wroblewski, V. J., Witcher, D. R., Becker, G. W., Davis, K. A., Dou, S., Micanovic, R., Newton, C. M., Noblitt, T. W., Richardson, J. M., Song, H. Y., & Hale, J. E. (2003). Decoy receptor 3 (DcR3) is proteolytically processed to a metabolic fragment having differential activities against Fas ligand and LIGHT. *Biochemical Pharmacology*, 65(4), 657–667. https://doi.org/10.1016/s0006-2952(02)01612-x

Wu, B., & Sun, Y. N. (2014). Pharmacokinetics of peptide-Fc fusion proteins. *Journal of Pharmaceutical Sciences*, 103(1), 53–64. https://doi.org/10.1002/jps.23783

Yao, M., Chen, B., Zhao, W., Mehl, J. T., Li, L., & Zhu, M. (2018). LC-MS differential analysis for fast and sensitive determination of biotransformation of therapeutic proteins. *Drug Metabolism and Disposition*, 46(4), 451–457. https://doi.org/10.1124/dmd.117.077792

Yin, S., Pastuskovas, C. V., Khawli, L. A., & Stults, J. T. (2013). Characterization of therapeutic monoclonal antibodies reveals differences between in vitro and in vivo time-course studies. *Pharmaceutical Research*, 30(1), 167–178. https://doi.org/10.1007/s11095-012-0860-z

Zell, M., Husser, C., Staack, R. F., Jordan, G., Richter, W. F., Schadt, S., & Pahler, A. (2016). In vivo biotransformation of the fusion protein tetranectin-apolipoprotein A1 analyzed by ligand-binding mass spectrometry combined with quantitation by ELISA. *Analytical Chemistry*, 88(23), 11670–11677. https://doi.org/10.1021/acs.analchem.6b03252

Zhang, Q., Schenauer, M. R., McCarter, J. D., & Flynn, G. C. (2013). IgG1 thioether bond formation in vivo. *Journal of Biological Chemistry*, 288(23), 16371–16382. https://doi.org/10.1074/jbc.M113.468397

12

New Modalities

Multidomain Therapeutics and Gene Therapy Programs

Kelly Colletti[a] *and Mark Ma*[b]

[a]*Laboratory Sciences, Charles River Laboratories (corresponding author)*
[b]*Translational Clinical Science, Rallybio*

Chapter Outline

- Definition of bispecifics, PEGylated, antibody–drug conjugates (ADCs), probody, and gene therapy therapeutics.
- Current bioanalytical practices in support of bispecifics, PEGylated, ADCs, probody, and gene therapy, including pharmacokinetics, pharmacodynamics, and immunogenicity.
- Challenges and solutions to address bioanalytical challenges.

12.1 Introduction

The body of scientific knowledge on the structure and function of large molecules in nature has progressed to a point where scientists are able to produce novel therapeutics by combining these natural components into new modalities, creating molecules with great therapeutic potential. Innovative techniques for cloning, recombination, protein purification, and conjugation have allowed biotechnology engineers to produce bispecific antibodies to target more than one antigen, PEGylated proteins to extend therapeutic half-life and

Bioanalytical Aspects in Biological Therapeutics, First Edition. Edited by Xiaohui Xu and Weifeng Xu.
© 2022 John Wiley & Sons, Inc. Published 2022 by John Wiley & Sons, Inc.

increase solubility, probodies that biotransform upon delivery to reduce off-target effects, antibody–drug conjugates (ADCs) that deliver toxins to specific cellular tumor targets, and gene therapies to genetically treat disease by either increasing or decreasing the production of specific proteins. This chapter will define and discuss each of these novel modalities and specifically describe the current bioanalytical strategies and challenges associated with the development of these molecules.

12.1.1 Multidomain Therapeutics (Bispecifics, PEGylated, Antibody–Drug Conjugates, and Probody Therapeutics)

12.1.1.1 Bispecifics

One of the new modalities that is currently under intensive investigation are bispecific therapeutics. Bispecific therapeutics include a diverse group of proteins utilizing both binding domains from antibody-based structures and non-antibody binding domain protein families. These molecules are able to bind to multiple therapeutic targets (cellular or soluble) to achieve the desired therapeutic effect. Specifically, a bispecific is defined as an antibody- or scaffold-based protein that contains more than one functional domain that binds two or more independent antigenic targets. Bispecifics are among the fastest growing new modalities with over 100 active clinical trials (www.clinicaltrials.gov). Two bispecific therapeutic products are currently approved for use, blinatumomab (targets: CD19 and CD3) and emicizumab (targets: factors IXa and X). A third molecule was approved for use in the European Union, catumaxomab (targets: EpCAM and CD3), but marketing authorization was withdrawn in 2017 for commercial reasons. Many of the bispecific molecules in development are oncology therapeutics with a small proportion of molecules targeting immunologic disorders such as psoriasis and autoimmune diseases. Although many bispecific antibodies and constructs may be active by binding to two cellular or soluble targets, they may also have an IgG Fc functional domain. In this chapter, the term bispecific is used for all multispecific protein therapeutics.

Many of the therapeutic targets that are being investigated as part of bispecific molecules were first successfully evaluated as monoclonal antibodies or protein therapeutics. Multiple targets are designed to work synergistically such that there is redundancy for treating specific immunologic diseases or targeting the therapeutic to the specific site of the disease (e.g., CD3 T cells targeted to cancer cells) through immune cell recruitment. Bispecific therapeutics are highly diverse in structure, which creates additional challenges to the development of a bioanalytical strategy for both nonclinical and clinical phases of drug development. Bispecific therapeutics are produced by combining various, single-chain variable fragments (scFv), Fabs, Fc domains, paratopes, SH3 domains, antibody structures, and fused peptides (Weidle et al., 2013; Xu et al., 2015). These molecules can be engineered to incorporate multiple protein structures resulting in molecules of various shapes and sizes with small nanobodies at one end of the spectrum and large antibody structures with up to four antigen-binding sites at the other (Brinkmann & Kontermann, 2017). The half-life of these molecules can be drastically different with some rapidly degrading following dose administration if they contain structures such as Fab or scFv to others having long half-lives that are more characteristic of monoclonal antibodies. Due to the diversity of these structures, the bioanalytical strategy including critical reagents required and the design of the bioanalytical assays must be carefully considered to support the pharmacokinetic (PK) and immunogenicity assessments.

12.1.1.1.1 Biotransformation of Bispecifics

Bispecific molecules have the potential to undergo biotransformation, or cleavage via proteases or other mechanisms into structurally distinct proteins that may retain all or some of their therapeutic effect (Kellie et al., 2020). They can also recombine in vivo to produce therapeutically active monoclonal antibodies or antibody fragments. Bioanalytical assays used for the detection of the bispecific must be thoughtfully designed to detect and characterize biotransformation ex vivo. The bioanalytical assay can be designed to measure the molecule based on one target binding domain or can be designed to require both binding domains to be intact for the assay to yield results.

12.1.1.1.2 Development of PK Assays for Bispecific Therapeutics

Often this requires highly specific critical reagents such as recombinant protein targets, or anti-idiotypic antibodies specific to a binding domain. For molecules with more than two target binding domains, this can pose a considerable challenge, which may require multiple assays to fully characterize the PK of the therapeutic (Ma et al., 2019). One of the assays often used to further characterize the PK of the bispecific antibody-based molecule is a generic assay that measures the Fc backbone of the antibody structure to determine the total antibody content in circulation (Figure 12.1). This can be accomplished with anti-human IgG specific capture and detection antibodies in the nonclinical phase using an immunoassay platform such as enzyme-linked immunosorbent assay (ELISA) or Gyrolab. In the clinical space this can be accomplished using an immunocapture liquid chromatography-mass spectrometry (LC-MS) method. The total antibody can be measured and compared to the data from the intact or domain-specific assays to determine if there is a significant amount of

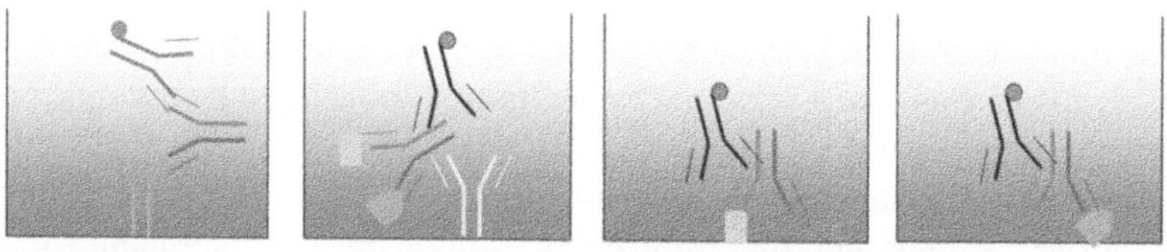

Figure 12.1 Intact, total, and free function domain methods for bispecifics.

biotransformation occurring in vivo to characterize the PK. Additional assays may be required to characterize biotransformation depending on the structure of the bispecific (Figure 12.2). For example, if the therapeutic has a fused peptide or cytokine, a specific assay may be needed to measure a smaller protein product. It will detect the biotransformation of the fused domain to further define the level of biotransformation observed in vivo. The structure of the bispecific drives the bioanalytical strategy chosen for both the nonclinical and clinical phases of analysis to detect and characterize any biotransformation in vivo.

12.1.1.1.3 Assay Sensitivity Requirements for Bispecific PK Assays

One of the major bioanalytical challenges for bispecific therapies is the demands of highly sensitive bioanalytical assays. Specifically, bispecific T cell engagers (BiTE) may be extremely potent and thus require dosing regimens in the microgram per kilogram range. They may also have short half-lives and rapidly degrade following dose administration. Bioanalytical assays with sensitivities in the picogram per milliliter to femtogram per milliliter range are often required to assess the PK of these molecules. These assays typically require highly specific reagents, such as affinity-matured anti-idiotypic antibodies or stable recombinant proteins. High-quality reagents of sufficient quantity are essential for the success of these assays and maintaining the assay life cycle. In order to produce an assay with the required sensitivity, an immunoassay platform such as single-molecule counting (SMCxPro) from Millipore-Sigma can be developed. This assay platform requires detection reagents that are

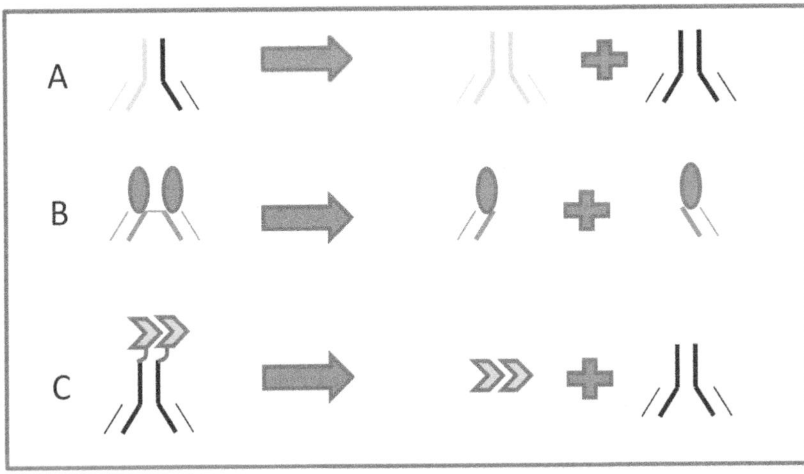

Figure 12.2 Examples of biotransformation. (A) Bispecific antibody: intact assay in conjunction with a total assay would detect biotransformation. (B) Fab fused to peptide: intact assay and a total assay would be needed to detect biotransformation. (C) Protein fused to antibody: intact assay in conjunction with a total antibody assay or functional domain assays would detect biotransformation.

conjugated to a fluorophore (Alexa Fluor) and can be developed in plate-based and bead-based formats. Unlike the first-generation single-molecule counting systems, the SMCxPro does not contain any fluidics and captures the single-molecule signal following elution from the immunocomplex. Although this platform still requires high-affinity and specific critical reagents, the sensitivity is generally at least 10-fold greater than a traditional immunoassay platform.

12.1.1.1.4 Bioanalytical Strategy for Bispecific Therapeutics

Given these challenges, the bioanalytical strategy for bispecifics may evolve over the course of drug development. For the nonclinical stage of development, there are two main objectives: (1) determine the PK to inform first-in-human dosing and (2) demonstrate exposure over the course of the safety assessment study. At the nonclinical stage, the characterization of biotransformation is only required based on any contribution to safety endpoints. For this reason, an intact assay that uses binding to two antigenic targets is recommended as the primary assay for nonclinical studies. If the exposure is lower than expected, the total assay (generic anti-human IgG) can be informative to determine if the loss of exposure is due to biotransformation or the formation of a blocking immune response. Although not recommended due to the inability to detect biotransformation, a generic anti-human IgG assay may be used as the primary assay for nonclinical studies. At the clinical stage it is much more important to fully characterize the PK/PD of all protein species. A primary assay, such as an intact assay, may be validated and then secondary assays that measure any biotransformed protein products may be qualified for further characterization. Measurement of pharmacodynamic (PD) effects are also important at this stage to understand the potential root causes of any loss of function. The design of these bioanalytical assays will be highly dependent on the structure and mechanism of action of the bispecific. For example, if the bispecific is an IgG with two soluble targets, an intact assay may be used as the primary assay, with free assays to each target used as characterization assays to ensure that the IgG did not undergo any biotransformation into single antigen species or that anti-drug antibodies (ADAs) are bound to a single epitope on the bispecific. For a bispecific IgG with two cellular targets, there may be no need for additional target binding assays, as any biotransformation would lead to loss of function, which could be detected with a PD assay. Additionally, characterization assays may be developed to detect fused peptides or cytokines that have been proteolytically cleaved from the parent protein product. Assays utilizing hybrid LC-MS platforms may be used to assess total therapeutic concentration to further define any potential concentration difference between the intact assay and the total therapeutic that could be indicative of biotransformation.

12.1.1.1.5 Immunogenicity Testing for Bispecific Therapeutics

Immunogenicity testing for bispecific antibodies is generally conducted using traditional ligand-binding assays. Although there is no specific evidence that bispecific therapeutics have a higher degree of immunogenicity other than anecdotal observations, there is a propensity of these therapeutics to have non-natural epitopes created at the sites of protein fusion. The immunogenicity testing strategy may evolve from the nonclinical stage to the clinical phase of drug development. For nonclinical studies, the immunogenicity testing results can explain any loss of exposure or toxicities associated with immune-complexing such as nephrotoxicity. Typically these assays can be developed using traditional bridging immunoassays utilizing ELISA or Meso Scale Discovery (MSD) platforms. A single positive control that has specificity and adequate sensitivity (between 100 and 500 ng/mL) is sufficient to develop and validate a nonclinical assay. The positive control does not need to be specific to any portion of the therapeutic at this stage. There is no requirement for further characterization of the immune response as it will not translate to the human immune response. These nonclinical assays can be streamlined to detect ADAs with minimal validation. At the clinical stage there is likely an additional need to assess specificity to novel linkers or moieties that are not natural structures (FDA, 2019). Also, bispecifics pose a unique challenge in that ADAs may neutralize binding to a single target or to multiple targets requiring further characterization of the immune response. Multiple positive controls and/or protein structures for immunodepletion may be required to fully characterize the immune response at the clinical stage (FDA, 2019). These assays will require full validation according to the US Food and Drug Administration (FDA) guidance and may also require additional characterization assays to further define the overall immune response.

12.1.1.2 Other Multidomain Biotherapeutics

Similar strategies are used for the other multidomain biotherapeutics including PEGylated molecules, ADCs, and probodies. These novel molecules currently have been used in the treatment of cancer, chronic kidney diseases, hepatitis, multiple sclerosis, hemophilia, and gastrointestinal disorders. Despite the different designs and constructs of these molecules, the bioanalytical approaches are similar to the bispecific drugs, which include generic assays, specific assays, and characterization assays to fully assess the PK, PD, immunogenicity, and potential biotransformation.

12.1.1.2.1 PEGylated Therapeutics

The PEGylation of proteins have been performed for 25 years to extend half-life and increase solubility (Swierczewska et al., 2015). Bioanalytical strategy for PEGylated molecules includes the use of anti-PEG antibodies in immunoassays for both PK and immunogenicity assays. The first-generation commercially available anti-PEG antibodies were rabbit monoclonal antibodies with relatively low reactivity. These reagents have been much improved and there are a number of commercially available anti-PEG antibodies with high affinity. For PK assays an anti-PEG antibody can be utilized as a specific detection reagent to measure PEGylated intact therapeutic. The immunogenicity strategy for PEGylated molecules should be carefully considered as there is a high degree of preexisting anti-PEG antibodies in the human population due to exposure to PEG in household and personal care items. Immunogenicity assays for PEGylated molecules can be designed to detect anti-PEG antibodies as a stand-alone assay or as a specificity assay in combination with the anti-therapeutic antibody assay. Anti-PEG and anti-protein therapeutic positive controls are generally required to fully validate the immunogenicity assay(s) for a PEGylated therapeutic.

12.1.1.2.2 Antibody–Drug Conjugates

The ADCs are complex molecules composed of an antibody linked to a biologically active cytotoxic (anti-cancer) payload or drug. There are currently nine approved ADCs (Table 12.1).

Table 12.1 Approved antibody–drug conjugates (ADCs).

ADC therapeutic	Molecular targets
Brentuximab vedotin	Lymphoma, Hodgkin lymphoma, and acute lymphoblastic leukaemia (ALL)
Trastuzumab emtansine	HER-2-positive breast cancer
Inotuzumab ozogamicin	CD22-positive acute myeloid leukaemia (AML)
Gemtuzumab ozogamicin	CD33-positive ALL
Polatuzumab vedotin-piiq	Relapsed or refractory diffuse large B cell lymphoma
Enfortumab vedotin	Urothelial cancer
Trastuzumab deruxtecan	HER-2-positive breast cancer
Sacituzumab govitecan	Metastatic triple-negative breast cancer
Belantamab mafodotin	Anti-CD38 multiple myeloma

Most ADCs are for the treatment of cancer by targeting tumor cells to deliver a cytotoxic molecule directly to kill the tumor cell. Like bispecifics, ADCs can also undergo biotranformation in vivo by cleavage of the drug. There are three drug species that require PK assay development including (1) free drug, (2) total antibody, and (3) conjugated antibody. Typically the free drug is measured using an LC-MS assay based on the size of the conjugated molecule. Total antibody can be measured using an immunoassay or a hybrid LC-MS assay. The conjugated drug is generally also measured using a specific immunoassay utilizing a high-affinity anti-payload antibody. Anti-payload antibodies can be difficult to procure due to their cytotoxic activity which makes mounting an immune response in an animal model problematic. Immunogenicity of ADCs is similar to bispecific antibodies with additional characterization assays considered based on unique linkers or payload characteristics.

12.1.1.2.3 Probodies

The newest modality to enter development is the probody, which is a monoclonal antibody that contains a light chain extension consisting of a masking peptide that blocks the antigen-binding site and a protease-specific substrate-containing linker. These modified antibody therapeutics may be bispecific antibodies, ADCs, or conventional antibody therapeutics. The masking peptide is designed to be cleaved by the aberrant protease activity in the tumor microenvironment. This masking of the active site reduces off-target toxicity and the adverse events that are associated with potent anti-tumor antibody-based therapeutics. As has been discussed in this chapter, the probody has the potential to be biotransformed in vivo outside of the tumor microenvironment. The same principles apply to the measurement of the intact, unmasked antibody, and cleaved peptide product that have been discussed for bispecifics and ADCs. Immunogenicity to all three protein products also must be investigated to fully characterize a clinical immune response (see more discussion in Chapter 8 for the neutralizing antibody assay strategy for probody).

12.1.2 Gene Therapy Programs

12.1.2.1 Introduction

The first successful gene therapy study for humans was conducted in May 1989 (Rosenberg et al., 1990). Over the past decade, advances in viral vector delivery systems, enabling the effective and safe development of gene therapies and immunotherapies, have enthused this already active field. According to clinicaltrial.gov, there are 642 gene therapy-related clinical trials that are ongoing or recently completed (clinicaltrials.gov, 2020). In recent years, an increasing number of gene- and nucleic acid-based products have advanced to late-stage clinical development. Gene therapy is an experimental technique that delivers therapeutic nucleic acid polymers into a patient's cells to modify gene expression at the DNA or RNA level to treat or prevent disease. With their unique ability to target "undruggable" targets, gene therapy has long formed the third major drug platform in addition to the traditional small molecule and large molecule therapeutics. Many different terms have been used across the industry, including nucleic acid-based therapy, oligonucleotide therapy, and DNA/mRNA therapy. The European Medicines Agency (EMA) guidelines published in 2015 provide a more comprehensive definition of gene therapy as: "Gene therapy medicinal products (GTMP) generally consist of a vector or delivery formulation/system containing a genetic construct engineered to express a specific therapeutic sequence or protein responsible for the regulation, repair, addition or deletion of a genetic sequence" (EMA, Committee for Advanced Therapies, 2015). Therefore, we use GTMP in this chapter to describe all products that modify gene expression at the DNA or RNA level in order to achieve therapeutic effects.

12.1.2.2 Structure

GTMPs have demonstrated patient benefits in specific areas of therapeutics to address unmet medical needs. Unlike other conventional therapeutic molecules, a GTMP contains both vector and transgene. Vectors used in GTMP can be designed and modified to target specific tissues or cells to avoid off-target effects or nonspecific toxicity. The modifications include deletion of genes associated with virulence, pathogenicity, or replication competence. Two delivery systems are commonly used: viral-based and nonviral-based delivery vectors (Wold & Toth, 2013; Yin et al., 2014). Viral-based transgene delivery systems illicit high transduction efficiencies and were used in the first generation of GTMPs. However, potential mutations, posttreatment recombination, potential oncogenic effect, and high cost have been the concerns in the employment of a viral system to deliver GTMP to patients. In contrast, nonviral vectors, such as lipid nanoparticles (LNP), have shown fewer safety concerns due to their relative simplicity, though nonspecific cytotoxicity associated with cationic liposomes has been observed. In addition to delivery systems, the expressed gene moieties also present unique challenges in PK assessment and bioanalytical method development.

12.1.2.3 Pharmacokinetics

Informed by cumulative scientific data over several decades, the conventional wisdom regards the systemic drug concentrations as the driver for the pharmacological effects. This general principle has to be altered as tissue concentration is the driver for the pharmacological effects for the majority of available gene therapies. Consequently, evaluating the PK of GTMP is an essential task in pharmaceutical discovery research as well as during nonclinical and clinical development. To GTMP, a prominent example of the utility of tissue exposure is the determination of exposure margin to facilitate the selection of the starting dose for the first-in-human study. Furthermore, due to the dependency of pharmacological response (or PD) on tissue drug concentration, a good understanding of the PK–PD relationship can be very useful in projecting the exposure and dose that would achieve the desired response. The importance of the bioanalytical methods that measure drug concentrations to enable the PK characterizations, therefore, cannot be overemphasized.

In order to substantiate the reliability and validity of the PK data, regulatory agencies expect those using validated methods to measure drug concentrations and provided their expectations in regulatory guidance/guidelines (EMA, 2011; FDA, 2018). However, these guidance/guideline documents do not clearly articulate the regulatory expectations concerning the assays used for support of GTMP nonclinical and clinical development, such as polymerase chain reaction (PCR), hybridization methods, and immunogenicity assays. Following the unique requirements of the tissue drug concentration being a driver for the pharmacological effects, it is generally recognized that drug concentrations representing analytes relevant to the pharmacological effects in both blood and tissue samples should be measured.

12.1.2.4 Bioanalytical Strategy and Platforms

Bioanalytical methods for quantification of DNA or RNA transgene, the vector, and the expressed gene moieties and assessment of their immunogenicity are essential for understanding PK, PD, and safety. Unlike most biotherapeutics, GTMPs may have specific tissue/cell tropism and can enter cells. Many products are administrated directly into the disease sites. For viral-based gene therapies, preexisting antibodies to the viral capsid due to prior infection/vaccination with related viruses may affect the safety and efficacy of the GTMP. All of these factors add to the complexity of the bioanalytical support for GTMP development. The sections that follow summarize the current practices and trends in the bioanalytical industry to illustrate the considerations and practices used in the design, development, qualification/validation, and implementation of bioanalytical methods as well as immunogenicity assessment for GTMP programs.

PK assessment employs several analytical techniques to quantify GTMP based on the requirement of assay sensitivity, assay specificity, analyte properties, and technical expertise in each bioanalytical laboratory. Hybridization-ELISA, quantitative PCR (qPCR), and MS have been the methods of choice in past years (Tremblay & Oldfield, 2009; Wang & Yang, 2010; Zhou et al., 2000).

Hybridization-ELISA, high-performance/pressure liquid chromatography (HPLC), and capillary gel electrophoresis (CGE)–ultraviolet (UV)/fluorescence methods have been the traditional methods to quantitatively determine the analyte of interest for support of PK/toxicokinetic (TK) evaluation of GTMP drugs. Recently, more sensitive and more specific qPCR, and MS-based methods including liquid chromatography tandem mass spectrometry (LC-MS/MS), high-resolution accurate mass (HR-AM), and hybridization-based LC-fluorescence have been added to the "toolbox."

Compared to PCR and MS methods, the traditional methods still offer adequate sensitivity and operational convenience. The hybridization-ELISA or the branched DNA (bDNA) assay employs traditional instrumentation, minimal sample processing, and is performed on a 96-well platform that enables high throughput and easy assay transfer to established bioanalytical laboratories. The bDNA assay, also known as the Quantigene assay, is commonly used for mRNA quantification in various types of samples. It is a single-step, probe-based ELISA assay that works through cooperative hybridization of probes to the target mRNA. A probe set usually consists of 20–40 oligonucleotides that bind specifically to the target mRNA regions. The probe set consists of capture extenders (CEs) to anchor the mRNA to the plate, label extenders (LEs) that amplify the signal of the target mRNA, and blocking probes (BL) to block the regions of mRNA that are not bound by probes within the region of probe design. Lyzed study samples are added to the 96-well assay plates in duplicate, and the ELISA-like assay protocol is followed. The luminescence signal obtained from each well is proportional to the RNA concentration in the samples, which is back-calculated against a calibration curve prepared in respective matrices. These methods can be validated following existing regulatory guidance for ligand binding assays. With the advanced amplification system of the luminescence signal, the sensitivity of the hybridization-ELISA is comparable to the qPCR assays.

qPCR assays have been recognized based on their high sensitivity, accuracy, and practical ease, while spectrometry assays have been considered to have the best specificity. Real-time PCR or/and qPCR is the current gold standard for quantifying gene expression, detection of viral shedding, and determination of virus copy number as the analyte of interest. However, it has been a task to design a suitable PCR assay for quantitative determination of GTMP with a short length of sequences. For example, the average length of an miRNA product is around 20 nucleotides. In addition, assay contamination and cross-reaction with other sources remain one of the main challenges.

For viral-based gene therapies, the US Food and Drug Administration (FDA) has specific requirements for PK assays with regard to sample collection and method sensitivity. For example, FDA 2006 guidelines state that to use a quantitative, sensitive PCR assay to analyze the samples for vector sequences: "You should submit data to your IND to demonstrate that your assay methodology is capable of specifically detecting vector sequence in both animal and human tissues." Moreover, "the assay should have a demonstrated limit of quantitation of < 50 copies of vector/1 μg genomic DNA so that your assay can detect this limit with 95% confidence" (FDA, 2006). The specific requirement of sensitivity poses challenges in PCR method qualification and validation. Challenges in outsourcing laboratory work have to be considered in selecting PCR-based assays for regulated study support since qPCR is not a traditional bioanalytical assay platform. Bioanalytical contract research organization (CRO) laboratories do not typically maintain instrumentation and expertise to validate qPCR methods and test samples.

LC-MS/MS-based methods have seen enormous growth in recent years. These types of assays provide higher analytical specificity than conventional HPLC and higher throughput than gas chromatography-mass spectrometry (GC-MS). Utilization of LC-MS/MS for GTMP development is at the vanguard of nonclinical candidate screening and nonviral-based vector quantification. For LNP quantification, multiple LC-MS methods can be developed and validated to ensure the intact LNP can be quantitatively determined from various matrices. However, the absolute sensitivity and specificity are highly dependent on the specific assay technique, assay reagents, and the structure of the target analyte. Several limitations of LC-MS/MS such as concerns of sensitivity, throughput, and transferability have been revealed. Table 12.2 provides a comparison of these methods.

For gene therapies entering clinical trials, their efficacy, safety, tissue distribution, and immunogenicity must be assessed in nonclinical animal models. Evaluation of biodistribution of gene therapy products focuses on localization of both the transgene and delivery vector, as well as the expressed protein. Since tissue exposure shall be the driver of clinical dose selection, the potentially deleterious effects of the gene therapy vector or protein product on normal healthy tissues need to be evaluated in nonclinical toxicology studies. It also requires an understanding of the probability of either entity reaching off-target organs.

Table 12.2 Comparison of conventional hybridization assays, polymerase chain reaction (PCR)-based assays, and liquid chromatography-mass spectrometry (LC-MS)-based assays for RNA/DNA quantification.

Methods	Hybridization assays	PCR-based assays	LC-MS assays
Sample preparation	miRNA, RNA, and DNA isolation may not be needed	miRNA, RNA, and DNA isolation is needed	Need sample extraction
Instruments	Traditional instrumentation, (e.g., plate washer, plate reader, incubators)	PCR machine	LC-MS
Throughput	Easier to automate (possibility to expand to 384-well formats)	Medium/high throughput – reverse transcription (RT) and PCR reactions are time consuming	Medium throughput
Sensitivity	High sensitivity	Highest sensitivity	Potentially acceptable
Interference/ specificity	Need to distinguish between modified RNA (modRNA) and endogenous mRNA signal (cross-reactivity)	Prone to contamination	High specificity
Reagent availability	Custom probe sets need to be designed for specific sequences	Combination of custom probes and commercially available reagents	Common reagents, no need of a specific probe
Method validation	Ligand binding assay (LBA) method validation guidance	No current guidance	LC-MS validation guidance

However, deciphering tissue exposure in clinical settings is very challenging since the collection of the human specimen is difficult and very limited. In particular, the consistent ratio of plasma/tissue from different animal species should be established to bridge the nonclinical and clinical dose selections. The correlation between systemic and tissue exposure will be established in the nonclinical phases. To obtain a consistent ratio of plasma/tissue is very critical for the clinical dose selection to support clinical studies. Also, multiple matrices shall be used for exposure characterization with sufficient analytical sensitivity. Therefore, multiple methods are often required to accomplish full PK characterization.

The immunogenicity strategy in support of GTMP development shall consider three potential challenges: the immunogenicity against DNA/RNA transgene, the immunogenicity to the delivery vectors/vehicles (e.g., viral capsid, liposome), and DNA/RNA-encoded proteins as effective molecules for mRNA-based enzyme replacement therapy.

Risk-based approaches shall be used to assess the potential immunogenicity of GTMP. Specifically, bioanalytical assays for determination of anti-DNA/RNA, anti-delivery vector/vehicles (e.g., anti-lipid component of LNP), and anti-protein antibodies shall be separately developed and validated. Considering the likelihood, clinical transformability, and potential impact, the immunogenicity assessment will not apply to the samples from the toxicity/TK studies unless it is deemed to be necessary. However, it must be noted that FDA guidelines on nonclinical assessment of investigational cellular and gene therapy products indicate that the likelihood of immunoresponses to viral vector is high and needs to be monitored. In the case of the GTMP using a viral vector, a preexisting antibody to the viral capsid can significantly hamper the effectiveness of the GTMP administration, and thus shall be tested for screening subjects or investigating the impact on PK/PD. For clinical studies, the immunogenicity strategy shall be implemented for all types of GTMPs. However, the strategies as to which assays are to be included in which development phases depend on the properties of each vector type. If immunogenicity is a concern (e.g., with viral capsids or allogeneic cellular products) then each subject's immune response to the product should be evaluated. This evaluation may include monitoring for evidence of both cellular and humoral immune responses. Also, other approaches to monitoring acute immune response shall be considered in the overall immunogenicity assessment.

Using mRNA therapies as an example to illustrate the consideration and strategy, the anti-mRNA immunogenicity assays seems less needed since the current data do not indicate that there is a significant induction of immunogenicity against mRNA itself in the absence of an autoimmune disease. Therefore, the anti-RNA antibody is considered as low risk and may not need to be monitored routinely. Anti-protein antibodies can develop against proteins expressed from mRNA, in particular, if repeat administration regimens are pursued. Anti-mRNA-encoded protein antibody shall be screened for in clinical studies of mRNA-mediated protein replacement. Anti-lipid nanoparticle (LNP) antibodies can be considered as low incident and may be determined to be of low impact on safety and efficacy.

12.1.2.5 Current Regulatory Landscape

The analytical methods have to be qualified or validated to enable assessment. However, to qualify or validate the matrix specific assay individually is often operationally difficult. Therefore, fit-for-purpose approaches have been applied to address the challenges, including only validating the method using one type of tissue and qualification of other tissue samples via the method selectivity test.

In previous years, many regulatory documents have touched upon the use of nontraditional bioanalytical methods to assess the safety or clinical endpoints. This guidance includes Guidance for Industry-Gene Therapy Clinical Trials – Observing Subjects for Delayed Adverse Events (FDA, 2006); ICH Considerations – General Principles to Address Virus and Vector Shedding (EMA ICH, 2009); Pathogen Safety Data Sheets and Risk Assessment (Public Health Agency of Canada, 2011); FDA Guidance for Industry – Clinical Considerations for Therapeutic Cancer Vaccines (FDA, 2011); FDA Guidance for Industry – Design and Analysis of Shedding Studies for Virus or Bacteria-Based Gene Therapy and Oncolytic Products (FDA, 2015); and the recent EMA guideline on the quality, nonclinical, and clinical aspects of gene therapy medicinal products (EMA, 2018). However, those regulatory guidelines still lack detail on how assay validations should be designed and performed including the appropriate acceptance criteria to apply. Therefore, many bioanalytical laboratories follow the general method validation guidance for bioanalytical assays for ligand binding or chromatography assays.

For support of regulated studies, a bioanalytical scientist has designed and conducted fit-for-purpose method validations to ensure method suitability and data integrity. The accuracy and precision, sensitivity, specificity, and stability of methods are tested to meet the requirements in the clinical setting. Of significance, the stability data shall be used to guide sample collection, shipping, and storage. Having validated stability in multiple storage temperatures and conditions enables the extension of the study design for the collection of more valuable data in both institutional and home settings. Very often the method shall be fully validated and implemented in a bioanalytical CRO as the in-house source can be a constraint for routine sample analysis.

12.1.2.6 Future Considerations

Looking forward, more gene therapy drugs are advancing to late-stage development. This increasing trend in genomic therapeutics opens the opportunity for bioanalytical scientists to work with regulatory agencies in establishing guidance for genomic and viral detection methods that are critical for clinical studies. Through close collaboration among pharmaceutical companies, bioanalytical CROs with expertise, and regulatory agencies we optimistically look forward to having a well-established set of industry practices and regulatory guidance shortly.

Although GTMP has demonstrated clinical benefits for several diseases, including inherited disorders, some types of cancer, and certain viral infections, this novel drug platform is still under investigation to ensure that it will be safe and effective.

As more genetic therapeutics are advancing to late-stage development, bioanalytical scientists are facing increasing technical challenges and regulatory rigor in method design, optimization, and validation. In addition, novel technologies in this area (e.g., Digital Droplet PCR) offer technical advantages but also make it difficult to validate these assays to meet regulatory requirements. Further discussions and input among industrial scientists, technology providers, and regulatory agencies will be valuable for providing the best practices for managing technical and regulatory challenges in the future.

References

Brinkmann, U., & Kontermann, R. (2017, February-March). The making of bispecific antibodies. *MAbs*, 9(2), 182–212. https://doi.org/10.1080/19420862.2016.1268307

clinicaltrials.gov. (2020). Gene therapy clinical trials worldwide database. https://www.clinicaltrials.gov

European Medicines Agency (EMA). (2011). Guideline on bioanalytical method validation. https://www.ema.europa.eu/en/documents/scientific-guideline/guideline-bioanalytical-method-validation_en.pdf

European Medicines Agency (EMA). (2018). Guideline on the quality, non-clinical and clinical aspects of gene therapy medicinal products. https://www.ema.europa.eu/en/documents/scientific-guideline/guideline-quality-non-clinical-clinical-aspects-gene-therapy-medicinal-products_en.pdf

European Medicines Agency (EMA), Committee for Advanced Therapies. (2015). Guideline on the quality, non-clinical and clinical aspects of gene therapy medicinal products. https://www.ema.europa.eu/en/documents/scientific-guideline/guideline-quality-non-clinical-clinical-aspects-gene-therapy-medicinal-products_en.pdf

Food and Drug Administration (FDA). (2006). Guidance for industry-gene therapy clinical trials – Observing subjects for delayed adverse events. https://www.ngvbcc.org/pdf/gtclin.pdf;jsessionid

European Medicines Agency (EMA) ICH. (2009). Considerations – General principles to address virus and vector shedding. https://www.ema.europa.eu/en/ich-considerations-general-principles-address-virus-vector-shedding

Food and Drug Administration (FDA). (2011). Clinical considerations for therapeutic cancer vaccines. https://www.fda.gov/regulatory-information/search-fda-guidance-documents/clinical-considerations-therapeutic-cancer-vaccines

Food and Drug Administration (FDA). (2015). Design and analysis of shedding studies for virus or bacteria-based gene therapy and oncolytic products. Guidance for industry. https://www.fda.gov/regulatory-information/search-fda-guidance-documents/design-and-analysis-shedding-studies-virus-or-bacteria-based-gene-therapy-and-oncolytic-products

Food and Drug Administration (FDA). (2018). Bioanalytical method validation. Guidance for industry. https://www.fda.gov/regulatory-information/search-fda-guidance-documents/bioanalytical-method-validation-guidance-industry

Food and Drug Administration (FDA). (2019). Immunogenicity testing of therapeutic protein products — Developing and validating assays for anti-drug antibody detection. https://www.fda.gov/regulatory-information/search-fda-guidance-documents/immunogenicity-testing-therapeutic-protein-products-developing-and-validating-assays-anti-drug

Kellie, J., Pannullo, K., Li, Y., Fraley, K., Mayer, A., Sychterz, C., Szapacs, M., & Karlinsey, M. (2020). Antibody subunit LC-MS analysis for pharmacokinetic and biotransformation determination from in-life studies for complex biotherapeutics. *Analytical Chemistry*, 92(12), 8268–8277. https://doi.org/10.1021/acs.analchem.0c00520

Ma, M., Colletti, K., Yang, T.-Y., Leung, S., Pederson, S., Hottenstein, C. S., Xu, X., Warrino, D., & Wakshull, E. (2019). Bioanalytical challenges and unique considerations to support pharmacokinetic characterization of bispecific molecules. *Bioanalysis*, 11(05), 427–435. https://doi.org/10.4155/bio-2018-0146

Public Health Agency of Canada. (2011). HERPES SIMPLEX VIRUS. Retrieved from pathogen safety data sheets and risk assessment. https://www.canada.ca/en/public-health/services/laboratory-biosafety-biosecurity/pathogen-safety-data-sheets-risk-assessment/herpes-simplex-virus.html

Rosenberg, S. A., Aebersold, P., Cornetta, K., Kasid, A., Morgan, R. A., Moen, R., Karson, E. M., Lotze, M. T., Yang, J. C., Topalian, S. L., Merino, M. J., Culver, K., Miller, A. D., Blaese, R. M., & Anderson, W. F. (1990). Gene transfer into humans–immunotherapy of patients with advanced melanoma, using tumor-infiltrating lymphocytes modified by retroviral gene transduction. *The New England Journal of Medicine*, 323(9), 570–578. https://doi.org/10.1056/NEJM199008303230904

Swierczewska, M., Lee, K. C., & Lee, S. (2015). What is the future of PEGylated therapies? *Expert Opinion on Emerging Drugs*, 20(4), 531–536. https://doi.org/10.1517/14728214.2015.1113254

Tremblay, G., & Oldfield, P. (2009). Bioanalysis of siRNA and oligonucleotide therapeutics in biological fludis and tissues. *Bioanalysis*, 1(3), 595–609. https://doi.org/10.4155/bio.09.66

Wang, Z., & Yang, B. (2010). End-point stem-loop real-time RT-PCR for miRNA quantification. Chapter: MicroRNA Expression Detection Methods MicroRNA Expression Detection Methods:131–140.

Weidle, U. H., Tiefenthaler, G., Weiss, E. H., Georges, G., & Brinkmann, U. (2013). The intriguing options of multispecific antibody formats for treatment of cancer. *Cancer Genomics and Proteomics*, 10(1), 1–18. https://cgp.iiarjournals.org/content/10/1/1.long

Wold, W. S., & Toth, K. (2013, December). Adenovirus vectors for gene therapy, vaccination and cancer gene therapy. *Current Gene Therapy*, 13(6), 421–433. https://doi.org/10.2174/1566523213666131125095046

Xu, Y., Lee, J., Tran, C., Heibeck, T. H., Wang, W. D., Yang, J., Stafford, R. L., Steiner, A. R., Sato, A. K., Hallam, T. J., & Yin, G. (2015, January-February). Production of bispecific antibodies in "knobs-into-holes" using cell-free expression system. *mAbs*, 7(1), 231–242. https://doi.org/10.4161/19420862.2015.989013

Yin, H., Kanasty, R. L., Eltoukhy, A. A., Vegas, A. J., Dorkin, J. R., & Anderson, D. G. (2014). Non-viral vectors for gene-based therapy. *Nature Reviews Genetics*, 15(8), 541–555. https://doi.org/10.1038/nrg3763

Zhou, L., Cryan, E. V., Minor, L. K., Gunnet, J. W., Demarest, K. T., Wang, L., & Ji, C. (2000, January 15). A branched DNA signal amplification assay to quantitate messenger RNA of human uncoupling proteins. *Analytical Biochemistry*, 282(1), 46–53. https://doi.org/10.1006/abio.2000.4587

13

Regulatory Aspects for Assay Development, Validation, and Sample Analysis

Amy Lavelle[a,b] *and Megan Wiberg*[a]

[a]*PPD*
[b]*Corresponding author*

13.1 Introduction

There are currently numerous documents in various stages of publication that guide the bioanalytical support of both clinical and nonclinical protocols necessary to submit data to regulatory authorities across the globe for approval of novel biological therapeutics. In this chapter, recommendations will be based on guidance documents published by the US Food and Drug Administration (FDA) and the European Medicines Agency (EMA). Other resources, including various technical publications by the International Council for Harmonisation of Technical Requirements for Pharmaceuticals for Human Use (ICH), are also pertinent for considering how to develop and validate robust methods to support bioanalytical studies and ensure the generation and reporting of high-quality and scientifically sound data.

There are primarily five documents that govern the development and validation of bioanalytical methods and subsequent analysis of samples for both pharmacokinetic (PK) (quantitative) measurements and immunogenicity

Bioanalytical Aspects in Biological Therapeutics, First Edition. Edited by Xiaohui Xu and Weifeng Xu.
© 2022 John Wiley & Sons, Inc. Published 2022 by John Wiley & Sons, Inc.

(qualitative or semiquantitative) assessments. Additionally, primary guidance for biomarkers is listed below. These documents are:

- "Bioanalytical Method Validation, Guidance for Industry" – published by FDA in May 2018 (FDA US DHHS, 2018).
- "Immunogenicity Testing of Therapeutic Protein Products – Developing and Validating Assays for Anti-Drug Antibody Detection Guidance for Industry" – published by FDA in January 2019 (FDA US DHHS, 2019).
- "Guideline on Bioanalytical Method Validation" – published by EMA to take effect in February 2012 (EMA CHMP, 2012).
- "Guideline on Immunogenicity Assessment of Biotechnology-Derived Therapeutic Proteins" – draft released for comments by EMA in October 2015, effective December 2017 (EMA CHMP, 2017).
- "Bioanalytical Method Validation M10 (ICH M10)" – draft distributed for comments by ICH in February 2019 (ICH, 2019).
- "Points to Consider Document: Scientific and Regulatory Considerations for the Analytical Validation of Assays Used in the Qualification of Biomarkers in Biological Matrices" – published by the Critical Path Institute (C-Path) in June 2019 (Piccoli & Sauer, 2019).

The recommendations for anti-drug antibody (ADA) method development, validation, and sample analysis in this chapter are based on the recently finalized FDA document, EMA documents, as well as current industry trends and expectations. Furthermore, while the titles of these documents suggest that they are pertinent only to the development and validation of the bioanalytical methods, each goes on to address implementation of the methods to support sample analysis and/or reporting of data from analytical studies. This chapter is not intended to substitute full review and understanding of the available guidance documents and regulations; however, the intent is to provide a summary of the salient points as well as highlight recommendations for practical consideration.

13.2 Regulatory Considerations During Ligand Binding Assay Development – PK

While assay development remains largely unregulated, FDA guidance provides a significant amount of detail to assist in the development of an optimized, robust assay (both PK and ADA) that can stand up to regulatory review. It is strongly recommended that guidance documents be reviewed and referenced when preparing a therapeutic protein product for regulatory submission and clinical use. The following sections highlight specifics recommended to focus on during development and validation but are not intended to replace guidance and are not all encompassing.

Large molecule ligand binding assay (LBA) development's main purpose is to define the procedure and its limitations while ensuring the method is optimized for validation and suited to its intended purpose (FDA, US Department of Health and Human Services, 2018). Although assay method development is generally regarded as not subject to audit, the work done is important to the method life cycle and therefore should be considered available to regulatory viewing. FDA guidance states that method development does not require Good Laboratory Practice (GLP) documentation; however, in our experience, good documentation is key to method life cycle maintenance and troubleshooting assistance and we suggest following "GLP-like" practices. The 2018 FDA BMV guidance specifically notes that issues, procedural changes, and resolutions be well documented to justify method changes. This particularly comes into play, for example, when a method fails validation and needs to go back to development for optimization, or new critical reagent lots are bridged in and perform differently from original lots.

In assay development, most parameters that will be evaluated in a validation are assessed, although in a limited fashion to the extent that it can provide confidence in a "validate-able" method. Guidance specifically states the following to be assessed prior to starting method validation: reference standards, critical reagents, calibration curve, quality controls (QCs), selectivity and specificity, sensitivity, accuracy, precision, recovery, and analyte stability (FDA, 2018). While not every parameter needs thorough evaluation, as that will be achieved in validation, there are a few key aspects of a method that must be assessed and are subject to regulatory review, often being noted in the validation report as they may not be evaluated specifically with a priori criteria in the validation. This includes a minimum required dilution (MRD) evaluation for both PK and ADA methods. This assessment can be paired with selectivity and specificity testing in the intended biological matrix including specific disease state(s) during method development or assessed as its own experiment compared against a buffer curve to assess interference in the intended matrix. Selectivity, also referred to as interference, in hemolyzed or lipemic matrix may also be evaluated in development, although it may or may not affect the optimization of the assay. For example, routinely, if hemolyzed matrix interference is identified, the assay procedure may not change to mitigate this interference, but it will be noted during sample analysis. Alternatively, matrix interference identified during disease state spiked selectivity would almost certainly result in method optimization to mitigate effects, by MRD optimization or quantitation range adjustment. Therefore, evaluation of both MRD and selectivity during assay development are imperative to be assessed and deemed acceptable prior to validation. Quantitation range of the assay, concentrations of calibration standards, and evaluation of the curve regression should also be established prior to moving into validation, although validation data may drive finalization of these method parameters. Freshly prepared QCs can be used during development to assess accuracy and precision across the range of the assay, while an evaluation of fresh versus frozen is helpful to ensure there are no freeze–thaw effects prior to validation. Additionally, FDA guidance recommends development data for evaluating analyte stability in matrix to include collection, handling, and storage temperatures. Whole blood stability and co-medication stability are also useful assessments to perform in development, so the data are available if needed but not necessarily validated, unless the study requires this information. The American Association of Pharmaceutical Scientists (AAPS) Ligand Binding Assay Bioanalytical Focus Group (LBABFG) (DeSilva et al., 2003) recommends the evaluation of dilutional linearity, which is a helpful assessment of the ability of the method to identify a hook effect (or prozone) as well as the ability to dilute samples with initial results above the quantitation range onto the curve. An unidentified hook effect could have serious implications in the assay's ability to accurately quantify samples and may affect curve range. Therefore, early identification in development and addition to the method parameters is recommended.

The 2018 FDA BMV guidance specifically notes several cases where development data play a significant role in subsequent validation regulations, with the ICH M10 (FDA, 2019, draft) briefly reiterating specific parameters for assay development. EMA (2012) guidelines are less specific regarding assay development but, in general, industry adheres to FDA guidance suggestions for development assessments. DeSilva et al. (2003) provides a clear outline of what should be assessed in assay development, including documentation, as well as mention of drafting the method prior to moving into validation, based on recommendations from the AAPS LBABFG.

13.3 Regulatory Considerations During LBA Development – Biomarkers

While biomarkers remain largely regarded as unregulated and focus more specifically on a "fit for purpose" or "context of use" qualification (Lee et al., 2006; Piccoli & Sauer, 2019), developing an assay to GLP (or "GLP-like") standards is preferable particularly depending upon phase of trial or endpoint need. The main area of focus during development is appropriate matrix selection. Ideally, a method validation is conducted in the same biological matrix expected in a study. Section V.A. of the 2018 FDA BMV guidance focuses on endogenous compounds, which can play a significant role in the ability to find an interference-free biological matrix appropriate for the assay. Matrix should have "no measurable analyte" and "no matrix effect or interference" when used to prepare calibration standards. Interferent- or endogenous compound-free matrix for a biomarker assay is often difficult to obtain. Therefore, the use of alternative matrices, or a surrogate matrix, is allowed with scientific justification. This is explored in development and documented appropriately for future potential audit review.

Assay QC concentrations must also account for endogenous levels in the biological matrix. These QCs can be evaluated and determined prior to validation during development and then qualified in validation, often as intact or endogenous QCs, derived from varying levels of endogenous biomarker in sample matrix – either by diluting or fortifying pooled samples or identifying individuals with endogenous levels spanning the quantitation range. If individuals are chosen, methods for bridging in new endogenous QCs must be clearly defined in the analytical method, as availability and volume of these specific matrices will be depleted often prior to study end.

One additional consideration when developing biomarker assays is the nature of the reference material used for such assays. Characterization may not be well explored and often lot-to-lot variability is inherent. Therefore, expectations that significant bridging work or re-optimization/re-qualification may be necessary if new lots are needed should be factored into timelines and resourcing. In general, the extent of method development should reflect the end goal of the assay, whether it be exploratory or used in a pivotal study.

13.4 Regulatory Considerations During LBA Development – ADA and Neutralizing Antibodies

ADA clinical method development and validation have historically been regarded as vague and somewhat "unregulated," following white papers and industry best practices for guidance on criteria and experimental design (Büttel et al., 2011; Koren et al., 2008; Mire-Sluis et al., 2004; Shankar et al., 2008). However, FDA finalized their guidance in 2019. The previous draft version was adhered to as auditable by regulatory authorities prior to this release.

While ADA assay parameters vary from PK, the general rules for development apply. FDA guidance (2019) and EMA draft guidance (2017) both specifically discuss development of ADA assays in Section V of the FDA guidance and Section 7 of the EMA draft guidance. Development is discussed in detail in both documents and it is strongly recommended to refer to these documents when developing an ADA assay. EMA provides helpful tables assessing different assay platforms, highlighting their advantages and disadvantages.

When developing an immunogenicity assay, the goal is to develop a sensitive assay that will detect both low- and high-level ADA across all tiers of testing that will be conducted. Therefore, it is recommended to assess parameters in all expected tiers including screening, confirmation, titer, and neutralizing/characterization. Previously, a focus was placed on Tier 1/screening; however, with the implementation of the newest FDA guidance and its emphasis on the confirmatory tier/Tier 2, evaluation of confirmatory testing of selectivity, sensitivity, and interference is strongly suggested. Positive control selection is imperative to determining an accurate representation of the assay's capabilities to detect ADA. The positive control can be either monoclonal or polyclonal affinity purified to be specific to the therapeutic. As with any assay development assay format, critical reagent selection, assay buffer and blocking buffer formulation, number of washes, and wash buffer composition all play a critical role in developing a sensitive and specific assay and should be clearly evaluated and documented during development. FDA guidance allows for development of a separate screening and confirmatory assay, both in method and platform; however, developing one assay is more time- and cost-efficient and avoids possible differences in data reporting and interpretation impacted by the platform or method.

MRD assessment should be documented during development comparing buffer to matrix, keeping in mind components of the assay diluent that may interfere in the assay's ability to detect ADA. While comparing normal controls to individuals, samples should also be prepared with ADA at varying levels to assess recovery in matrix versus buffer, targeting the highest signal-to-noise ratio and signal close to that of the diluent (FDA, 2019). FDA guidance recommends not exceeding a minimum dilution of 100.

For both screening and confirmatory tiers, it is important to consider what may interfere and affect the cutpoint of the assay. This could include preexisting antibodies, rheumatoid factors, human anti-mouse antibodies, and the dosing level of therapeutic that is expected (see Tate & Ward, 2004, review article for more detailed descriptions of interferents in immunoassays). Additionally, cross-reactivity or target interference can play a role in the choice of assay format and procedural steps aiming to mitigate the interference, since endogenous drug target or similar protein products could affect the assay's ability to specifically detect ADA. A preliminary cutpoint should be determined in development to allow for early evaluation of additional parameters, including drug tolerance and sensitivity. FDA guidance allows for as few as 5–10 population-specific treatment-naïve individuals (or pooled individuals, depending on matrix type). If a greater sample size is evaluated, a more representative preliminary cutpoint may be determined. Additionally, more individuals would also give an idea of outliers, which could point to expected preexisting antibodies or other interfering factors and provide a path forward for cutpoint determination in validation. When creating the negative control pool, the signal of the negative control pool should closely match treatment-naïve individuals, in the same matrix where possible to ensure the control is below the cutpoint but not so different that the assay acceptability is affected. In situations where controls do not match or study-specific matrix is not available for early validation work, a matrix-specific cutpoint may need to be evaluated once the study has enough samples available.

When developing the assay, two keys elements are the focus in determination of whether an assay is ready for validation and should therefore be assessed early in development: sensitivity and drug tolerance. These will have the most significant impact on the assay format and procedure. The assay should be optimized to ensure sensitivity targets are met (100 ng/mL for clinical assays, 250–500 ng/mL for nonclinical) and be evaluated in pooled and individual matrix to ensure the negative control behaves as expected in the samples. A preliminary low positive control (LPC) can be set from the development data to be used in validation cutpoint analysis. Since this is determined from a cutpoint often from a much smaller set of individuals than will be used for validation, it is not necessary to determine an LPC at 1% failure rate, but instead something low enough to monitor the low end of the assay. Alternatively, a sensitivity curve can be utilized during validation cutpoint screening from which the sensitivity, titer precision, and the 1% LPC calculation can be analyzed.

Drug tolerance is targeted to minimize the interference of the therapeutic and the assay's ability to detect ADA. Prior knowledge of expected levels of drug dosing in patients helps determine the assay's target although commonly the achievable target is based on the sensitivity of the assay (i.e., "best possible"). Sample pretreatment can disrupt the ADA–drug complexes to allow for more accurate detection in the assay. If an approach such as this is needed, it is best to determine and document this early on in assay development to ensure the rest of the assay is optimized with a pretreatment included. Possible assay format optimizations for increasing drug tolerance include acid dissociation, utilizing coated beads, solid phase extraction, or a precipitation and acid dissociation (PandA) assay. An alternative approach would be to adjust the study protocol to collect patient samples when drug product levels are lowest.

Additional parameters recommended for evaluation in development include robustness, specificity and selectivity, and precision. The concentration of confirmatory drug used during Tier 2 should be optimized with documentation. This is generally done during sensitivity testing to demonstrate sufficient inhibition across the range of the assay.

For neutralizing antibody (NAb) assays, these recommended levels of sensitivity and/or drug tolerance may not be achievable; however, development data should be available to show that all attempts were made to develop the most sensitive and tolerant assay. This includes moving from cell-based NAb to plate-based NAb. Both EMA (2017) and FDA (2019) guidance recommend determining the appropriate format based on the mechanism of action of the therapeutic, and cell-based is recommended since it more closely mimics in vivo mechanisms. The in vivo mechanisms can create their own issues in assay development, such as increased assay variability, smaller range, and limitations on optimization. Thus, documentation and justification demonstrating due diligence in developing the assay is required to instead proceed with a plate-based NAb format. The dynamic range of the assay is important early in development to determine a therapeutic product concentration in the linear range so that the activity is sensitive to inhibition (FDA). Other considerations for NAb assays include matrix interference effects, much as in ADA development (refer to Chapter 8 for further details).

13.5 Regulatory Considerations During LBA Validation – PK

The validation of an assay is not started until there is a fully developed and optimized method available. However, that method is still considered a living document, and therefore the method is not final or ready for use in sample analysis until the validation has been successfully completed and all a priori criteria met. It may be determined during validation that the method needs adjustments or even re-optimization and necessitates going back to development. Changes to the method such as incubation times or temperatures, quantitation range or positive control concentration adjustments, or minor diluent changes can be done and implemented during validation, provided all is documented. Changes required that affect the entire assay, such as MRD adjustment, sample pretreatments, or similar major changes to procedure that may affect quantification of samples would require a validation restart, as they would affect the calculation of acceptable data.

FDA (2018), EMA (2012), and ICH M10 (2019, draft) all clearly outline expectations of a PK validation for LBA. The purity and quality of the reference standard and critical reagents can have a large impact not only on an assay's ability to accurately quantitate samples, but on its ability to perform robustly, and therefore should be considered carefully prior to starting a validation. Once in validation, a priori criteria, outlined in an approved validation plan, written per guidance and specific standard operating procedures (SOPs), are expected to be met. If they are not met, the method either requires re-optimization, or strong scientific justification is necessary for moving forward into analyzing samples. Circumstances such as these should be discussed in depth in the validation report. For example, a failed criterion may have little impact on the study, depending on the purpose of the study and the phase of the trial. This could include hemolytic interference or freeze–thaw stability beyond a certain number of cycles (five to six cycles tend to be the standard optimal number, but often three cycles are sufficient). Parameters that must be met for a method to be deemed appropriate and acceptable for use include quantification range determination, selectivity at the lower limit of quantitation (LLOQ), specificity (including interference testing), and dilutional linearity. Selectivity should be carried out in the same biological matrix as the study samples. Additionally, it is recommended that assay controls be prepared in the same biological matrix. However, in some studies the availability of disease-specific matrix or other matrices such as pediatric or cerebral spinal fluid make it difficult to prepare enough controls to cover an entire study. Therefore, it is acceptable to prepare controls in normal matrix, provided study-specific selectivity/recovery samples are demonstrated to be acceptable at the same LLOQ as determined in the normal/healthy matrix and infer that the disease matrix behaves similarly to normal in the assay. Stability must be validated mimicking sample handling. Carryover and parallelism are also parameters to be evaluated per guidance, where appropriate either by platform or when samples are available.

In general, a full method validation as outlined in guidance should be carried out under most circumstances, particularly for clinical/GLP studies always with the endpoint in mind and focusing on evaluating validation samples as one would study samples. The 2019 draft ICH M10 and 2018 FDA BMV are excellent resources for validation expectations.

13.6 Regulatory Considerations During LBA Validation – Biomarkers

The C-Path Points to Consider Document (Piccoli & Sauer, 2019) and ICH M10 (ICH, 2019) define paths for qualification and validation in assays with biomarkers and endogenous compounds. It is strongly recommended to qualify a biomarker assay as closely to guidance as possible, designing the strategy around the endpoint of the data. Therefore, context of use approaches for method validation are acceptable when clinically justifiable (Lee et al., 2006;

Piccoli & Sauer, 2019). If the assay will be used for regulatory decision-making, then the assay should be fully validated, per GLP guidance. This includes accuracy, precision, sensitivity, selectivity, parallelism, range, reproducibility, and stability. The Workshop report from Crystal City VI (Arnold et al., 2016) and now the C-Path document provide detailed descriptions of the assay parameters to be assessed, highlighting the unique experiments that are indicative of a biomarker assay. Specifically, parallelism in place of quantitative spiked recovery, appropriate matrix selection for calibrators (and QCs), endogenous QCs, precision versus accuracy, and stability. The ICH M10 outlines multiple options for assay control selection and determination. Parallelism is critical to demonstrating the assay's ability to quantitate and dilute samples reproducibly. As such, it is used in place of spiked selectivity, which is often difficult to reliably capture or apply criteria to, given the nature of the endogenous protein concentrations in individuals.

Endogenous QC concentrations are generally set during validation for the method, to provide a large enough data set to apply a meaningful concentration or concentration range in a regulated setting. This can be determined either by the average value over a set number of runs, a range based off the mean value factoring in standard deviation, or the median value. The qualification or validation plan defines criteria a priori for setting initial QC values. The method or bioanalysis plan describes bridging in or creating a new lot of endogenous QCs, as they will vary lot to lot (or pool to pool) when in sample matrix due to individual patient variation. The levels of endogenous QCs should target expected sample levels, spanning a low, mid, and high range when possible.

Stability should be determined both for the assay controls as well as matrix controls or endogenous QCs. For matrix controls, stability is based on the T0 for pooled endogenous QCs (although individuals could be used), as it is nearly impossible to set a standard stability starting point, particularly if the endogenous protein levels are expected to change over time. Therefore, endogenous QCs or study samples are then compared back to freshly thawed/T0 samples as a reference. Sample collection and handling can create artificial effects on stability and should be taken into consideration when assessing stability parameters in a biomarker validation or qualification.

The nature of the biomarker creates a question of how to assess true accuracy in the assay. Relative accuracy is expected to be part of a biomarker validation. Precision is also expected in the validation, with the understanding that intra-subject precision is more meaningful than inter-subject. Alternatively, precision and accuracy can be assessed using surrogate calibrators and endogenous QCs. Total error is often added as part of the analysis of variance testing criteria (Arnold et al., 2016). It is generally expected that a robust biomarker assay, with its known caveats, should meet the majority of standard PK validation acceptance criteria, again, keeping the use of the resulting data and the endpoint of the study in mind and scientifically justifying any deviation from those limits.

13.7 Regulatory Considerations During LBA Validation – ADA and NAb

The guidance for immunogenicity is far less detailed and specific for ADA method validation than for PK. The main criteria outlined in the FDA guidance are described in the sections that follow.

13.7.1 Screening

The screening cutpoint should be carried out in at least 50 treatment-naïve samples from the appropriate patient population if available, over 3 days by at least two analysts. If not available, commercial individuals are acceptable for validation, but the study would then require reassessment once samples are available, as an in-study cutpoint. Balanced design is strongly recommended, and a few papers describe statistically the best way to carry this out (Devanarayan et al., 2017; Shen et al., 2015; Zhang et al., 2013, for example), although various methods are accepted across industry. FDA guidance outlines a six-panel balanced design based on Shen et al. (2015). A 5% false-positive rate (FPR) cutpoint is determined from the screening data and 1% FPR for the confirmatory cutpoint. Since development gives limited information as to how the samples will actually perform in cutpoint analysis, it is suggested that determination of the cutpoint be data driven and not a priori, although the specific mathematical methods for determining the cutpoint can be outlined as options, in the validation plan or SOP. Additional parameters expected for the screening tier of the validation include sensitivity determination, drug tolerance, specificity and selectivity, precision, reproducibility, and robustness.

Current industry practice and guidance recommends an LPC determined to have a 1% FPR rate at the assay low end to monitor those samples very close to the cutpoint. FDA is clear that sensitivity levels should be at least 100 ng/mL for clinical studies, both for screening and confirmatory tiers. Drug tolerance target levels are less clear and are generally based on the expected sample concentrations. Spiked selectivity at this low level can be evaluated either as calculated percent recovery or as positive/negative in disease-state individuals, based on the study protocol. Stability is not discussed in the FDA guidance apart from in-use stability of critical reagents, but it is suggested at minimum to demonstrate process sample handling (refrigerator and/or room temperature) and freeze–thaw stability at the low and high positive control concentrations. Long-term stability is under debate in the industry, with the consensus that antibodies are stable and therefore it is unnecessary to test as one would for a quantitative assay.

13.7.2 Confirmatory

FDA draft guidance released in 2016 included the recommendation that confirmatory assays should be fully validated to screening assay standards, as the inhibition data could prove critical to assessing patient response. This was a change from the previous draft guidance and is part of the final guidance released in 2019. Methods that have been online prior to this new recommendation should be re-validated or "up-validated" under these conditions to ensure compliance. A 1% FPR is recommended for confirmatory assays, using a cutpoint determined in the same individuals as assessed during screening, comparing treatment-naïve to individuals incubated with a known concentration of drug (optimal concentration determined in development).

13.7.3 Titration

The titer cutpoint may be the same as the screening cutpoint or the assay may necessitate a cutpoint specific to titer if, for example, the assay diluent or matrix results in a higher response than the screening cutpoint and therefore the samples cannot be diluted below the cutpoint. A titer cutpoint is often set at the screening 1% FPR (as opposed to the 5% for screening), or even 0.1%, based on the theory that it goes beyond the confirmatory's 1%. However, it is not often necessary to use such a tight parameter at the point of titer evaluation. The latest FDA guidance requires inclusion of the assay MRD when reporting sample titers, which is a change from how many of those in industry were previously reporting titer.

13.7.4 Neutralizing Antibodies (NAb)

A NAb assay cutpoint does not require as many individuals for assessment, citing 30 instead of 50 individuals (FDA, 2019). While it is acknowledged that these assays may not achieve the sensitivity of an ADA assay, given the ability of the neutralizing activity of the antibodies used for the positive control, it is still important to validate the sensitivity. For cell-based assays, specificity should focus on the assay's ability to inhibit the therapeutic and not due to other external stimuli. Also, due to the nature of the cell-based assay, precision should be assessed more extensively than is recommended for a screening assay. Cell-specific robustness is one additional unique parameter recommended for validation by the FDA guidance.

13.8 Additional Considerations

Novel modalities have become a significant part of drug discovery and include therapeutics that may need special consideration during assay development and validation. These include antibody–drug conjugates (ADCs), multi-domain protein products, fusion and bispecific proteins, enzyme therapeutics, PEGylated proteins, peptides, and adeno-associated viral vectors. Each pose their own special characteristics that may affect both the development and validation of the assay but are beyond the scope of this chapter. There are a number of white papers focusing on recommendations for these modalities. EMA also recommends developing additional assays for relevant biomarkers or PK measurements to evaluate clinical impact. Isotyping, affinity, and other characterization assays may also be necessary depending on clinical impact as well as induced antibody response against impurities, either product-related or process-related (EMA, 2017).

13.9 Conclusions

Development and validation of an LBA follows a combination of current guidance, industry recommendations, and white papers. The results from a validated method provide critical information to support new therapeutics, demonstrating safety and efficacy. Plans, SOPs, and reports should assess the recommended parameters and meet the recommended a priori acceptance limits as outlined in guidance following ALCOA (attributable, legible, contemporaneous, original, accurate) guidelines for documentation of all relevant setup and results. Outlined earlier are highlights from guidance and suggestions based on experience and feedback from regulators. The data are only as good as the validated method and therefore significant focus should be on development of a specific, robust, and optimized assay.

13.10 Bioanalytical Sample Analysis

Once the methods are validated, they can be used to support regulated bioanalysis. However, analyte- and matrix-specific methods are not sufficient on their own. There are specific regulations detailing requirements for the conduct of bioanalytical studies. The Good Laboratory Practice (GLP) regulations, which apply only to nonclinical (animal) studies to support applications for drug products regulated by a governmental body such as the FDA, were put in place to ensure the scientific integrity and quality of the data ultimately used to gain approval to move into human test subjects (FDA GLP). The Good Clinical

Practice (GCP) guidelines, which are applicable to the conduct of clinical trials to evaluate the safety and efficacy of new drug products on consenting human subjects, were put in place to protect the rights and safety of those subjects (FDA GCP).

A nonclinical study for which the data are intended to be submitted to a regulatory agency must adhere to one or more of the three different GLP regulations:

1) FDA GLPs for Nonclinical Laboratory Studies, also known as FDA 21 Code of Federal Regulation, Part 58 (FDA, 2021).
2) The Organisation for Economic Co-operation and Development (OECD) Principles for Good Laboratory Practice (OECD, 1998).
3) Ministry of Health, Labor and Welfare (MHLW) Ordinance on the GLP Standard for Conduct of Nonclinical Safety Studies of Drugs, also known as MHLW Ordinance No. 114 (MHLW, 2008).

The OECD is an international body made up of 36 member countries and provides support to nonmember countries with emerging or developing economies. MHLW is the regulatory authority within the government of Japan. The requirements of the three different GLPs are largely aligned; however, it is critical that the correct regulations detailed in the GLP study protocol are referenced and adhered to in the generation of bioanalytical data. The sponsor of the study will need to choose the appropriate GLPs to follow based upon which regulatory authority the data will ultimately be submitted to and ensure these specific details are included in the compliance section of the protocol. In many cases, work performed within the US on OECD- or MHLW-compliant GLP studies may need only comply with the FDA GLP requirements as they are often deemed an acceptable substitution. All GLPs lay out the requirements for the conduct of nonclinical (animal) studies and include provisions for the management of studies, requirements for the facilities in which any part of the study is carried out, the equipment used, the documentation of all aspects of study conduct and data collection, as well as the reporting and archival of data, documents, and samples (FDA GLP).

For GLP studies, raw data collection must be done according to very specific requirements for regulated studies. All recordings are made with indelible ink and must be attributable, legible, contemporaneous, original, accurate (ALCOA) and complete, consistent, enduring, and available (CCEA). Changes or corrections to raw data/documentation must be done in a fashion that indicates who made the change, when it was made, why it was made, and without obliteration of the original documentation (FDA GLP).

Generally, from a bioanalytical standpoint, adherence to GLP requirements will ensure that the sample analysis conducted to support a clinical study will be compliant with the GCPs as well. Table 13.1 summarizes some of the differences between the GLPs and the GCPs. To summarize, the basic themes of both sets of guidelines are as follows:

- Properly trained and experienced personnel are assigned to manage and carry out the studies with sufficient resources, suitable and properly calibrated equipment, and within suitable facilities.
- A standard operating procedure (SOP) system is in place and followed to ensure the quality and integrity of the data being generated and reported.
- Specific expectations for documentation of raw data and reporting of results.

Table 13.1 Comparison of Good Laboratory Practice (GLP) and Good Clinical Practice (GCP) requirements.

Parameter	GLP	GCP
Scope and Purpose	Applicable to nonclinical (animal) studies from which data will be used to support regulatory filing. Designed to protect the quality and integrity of the data.	Applicable to all clinical studies to protect the rights and safety of the human trial volunteers.
Responsible Party	The Study Director (SD) is the single point of control for the study. The SD has overall responsibility for the study and ensures delegated contributors (e.g., bioanalytical laboratory) adhere to the protocol and applicable regulations. The SD is assigned by management at the test facility (i.e., where the in-life phase of the study is performed). Oversight of portions of the study, such as the bioanalytical phase, may be delegated to appropriately trained/qualified Principal Investigators (PIs) who must agree to comply with the protocol and applicable regulations and ensure direct communication with the SD.	Responsibility for the study lies with the Study Sponsor. Typically, a Medical Director fills this role. The Sponsor may employ delegates, for example, through outsourcing to a Contract Research Organization (CRO), to conduct various aspects in support of the clinical trial, but the overall responsibility for the study resides with the Sponsor.

(Continued)

Table 13.1 (Continued)

Parameter	GLP	GCP
Quality Assurance	The Lead Quality Assurance (QA) resides at the test facility and inspects data and processes for compliance with the appropriate GLPs and reports findings to the SD and test facility management. QA at any test sites (e.g., the bioanalytical laboratory) inspects data and processes at the site and reports findings to the SD and test site management. QA personnel are separate from the personnel engaged in the conduct or management of the study. QA's role is oversight only; management ensures quality systems and controls are in place and followed.	The Study Sponsor is responsible for implementing and maintaining quality assurance and quality control systems to ensure that studies are conducted, and data are generated, documented, and reported in compliance with the protocol, SOPs, and GCPs.
Study Management	Management at the test facility is responsible for assigning a trained and qualified SD for each GLP study. Is responsible for the design and implementation of SOPs and other quality systems. Designates a QA Unit that is managed and operates separately from the study personnel. Also ensures availability of resources including suitable facilities, qualified personnel, and appropriate equipment.	The Study Sponsor is responsible for ensuring the clinical study is properly managed in compliance with GCPs.
Facilities	Each testing facility is required to be of suitable size and construction to facilitate the proper conduct of nonclinical laboratory studies and with sufficient separation of operations that no activity will have a negative impact on the study.	Each study must be carried out in facilities deemed adequate for the foreseen duration of the trial to ensure proper and safe conduct.
Equipment	Equipment must be of appropriate design and adequate capacity to support the protocol. Equipment must be inspected, cleaned, maintained, and when used to generate or measure data, properly calibrated.	Each study must utilize equipment deemed adequate for the foreseen duration of the trial to ensure proper and safe conduct.

Parameter	GLP	GCP
Standard Operating Procedures (SOPs)	Management-approved SOPs must be in place to ensure the quality and integrity of the data generated in the course of a study. Any deviations from SOPs are communicated to and approved by the SD.	The Sponsor must implement a quality system that includes written SOPs to ensure trials are conducted and data are generated, documented, and reported in compliance with the protocol, GCPs, and applicable regulations.
Protocol	Each study shall have an approved written protocol that indicates the objectives for the study and all methods to be followed during the conduct of the study. The SD is responsible for approving the study protocol.	Each study must follow a written and approved protocol which describes the study objectives, design, methodology, statistical considerations, and organization of the trial. Protocols frequently provide background on the drug product and rationale for the study. Clinical protocols must receive approval from an Institutional Review Board (IRB)/Independent Ethics Committee (IEC) to ensure the rights and safety of trial subjects are protected. The Sponsor and Investigator sign the protocol to confirm agreement with conducting the study in compliance with the document.
Records and Reports	A final report is required for each study. Study records are archived in a secure location for at least 5 years following the date of registration if used to support a marketing permit or 2 years following study completion/ termination. The archivist is responsible for security of the archived records.	Clinical study reports are required for each trial. Retention time for records and data vary depending on the requirements of the regulatory authority to which the data are submitted for approval of the drug product. Many investigators/sponsors maintain records for 15 + years.

Table adapted from Eurofins (2019): http://vivotesting.com/what_is_the_difference_between_GLP_and_GMP_lab_testing_regulations

For clinical studies, it is particularly important to protect the identity of trial volunteers. For blinded studies, the treatment group assignment of each patient must be protected until the blind has been lifted. If the bioanalytical laboratory has been provided with a key to allow analysis of only those samples collected from treated patients, data cannot be provided to any blinded contacts until the blind has been lifted. If the study blind is compromised in any way prior to the unblinding, the integrity of the study may be questioned and all or part may have to be repeated.

13.11 Analysis of PK Samples

13.11.1 Initial Analysis

Along with study samples, it is critical to include a calibration curve and appropriate QCs on each analytical run. The acceptable performance of the calibration curve and QCs will validate the acceptability of the sample results from the same run. For efficiency of analysis, an appropriate dilution of the study samples, either at or beyond the assay MRD, can be selected prior to initial analysis based on the validated curve range, known dosing concentration, as well as the visit and time point at the time of sample collection. If this information is limited or unknown, it is suggested that a small subset of samples from subjects at different dose levels and time points be analyzed first to identify the expected concentrations and be able to move forward with analysis in the most efficient manner. When there is a demonstrated prozone, or hook effect, during method validation it is important to ensure that samples with concentrations expected to fall in or near the hook are analyzed at appropriate dilutions to ensure accurate data reporting. A common approach to analysis of samples in an assay in which a hook is demonstrated is to analyze post-dose samples at multiple dilutions. By evaluating the raw signal compared to the dilution factor, it can be determined if the sample has been diluted sufficiently to avoid the prozone. Signal should decrease as the dilution factor increases. If the signal increases with increased dilution, the result is not reliable, and the sample should be repeated at higher dilution factors. When reviewing the data generated from study samples it is important to look at whether the data generated make sense (are plausible).

13.11.2 Reanalysis of Study Samples

During the course of the study there will be samples that require analysis more than once. Reanalysis of study samples may be either planned or unplanned. Unplanned reanalysis of study samples may be required for a number of reasons including but not limited to: if there is unacceptable precision between replicates used to determine a mean result (e.g., two wells for an LBA), if the result is above the validated upper limit of quantitation (ULOQ) of the calibration curve, if a sample has been diluted beyond the MRD and the result is below the validated lower limit of quantitation (LLOQ), if the result is above the LLOQ and is known to be a true pre-dose (or baseline) sample, if the results for a sample analyzed at multiple dilutions due to a known prozone indicate that the sample is within the hook, or inadvertent repeat analysis. It may be necessary to generate additional results in cases where multiple results for a single sample do not agree. In all cases, SOPs or other documented a priori criteria should be in place to guide the selection of the reportable sample result. All results generated should be evaluated together to ensure accurate data reporting.

An example of deliberately repeating samples is incurred sample reanalysis (ISR) and is a requirement for both nonclinical and clinical studies. ISR is carried out by selecting a subset of samples that have been analyzed and have reportable results and reanalyzing them in a subsequent independent analytical run to compare the results to the original values obtained. Results from ISR are not used to select a final reportable value for the sample but are only for verifying acceptable reproducibility of the method within a study. Typically, approximately 10% of samples are selected from across the study from different subjects/patients who received various doses and at various visits/time points following dosing. This ensures that samples are selected across the full range of reported concentrations, including C_{max} and during the elimination phase. Guidance states that ISR samples must be repeated using the same method conditions and at the same dilution used to obtain the original result. The relative percent difference between the original and ISR result is calculated. At least two-thirds of the ISR results must be within 20% or 30% of the original result for chromatographic and LBAs, respectively. Should ISR fail, a plan to investigate the cause of the failure must be written and followed. Once complete, the results of the investigation and impact of the failure on the study must be documented as well. It is important to state that individual ISR results that do not agree with the original reported values should not be the sole basis for further reanalysis of those samples. However, ISR data may be helpful in identifying analytical errors that may prompt further investigation into a cause for variability of results. Any investigation done as a result of ISR data review must be well documented. The regulations specify which studies require ISR be performed; however, many study sponsors simply perform ISR during each study to help ensure quality and reproducibility of data (FDA, 2018).

Occasionally, there may be a desire to reanalyze study samples based on the overall study results. Such requests for reanalysis of samples, referred to as PK repeats as they are often identified as unexpected based on the PK profiles, must be based on a priori criteria and have clear justification. Reanalysis of study samples on this basis should be a last resort following investigation into all other potential causes for the unexpected results (e.g., mislabeling of samples, switching of samples during analysis, missed dosing, etc.). No PK repeats should be performed on studies to evaluate bioequivalence or bioavailability (FDA, 2018). PK repeats may also be prohibited for studies that will be submitted to certain regulatory authorities, such as Agência Nacional de Vigilância Sanitária (the Brazilian Health Regulatory Agency, or ANVISA).

13.12 Analysis of ADA Samples

In contrast to the quantitative results obtained from PK assays, ADA results are based on the evaluation of raw signals generated in the assay compared to established thresholds for classification as either positive or negative. A multi-tiered, risk-based approach is almost universally employed for the analysis of samples for ADA. In this tiered approach, samples are first screened against a cutpoint statistically determined to identify approximately 5% false-positive samples. Any samples that screen positive against this screening cutpoint are then moved on to a second tier in which the specificity of the positive signal is confirmed for the biologic therapeutic. Confirmed positive samples are then diluted serially until the signal response crosses the cutpoint to determine a titer. Further characterization of the antibody response, such as identifying if the ADA is neutralizing or determining the specific epitope to which the ADA is binding (e.g., in the case of bispecific antibodies, ADCs, or probodies, etc.) may also be required. Alternative approaches, while allowed, must be justified (FDA, 2019).

Inclusion of negative controls (NCs) and appropriate positive controls (PCs) in each analytical run is necessary to ensure the consistent performance of the ADA method. Run-specific screening and titration cutpoints are calculated by applying an additive or multiplicative factor to the response of the NC on each run. The confirmatory cutpoint is based on the change in signal when the sample is analyzed with and without a set concentration of drug during a single run. These cutpoints are determined either during validation or as part of the analytical study (see next section on In-Study Cutpoints). Because these assays provide qualitative assessments, monitoring the signal of these controls in the assays is critical for maintaining the sensitivity and overall consistent performance of the assay.

13.12.1 In-Study Cutpoints

As clinical development proceeds and studies are carried out in populations of patients with the targeted disease(s), being dosed with various other comedications, or with other sources of population variability, it is important to ensure that the cutpoints applied to the study samples are appropriate. Validated cutpoints may be suitable for nonclinical and early phase I studies. For phase II, and certainly any pivotal trials from which data will be used to support filing applications, the need for calculation of in-study cutpoints should be evaluated. A common and generally accepted approach to generating an in-study cutpoint involves the analysis of a subset of pre-first dose samples (baseline) in the screening and confirmatory tiers to evaluate an FPR. For example, for a clinical study, targeting approximately 50 pre-dose samples analyzed in two panels should be sufficient for this evaluation. If the FPR is in the range 2%–11%, the existing validated cutpoint may be considered appropriate for the study (Devanarayan et al., 2017). If the FPR is outside this range, the data generated from the pre-dose samples can be used to calculate in-study cutpoints following statistical approaches that are similar to the original validation. There are alternative approaches to calculating in-study cutpoints as well. For example, data generated from predose samples throughout the course of the study may be used to calculate a screening cutpoint that can be applied to the data retrospectively. In this scenario, there is no need for multiple panels to introduce analytical variability due to the fact that the data were generated over many runs by more than one analyst across months or even years. In either scenario, the design for how the in-study cutpoint will be calculated should be documented a priori and detailed in the bioanalytical study report (Devanarayan et al., 2017).

13.12.2 Reanalysis of ADA Samples

There are times when ADA samples will require reanalysis. As required for PK samples, SOP or other a priori criteria must be in place for reanalysis of samples for ADA to guide the selection of the reportable sample result. All results generated should be evaluated together to ensure accurate data reporting.

Reasons for reanalysis may include but are not limited to unacceptable precision when more than one replicate is used to generate a mean result as in the case of traditional LBAs, inadvertent reassays that demonstrate inconsistent results, or when positive samples with elevated raw signals (e.g., higher than the high positive control) do not confirm positive. For this latter scenario, the result may be falsely negative because there is insufficient drug added in the confirmatory tier to inhibit the signal. A strategy for this, such as titering the sample and then selecting an appropriate dilution to repeat the confirmatory tier of analysis, should be outlined. There is no ISR for ADA testing.

13.12.3 Other Cutpoint Considerations

Due to the fact that it is important to evaluate cutpoints for each study, especially at later stages of clinical development, it may be useful to establish assay cutpoints in normal matrix during initial method validation. This is not to say that other important method parameters, such as specificity and selectivity in the disease population of interest, should not be done during initial validation. However, the commercial availability of samples from patients with the disease of interest may be difficult to obtain, may be cost-prohibitive, or may require more time than is available to collect. It is also important to remember to evaluate the consent form that trial volunteers will sign to ensure it covers the use of their samples for evaluations such as determination of in-study cutpoints.

13.13 Final Data and Reports

For each nonclinical or clinical study that is conducted, data are generally provided in two formats: (1) an electronic summary of data in a tabular form (e.g., Excel, SAS, etc.) and (2) a detailed bioanalytical report summarizing data generated under the applicable study protocol. Generally, the more information included in reports, the less likely a reviewer will need to visit a site to audit a study. Both the FDA BMV Guidance as well as the ICH M10 draft include very specific requirements for information to include in validation and bioanalytical study reports. The summary report typically includes a tabular presentation of the sample results. Additionally, the report includes text to summarize the details pertaining to the conduct of the study and any information that is critical to the interpretation of the data presented. The bioanalytical report is incorporated into the overall GLP study report or clinical study report as appropriate. Each report should include all results generated as part of the study. For example, all results for samples that were reanalyzed for any reason must be included in the report. Data generated for a subject that withdrew consent, while no longer considered part of the study, must also be included in a table in the report (e.g., this may be in an appendix). Documentation of any GLP, protocol, SOP, or method deviations that occur during the study must be included in the report along with any impact.

13.14 Bioanalytical Method Life Cycle Management

Depending on the stage of drug development as well as the success of the novel biotherapeutic, it is likely that the bioanalytical method, once validated, will need to be functional to support nonclinical and/or clinical trials for several years. It is therefore necessary to ensure proper method life cycle management procedures are in place to maintain consistent results across studies. The most important piece of managing the life cycle of a method is planning for inevitable changes to the critical reagents. Any reagent that has a direct impact on the results of the assay is considered a critical reagent. For LBAs these include but are not limited to binding proteins, antibodies, conjugated antibodies, and enzymes. The performance of new critical reagents should be evaluated directly in the assay along with the old reagents for comparison whenever possible. When an older reagent is not available, data generated with the new reagent may be compared to relevant historical data. It is important to have a priori criteria defined for accepting the performance of a new reagent in a method. Retesting of critical reagents in the assay is also necessary to support their stability and to extend their continued use beyond initially assigned expiration dates when performing well. In all cases, the data to support successful performance of new reagents as well as the extension of the use of reagents beyond initial documented expiration should be well documented and maintained (ICH M10).

For ADA methods specifically, and for commonly used bridging formats especially, changes to critical reagents can have a significant impact on assay sensitivity as well as drug tolerance. Both should be taken into consideration when testing new critical reagents in ADA methods. If the sensitivity of the method is impacted by a new critical reagent, it may be necessary to adjust the LPC to ensure proper monitoring of the cutpoint during sample analysis. It is important to document and report any change in drug tolerance so that study data can be interpreted appropriately (FDA, 2019).

For quantitative assays, evaluation of analyte stability should also be part of life cycle management. If frozen calibrators and/or QCs are used for acceptance of analytical runs, the stability of the analyte in the same matrix the calibrators and QCs are prepared in must be demonstrated. Ideally, samples are analyzed within established stability as well.

13.15 Other Considerations During Bioanalysis

There are many other factors to consider throughout the use of a validated bioanalytical method. As the drug progresses through the development process, it is possible that indications other than the initial disease the drug was developed to treat will be identified. It is necessary to ensure that the validated method performs acceptably in matrix from patients of all disease populations tested. When a study enrolls patients from a disease population other than that which was validated, it may be necessary to test selectivity and parallelism in quantitative assays using samples from individuals diagnosed with the disease of interest. Samples may be obtained commercially, or if consent has been given, samples may be obtained from the study using those collected prior to the first dose of study drug. If selectivity and/or parallelism is unacceptable in the new disease state, modification to the method identified to overcome the issue will need to be validated appropriately.

During bioanalysis, samples requiring dilution to obtain a reportable result within the validated curve range must be diluted within the demonstrated assay linearity. If a highly concentrated sample requires dilution beyond the demonstrated linearity, additional validation of those dilutions is required. If the reference standard is not at a high enough concentration to prepare a QC that can be diluted to the necessary factor, incurred sample linearity or parallelism may be necessary.

13.16 Conclusions

Method validation does not end once sample analysis begins. It is an ongoing exercise throughout the life of a method to ensure the highest quality and scientific integrity of the bioanalytical data. Regulated bioanalysis is governed by numerous guidance documents, government regulations, and industry white papers. While not a replacement for carefully reviewing and understanding these documents, this chapter has highlighted some of the important topics to consider when planning to undertake support of a regulated study. Knowing what regulations apply, maintaining proper documentation throughout the course of the study, providing thorough reporting of the data, and planning properly to manage the life cycle of methods will help ensure success.

References

Arnold, M. E., Booth, B., King, L., & Ray, C. (2016). Workshop report: Crystal City VI – bioanalytical method validation for biomarkers. *The AAPS Journal*, 18(6), 1366–1372. https://doi.org/10.1208/s12248-016-9946-6

Büttel, I. C., Chamberlain, P., Chowers, Y., Ehmann, F., Greinacher, A., Jefferis, R., Kramer, D., Kropshofer, H., Lloyd, P., Lubiniecki, A., Krause, R., Mire-Sluis, A., Platts-Mills, T., Ragheb, J.A., Reipert, B.M., Schellekens, H., Seitz, R., Stas, P., Subramanyam, M., Thorpe, R., Trouvin, J.H., Weise, M., Windisch,

J., Schneider, C.K. (2011). Taking immunogenicity assessment of therapeutic proteins to the next level. *Biologicals*, 39(2), 100–109. https://doi.org/10.1016/j.biologicals.2011.01.006

DeSilva, B., Smith, W., Weiner, R., Kelley, M., Smolec, J., Lee, B., Khan, M., Tacey, R., Hill, H., & Celniker, A. (2003). Recommendations for the bioanalytical method validation of ligand-binding assays to support pharmacokinetic assessments of macromolecules. *Pharmaceutical Research*, 20(11), 1885–1900. https://doi.org/10.1023/B:PHAM.0000003390.51761.3d

Devanarayan, V., Smith, W., Brunelle, R., Seger, M. E., Krug, K., & Bowsher, R. B. (2017). Recommendations for systematic statistical computation of immunogenicity cut points. *The AAPS Journal*, 19(5), 1487–1498. https://doi.org/10.1208/s12248-017-0107-3

Eurofins CRL. (2019). http://vivotesting.com/what_is_the_difference_between_GLP_and_GMP_lab_testing_regulations (accessed January 2019)

European Medicines Agency (EMA), Committee for Medicinal Products for Human Use. (2012). Guideline on bioanalytical method validation. https://www.ema.europa.eu/en/documents/scientific-guideline/guideline-bioanalytical-method-validation_en.pdfAdopted by CHMP 2011.

European Medicines Agency (EMA), Committee for Medicinal Products for Human Use. (2017). Guideline on immunogenicity assessment of biotechnology-derived therapeutic proteins. https://www.ema.europa.eu/en/documents/scientific-guideline/guideline-immunogenicity-assessment-therapeutic-proteins-revision-1_en.pdfAdopted by CHMP 2017.

Food and Drug Administration (FDA) 21 Code of Federal Regulations Part 58, Good Laboratory Practices for Nonclinical Laboratory Studies. (2021). https://www.accessdata.fda.gov/scripts/cdrh/cfdocs/cfcfr/CFRSearch.cfm?CFRPart=58.

Food and Drug Administration (FDA), US Department of Health and Human Services. (2018). Bioanalytical method validation, guidance for industry 2018. https://www.fda.gov/files/drugs/published/Bioanalytical-Method-Validation-Guidance-for-Industry.pdf

Food and Drug Administration (FDA), US Department of Health and Human Services. (2019). Immunogenicity testing of therapeutic protein products – developing and validating assays for anti-drug antibody detection, guidance for industry. https://www.fda.gov/media/119788/download

International Council for Harmonisation of Technical Requirements for Pharmaceuticals for Human Use (2019). ICH Harmonised Guideline: Bioanalytical Method Validation M10. Draft, 2019.https://www.ema.europa.eu/en/documents/scientific-guideline/draft-ich-guideline-m10-bioanalytical-method-validation-step-2b_en.pdf

Koren, E., Smith, H. W., Shores, E., Shankar, G., Finco-Kent, D., Rup, B., Barrett, Y. C., Devanarayan, V., Gorovits, B., Gupta, S., Parish, T., Quarmby, V., Moxness, M., Swanson, S.J., Taniguchi, G., Zuckerman, L.A., Stebbins,C.C., Mire-Sluis, A. (2008). Recommendations on risk-based strategies for detection and characterization of antibodies against biotechnology products. *Journal of Immunological Methods*, 333(1–2), 1–9. https://doi.org/10.1016/j.jim.2008.01.001

Lee, J. W., Devanarayan, V., Barrett, Y. C., Weiner, R., Allinson, J., Fountain, S., Keller, S., Weinryb, I., Green, M., Duan, L., Rogers, .A., Millham, R., O'Brien, P.J., Sailstad, J., Khan, M., Ray, C., Wagner, J.A., (2006). Fit-for-purpose method development and validation for successful biomarker measurement. *Pharmaceutical Research*, 23(2), 312–328. https://doi.org/10.1007/s11095-005-9045-3

Ministry of Health, Labor and Welfare (MHLW) Ordinance on the GLP Standard for Conduct of Nonclinical Safety Studies of Drugs, also known as MHLW Ordinance No.114 (2008). https://www.pmda.go.jp/files/000153713.pdf

Mire-Sluis, A. R., Barrett, Y. C., Devanarayan, V., Koren, E., Maia, M., Parish, T., Scott, G., Shankar, G., Shores, E., Swanson, S.J., Taniguchi, G., Wierda, D., Zuckerman, L.A, (2004). Recommendations for the design and optimization of immunoassays use in the detection of host antibodies against biotechnology products. *Journal of Immunological Methods*, 289(1–2), 1–16. https://doi.org/10.1016/j.jim.2004.06.002

Organisation for Economic Co-operation and Development (OECD) Principles for Good Laboratory Practice (1998). https://www.oecd.org/officialdocuments/publicdisplaydocumentpdf/?cote=env/mc/chem(98)17&doclanguage=en

Piccoli, S. P., & Sauer, J. M. (2019). Points to consider document: Scientific and regulatory considerations for the analytical validation of assays used in the qualification of biomarkers in biological matrices. Biomarker Assay Collaborative Evidentiary Considerations Writing Group, Critical Path Institute (C-Path). https://c-path.org/wp-content/uploads/2019/06/evidconsid-whitepaper-analyticalsectionv20190613.pdf

Shankar, G., Devanarayan, V., Amaravadi, L., Barrett, Y. C., Bowsher, R., Finco-Kent, D., Fiscella, M., Gorovits, B., Kirschner, S., Moxness, M., Parish, T., Quarmby, V., Smith, H., Smith, W., Zuckerman, L.A., Koren, E. (2008). Recommendations for the validation of immunoassays used for detection of host antibodies against biotechnology products. *Journal of Pharmaceutical and Biomedical Analysis*, 48(5), 1267–1281. doi.org/10.1016/j.jpba.2008.09.020

Shen, M., Dong, X., & Tsong, Y. (2015). Statistical evaluation of several methods for cut-point determination of immunogenicity screening assay. *Journal of Biopharmaceutical Statistics*, 25(2), 269–279. https://doi.org/10.1080/10543406.2014.979196

Tate, J., & Ward, G. (2004). Interferences in immunoassay. *Clinical Biochemistry Review*, 25(2), 105–120.https://www.ncbi.nlm.nih.gov/pmc/articles/PMC1904417

Zhang, L., Zhang, J. J., Kubiak, R. J., & Yang, H. (2013). Statistical methods and tool for cut point analysis in immunogenicity assays. *Journal of Immunological Methods*, 389(1–2), 79–87. doi.org/10.1016/j.jim.2012.12.008

Index

Bioanalytical Aspects in Biological Therapeutics, First Edition. Edited by Xiaohui Xu and Weifeng Xu.
© 2022 John Wiley & Sons, Inc. Published 2022 by John Wiley & Sons, Inc.

b

d

e

i

s

Printed and bound by CPI Group (UK) Ltd, Croydon, CR0 4YY
26/10/2022
03157716-0004